The Math of Ancient History

The Math of Ancient History

Integrating Mathematical Problem Solving Strategies with History

Max W. Fischer

Teacher's Choice Press
San Jose New York Lincoln Shanghai

The Math of Ancient History
Integrating Mathematical Problem Solving Strategies with History

All Rights Reserved © 2002 by Max W. Fischer

The classroom teacher may reproduce copies of materials in this book for classroom instructional use only.

No part of this book may be reproduced or transmitted in any form or by any means, graphic, electronic, or mechanical, including photocopying, recording, taping, or by any information storage retrieval system, without the permission in writing from the publisher.

Teacher's Choice Press
an imprint of iUniverse, Inc.

For information address:
iUniverse, Inc.
5220 S. 16th St., Suite 200
Lincoln, NE 68512
www.iuniverse.com

ISBN: 0-595-23465-8

Printed in the United States of America

Contents

Problems dealing with the civilizations of... **Page Number**

List of Tables

CROSS REFERENCE OF PROBLEM-SOLVING STRATEGIES IN *THE MATH OF ANCIENT HISTORY*				
Strategies>	Use Objects	Make a Picture	Use a List or Table	Trial & Error
Civilizations		**Problem**	**Numbers**	
Mesopotamia		10	1,3,11,12	3,15A
Egypt	12	1,12	3,4	2A,4,13A,14A
Near East	2,6	2,6,8	3,4,9,12	5A,11A
Greece	1	12	11,14,15	4
Rome		5,6A,14	4,7,11	13
Mayan	2,3	4,13, 14,16	1,9,11	5A,6A,10A
Mound Builders		8	3,9,11	2A,5A,7A

"A" denotes problems that may utilize basic algebra as one possible method for problem resolution.

"C" denotes problems that involve calculator assistance.

CROSS REFERENCE OF PROBLEM-SOLVING STRATEGIES
IN *THE MATH OF ANCIENT HISTORY*

Strategies>	Find a	Brain-	Work	Logical
	Pattern	Storming	Backwards	Reasoning
Civilizations		**Problem**	**Numbers**	
Mesopotamia	7,8	2	13	4A,5,6A,9A,14A
Egypt	1,5	11	10	6C,7C,8C,9
Near East	10,14		7,15A	1,8,13
Greece	7,13		6A	1,2,3,5,
				8A, 9A,10C
Rome	9,14		8,10,13	1,2A,3A,12
Mayan	9,15		7A	8,12,16
Mound Builders	1A,9			3,4,6,8,10A

"A" denotes problems that may utilize basic algebra as one possible method for problem resolution.

"C" denotes problems that involve calculator assistance.

Introduction

As more and more junior high schools evolve into middle schools, interdisciplinary teaching becomes a critical aspect of that evolution. Since proper integration of the curriculum respects the integrity of each subject matter, one discipline becomes the most challenging subject to integrate consistently. Mathematics, with its scope inherently hinged upon sequential skills cannot be arbitrarily broken apart in order to fit any given thematic unit.

The one consistent skill math teachers will repeat periodically throughout the year will be problem solving. *The Math of Ancient History* presents one hundred realistic word problems based upon life in various ancient civilizations. These problems emphasize the use of numerous problem solving strategies by students such as "Find a Pattern", "Use an Object", "Draw a Picture" and many more as suggested by the National Council of the Teachers of Mathematics. *The Math of Ancient History* allows math instructors to integrate their subject with a middle school ancient history curriculum without interrupting the continuity of their own curriculum, a characteristic most math instructors will welcome as they prepare their students to meet state proficiency standards.

Allotting one problem per page, the reproducible text allows students room to work their solutions beneath each problem.

Chapter 1

Ancient Mesopotamia

Problem #1

Part of an ancient Sumerian caravan traveling north to the Anatolia Plateau region (modern Turkey) was carrying ingots of tin and copper that would be smelted into bronze. There were three times as many five pound tin ingots as there were ten pound tin ingots. There were six times as many five pound copper ingots as there were ten pound copper ingots. If the total weight of the copper and tin ingots totaled 780 pounds, how many ingots of each kind (five and ten pound tin; five and ten pound copper) were in the caravan?

Problem #2

Akad and Rinab were two mud brick makers in ancient Mesopotamia. They kept track of their daily brick count by laying out a system of straws as they progressed similar to the six lines shown below. Their system of straws showed them when to stop when eleven straws were properly placed. Add five more straight lines to the six shown to see how Akad and Rinab knew when to stop.

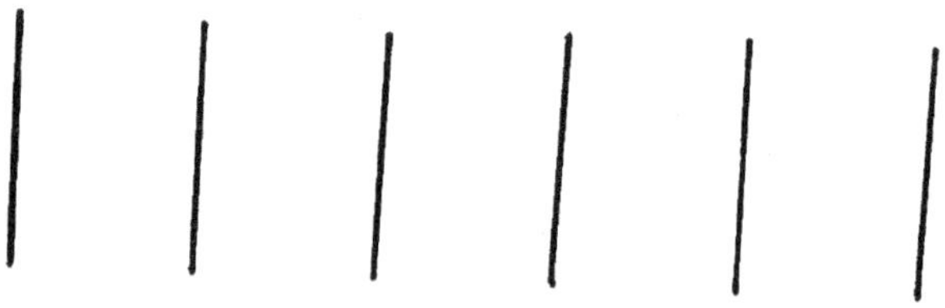

Problem #3

Cavu, Ramu and Sheck, three brick masons in the ancient Sumerian city of Ur, were building a merchant's new trading store. They were paid by their employer in measures of barley based on their experience and their quality of work. Cavu earned twice as much as Sheck while Ramu made one-third the amount of Cavu. If their employer paid out a total of 88 measures of barley to the three of them over the length of this job, how much did each man receive?

Problem #4

Cavu, Ramu and Sheck used their pay of barley measures to support their families. Cavu had seven mouths to feed in his family including himself. Sheck had four (including himself) while Ramu only had himself and his wife. If a measure of barley could produce two loaves of bread, which man was able to provide the most bread for each member of his family? Which provided the least?

Problem #5

In a Babylonian tavern a farmer, weaver, metal worker, carpenter, potter and stone worker were discussing the challenges, joys and hardships of their work. Obi was the first to speak, "My world keeps spinning past me each and every day at work."

Mek replied, "The heat's always on me in my work."

Kur quickly retorted, "That's nothing. You should see how many crack up in my job."

Ban countered by saying, "I rather enjoy my work, I understand all its angles."

Ibin, somewhat philosophical about the entire discussion, stated, "Just as in my line of work, we're all interrelated and part of the whole."

Shur put an abrupt end to the entire discussion with a harsh complaint, "You guys have it *so* tough! I support all of you, and you rarely give me any respect."

Identify the jobs of each man in this situation.

Problem #6

A Babylonian trader purchased 40 large wool blankets from a sheepherder for 10 pieces of silver each. He took them by caravan to Anatolia (central Turkey) and sold them for 15 pieces of gold each. The value of two silver pieces equaled that of one gold piece. If the trader had to pay a total of three silver pieces for each blanket to cover import/export taxes and other expenses, how much profit did he make from this deal? Give your answer in gold pieces.

Problem #7

In Turkey hundreds of ancient clay tablets that recorded business transactions and other information of the time have been uncovered from several archaeological sites. As of 1996, there were 1400 of these tablets. In the five preceding years the following number of tablets were uncovered: 100 (1992), 200 (1993), 110 (1994), 220 (1995), and 121 (1996). More tablets are expected to be unearthed. If the same recovery rate of tablets continues, how many tablets will there be by the year 2000? (When encountering decimals, round to the nearest whole number.)

Problem #8

Kali (Ka-le), a donkey herder in southern Babylon along the Tigris River, raises donkeys specifically to sell to caravan traders. In the past four years his female donkeys had given birth to 30, 35, 33 and 38 foals respectively. Meanwhile, Kali had sold 50, 54, 62, and 74 donkeys in those same four years. At that point he had 500 donkeys in his herd. Kali believed that a successful herder/seller of donkeys must always replenish his herd with new stock when the herd falls under 200 head. If the current trend of reproduction within his herd and sales of his herd would continue, how many years would it take before Kali had to buy more donkeys to add to his herd?

Problem #9

Referring back to Problem #8, Kali sold his donkeys to caravan traders for 25 silver pieces each and bought new donkeys to replenish his herd at 10 silver pieces a piece. After selling his herd to a point where he had less than 200 head left and buying new donkeys to replenish his herd, Kali made a profit of 4100 silver pieces. How many donkeys did Kali buy to add to his herd?

Problem #10

Donkeys used in caravan trade of ancient Mesopotamia could carry as much as 220 pounds provided the weight was distributed properly. One donkey was known to have carried the following trade goods:

(6) six pound ingots of tin
(7) twelve pound ingots of tin
(1) thirty-six pound pouch of diorite
(1) jug of sesame oil weighing 24 pounds
(8) large wool blankets at five pounds each
(**NOTE:** When folded these large blankets were 3'X7')

Use the drawing of the donkey below to decide how the load would be evenly distributed upon the animal's back.

Problem #11

Rimuk's army has just ransacked a village in northern Mesopotamia believing no challenging force is anywhere near them. Scouts for the army of Amal, whose has sworn vengeance on Rimuk for destroying this village, tell him Rimuk's army is 55 miles to the northwest. Rimuk's army is unconcerned about an attack and is traveling 3 miles per hour for twelve hours a day. Amal decides to set out after Rimuk by having his army undergo a forced march of 4 miles an hour for fourteen hours a day. How many days will it take for Amal's forces to overtake Rimuk's army?

Problem #12

Gilgu, clay sculptor in the city of Ur, makes praying statues that people use to petition their numerous Sumerian gods. He has decided to make a different praying statue to each of the various gods of Sumer. It has been an arduous task. He started eight years ago. Each of the first four years he made one third as many as he did this year. During the second four year period, each year he made two thirds of the number he made this year. This year Gilgu believes he made as many as he can in a year with 375 praying statues. He thinks he can finish in three more years. How many Sumerian gods are there?

Problem #13

Hammurabi, king of Babylon in the eighteenth century B.C., authored the famous law code that bears his name. Suppose Hammurabi took six months to codify Babylon's laws. He wrote twice as many laws in the second month as he did in the first month. In the third month he produced one-and-a-half as many laws as he did in the second month. During the fourth month, he wrote one-and-a-third as many laws as he did in the third month. In the fifth month he codified one-and-a-quarter as many laws as he did in the fourth month. During the last month he produced 66 laws which was 1.2 times as many as in the fifth month. How many separate laws were part of his code?

Problem #14

Sakid, Tela, Narum, and Akad were four young Sumerian students in a private school in the city of Nippur. They were working on their letters. Each of them practiced writing separate cuneiform letters. Some worked harder than others in their half-hour lesson. Sakid only wrote 30 letters while Tela produced five times that amount. Akad wrote ten more than three times as many as Sakid. Lazy Narum only wrote two-thirds as many as Sakid. Altogether they wrote one half of the Sumerian alphabet. How many characters were in the Sumerian alphabet?

Problem #15

The great emperor of Mesopotamia, Sargon of Akkad, was viewing a demonstration of a powerful weapon some 4300 years ago. An archer, Aziz, was told to shoot an arrow with his common bow as far as he could to the east. A second archer, Shulgi, using a composite bow was then told to shoot an arrow as far as he could to the east as well. Shulgi's arrow traveled three times as far as the one shot by Aziz. If the combined total of the two arrows was 3600 feet, Shulgi's arrow traveled the length of how many football fields?

Chapter 1 Solutions

Solution to Problem #1

Unless students understand advanced algebra, making an organized list of the possibilities with the various ingots will be their best option.

# of Sets	10# tin	5# tin	Total tin wgt.	5# copper	10# copper	Total Copper Wgt.	Total Wgt.
1	10	15	25	10	30	40	65
2	20	30	50	20	60	80	130
3	30	45	75	30	90	120	195
4	40	60	100	40	120	160	260
5	50	75	125	50	150	200	325
6	60	90	150	60	180	240	390
7	70	105	175	70	210	280	455
8	80	120	200	80	240	320	520
9	90	135	225	90	270	360	585
10	100	150	250	100	300	400	650
11	110	165	275	110	330	440	715
12	**120**	**180**	**300**	**120**	**360**	**480**	**780**

The caravan had 12 ten pound ingots of tin, 36 five pound tin ingots, 12 ten pound copper ingots and 72 five pound copper ingots.

Interesting note: Present-day Turkey was a major smelting region of the ancient Middle East because it had enough timber to support the fires required for that activity.

Solution for Problem #2

With some hard thinking and brainstorming amongst fellow classmates, students may be able to deduce the following:

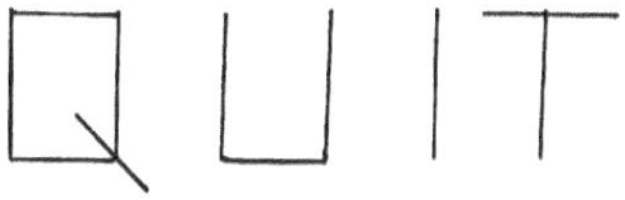

Interesting note: Sun-dried brick was the major building material of Mesopotamia. Stone and timber had to be imported from such areas as present-day Lebanon and the mountainous regions of Iran.

Solution for Problem #3

Dealing with a relatively small number such as 88 will afford students to make an organized trial and error list as follows:

Ramu's pay (1/3 of C)	Cavu's pay (3xR)	Sheck's Pay (1/2xC)	Total
10	30	15	55
12	36	18	66
14	42	21	77
16	**48**	**24**	88

Of course, students could apply basic algebra, with Ramu's earnings the basic variable (r).

r + 3r (*Cavu's earnings*)	+ 1.5r (*Sheck's earnings*)	= 88
5.5r = 88		
r = 16		
Ramu earned 16 measures of barley		
Cavu earned 48 measures of barley		
Sheck earned 24 measures of barley		

Interesting note: Barley was like a currency in ancient Sumeria. Not only was it used to pay wages, it was the staple source of rations for the military.

Solution for Problem #4

Using the measures that the workers earned in Problem #3, each worker's pay is multiplied by two to represent the loaves of bread that could be made from the barley. The number of loaves is then divided by the number of people in each family.

For Cavu, 48 X 2 =	96 loaves / 7 people =	13.7 loaves per person
For Sheck 24 X 2 =	48 loaves / 4 people =	12 loaves per person
For Ramu 16 X 2 =	32 loaves / 2 people =	16 loaves per person

Therefore, Ramu had the greatest number of loaves per person while Sheck had the least.

Interesting note: Sumerian barley was also the key ingredient in the world's first organized brewing of beer.

Solution for Problem #5

This problem dictates that students will have had some exposure to the various labor specialties alluded to within it. From that point logic and some hard thinking should deduce the following answer:

Obi is the potter. (A potter's wheel is constantly "spinning".)

Mek is the metal worker. (The "heat" is always on . . near the furnace.)
This is a Kur is the stone worker. (Rocks are being "cracked".)

Ban is the carpenter. (He knows the "angles".)

Ibin is the weaver. (Like the fabric in weaving, everything is "interrelated".

Shur is the farmer. (The development of farming allowed the others to specialize because one man could feed so many others.)

Interesting note: Other early vocations included beer brewing, baking, brick masonry, building, and writing (the keeping of records).

Solution for Problem #6

The trader laid out 400 silver pieces for the blankets (40 X 10) and 120 silver pieces for taxes and expenses (40 X 3) for a total expense of 520 silver pieces. Since he sold them for 15 gold pieces each, his gross income would have been 600 gold coins. However the expenses must be deducted to give the net income. To do this, the 600 gold pieces must be converted to a value of 1200 silver pieces. Subtracting 520 from 1200 leaves a net profit of 680 silver pieces or **340 gold pieces.**

Interesting fact: Babylonia was renowned for its wool industry. Wool blankets measuring approximately 13' X 15' were highly prized exports for that region.

Solution for Problem #7

Students need to identify the pattern to solve this one. Beginning with the 100 unearthed tablets in 1992 and the 200 recovered tablets in 1993 each subsequent even and odd year multiplies that original figure by one tenth. So 1994's total of 110 resulted from 100 X 1.1. 1995's total of 220 came about as 200 was multiplied by 1.1. Continuing this pattern goes as follows:

1996 = 121 (110 x 1.1)	1997 = 242 (220 X 1.1)
1998 = 133 (121 X 1.1)	1999 = 266 (242 X 1.1)
2000 = 146 (133 X 1.1)	
	1400 + 242 + 133 + 266 + 146 = **2187 tablets**

Interesting note: Half of the tablets found have been business letters while the other half has been constituted of legal documents. These clay tablets were sun-dried and then fire-baked expertly preserving them.

Solution for Problem #8

The pattern set for the previous four years is the key to this item's solution:

donkeys gained	*pattern*	*donkeys sold*	*pattern*	
30		50		
35	+5	54	+4	
33	-2	62	+8	
38	+5	74	+12	
From this point the pattern will continue for four years				
				Net Loss Head
36	-2	90	+16	-54
41	+5	110	+20	-69
39	-2	134	+24	-95
44	+5	162	+28	-118
	Total head loss via sales =			336

After four years Kali would be below 200 head with a total of 164 donkeys (500-336).

Interesting Note: Donkeys were the preferred beast of burden for this region especially when traveling through the rocky passes of the Anatolia Plateau of present-day Turkey and the Taurus Mountains of present-day Iran.

Solution for Problem #9

Using information garnered from Problem #8 is essential in this situation. Kali would have sold 336 donkeys over the next four years at 25 silver pieces each totaling 8400 silver pieces. If he made 4100 pieces in profit, it would be logical to deduce he had spent 4300 silver pieces for a new herd. At the price of 10 silver pieces a head (mention in Problem #8) he would have purchased **430 new donkeys.**

Interesting Note: Some Mesopotamia caravans had as many as 200 donkeys in them.

Solution for Problem #10

Students could cut out small geometric shapes out of construction paper—-rectangles for the tin ingots, an oval for the pouch of diorite, a circle for the jug of oil, an elongated curved rectangle for the blankets—- label their individual weights and use trial and error to approach this task. While the cut-outs may be helpful, securing the numerals that will offer a balanced load will be equally important.

The size of the blankets and their flexibility will dictate that they will be draped over the top of the remaining load equally distributing their weight in so doing. As the blankets make up 40 pounds, this leave 180 pounds to be equally divided on either side (90 pounds per side) of the animal.

One side will hold the 36 pound pouch of diorite, three 12 pound tin ingots, and three 6 pound tin ingots. The other side will hold the 24 pound jug of oil, four 12 pound tin ingots, and three 6 pound ingots.

Interesting note: Diorite is a dark rock of volcanic origin.

Solution for Problem #11

Students will find a table helpful in organizing the data behind this item's solution:

	MILES TRAVELED		
Day	*Amal*	*Rimuk*	*Distance Rimuk is ahead*
0	*0*	*0*	*55 miles*
1	*56*	*36*	*35 miles*
2	*56*	*36*	*15 miles*
3	*56*	*36*	*caught*

Amal's army will catch up to Rimuk's towards the end of the third day after the pursuit began.

Interesting note: Besides trade, the most common way to extend a group's resources in ancient Mesopotamia was through the use of military force. Exacting tribute from weaker neighbors was a worthy pursuit if the neighboring group had valuable resources and the aggressor was reasonably sure the cost in lives would not be too great. The geography of the Middle East, large stretches of flat land with few natural barriers, helped perpetuate this violent practice.

Solution for Problem #12

Students could construct a table to organize Gilgu's progress on a task that will span twelve years. Basically, the 375 statues made during the current year is the key figure. Each of the first four years he made 125 (1/3 of the current number); each of the second set of four years he made 250 (2/3 of the current number) and during the final four years he will make 375 yearly for a subtotal of 1500 and a grand total of **3000 Sumerian gods.**

Years at work	statues made	*subtotal*
1	*1/3 of 375 = 125*	*125*
2	" = *125*	*250*
3	" = *125*	*375*
4	" = *125*	*500*
5	*2/3 of 375 = 250*	*750*
6	" = *250*	*1000*
7	" = *250*	*1250*
8	" *250*	*1500*
9	*375*	*1875*
10	*375*	*2250*
11	*375*	*2625*
12	*375*	***3000***

Interesting Note: Praying statues were likenesses of individuals that could be placed at altars to represent that person's worship for a particular god.

Solution for Problem #13

Students need to start with the last month's total of 66 and work back towards the first month to obtain monthly totals which will offer the sum of Hammurabi's laws.

L x 1.2 = 66	1.2L = 66	L = 66/1.2	L = 55
L x 1.25 = 55	1.25L = 55	L = 55/1.25	L = 44
L x 4/3 = 44	4/3L = 44	L = 44/4/3	L = 33
L x 1.5 = 33	1.5L = 33	L = 33/1.5	L = 22
L x 2 = 22	2L = 22	L = 22/2	L = 11

11 + 22 + 33 + 44 + 55 + 66 = **231 laws**

Interesting note: Punishment under Hammurabi's Code ranged from fines to the loss of limbs to death. However, there were no prison sentences.

<u>*Solution for Problem #14*</u>

While students may logically surmise the total of the various students' efforts, a basic equation will be more than adequate solving this problem.

(30 + 5(30) + (3(30)+10) + 2/3(30))2 =	characters of alphabet
(30 + 150 + 100 + 20)2 = c	
(300)2 = c	
600 = c	

Interesting note: Only boys received formal education which would ready them to become scribes (official record keepers) or priests.

<u>*Solution to Problem #15*</u>

Students may use trial and error, however basic algebra will produce a simple equation:

d(distance of Aziz's arrow)	+ 3d(Shulgi's arrow)	= 3600 feet
4d = 3600 feet		
d = 900 feet		

Therefore, Shulgi's arrow traveled 2700 feet or 900 yards or 9 football fields in length

Interesting note: The composite bows of central and southwestern Asia were made of layers of bone, wood and animal sinew. The fact they could shoot arrows a half-mile could lead one to say they were the first war-tested missiles.

Chapter 2

Ancient Egypt

Problem #1

A farmer of ancient Egypt owned a triangular piece of land along the Nile River. The land was shaped like an isosceles triangle, 500 feet long on each of two sides with a 600 foot base. The base was parallel to the banks of the Nile. One year the flood waters just barely covered the entire triangle. During the first week that the waters started to recede, five feet of land (from the height of the triangle) became exposed the first day, seven feet the second day, nine feet on each of the third and fourth days, twelve feet the fifth day and fifteen feet each on the sixth and seventh day. At this rate, how many days will it take before the entire land surface of the triangle reappears?

Problem #2

Amut, an Egyptian peasant, leads an exhausting life. Working for a wealthy master, he has been manning a shadoof over the past five days emptying water from a small reservoir into irrigation canals for the master's fields. On the first day he poured twice as many gallons of water into the canal as he did on the second day. On the third day he emptied 125% more than the second day. On the fourth day he emptied one fourth more water than he did the first day while on the fifth day, Amut poured one and a half times the water of the second day into the canal. If Amut's five day total equaled 24,750 gallons of water, how many gallons of water did he divert each day into the irrigation canal with the shadoof?

Problem #3

An Egyptian trading mission has traveled to the land of Punt to barter with Chief Mombo. On their last trip, the Egyptians had successfully traded 150 copper spears for 100 leopard skins and 80 copper knives for 20 pair of elephant tusks. This time Chief Mombo demands more since leopard and elephant hunting has become increasingly dangerous for his men. He says the prices for leopard skins have doubled while tusk prices have tripled. If the Egyptians had planned on buying 150 leopard skins and 30 pairs of elephant tusks, how much will it cost them?

Problem #4

Marmez, an ancient Egyptian cosmetic merchant, has just received 308 pounds of cosmetic material. Three sizes of pottery jars contained new materials. There were jars with 20 pounds of animal fat; jars with 3 pounds of black mineral powder and jars with 5 pounds of green mineral powder. How many jars of each material made up the cosmetic order.

Problem #5

A wall painter was depicting the life and times of the Pharaoh Cheops within his tomb deep inside the Great Pyramid at Giza. The wall on which the man was painting was 17 feet wide and 35 feet high. On the first day of his work, he painted an area 3 feet by 3 feet; on the second day, an area 2 feet by 5 feet; on the third day, an area 3 feet by 4 feet. On the fourth day, once again, he completed an area 3 feet by 3 feet. This rate of completion remained the same for the entire time it took the painter to illustrate the wall. How many days did it take to finish painting the wall within Cheops' tomb?

Problem #6

Cheops' final monument, the Great Pyramid at Giza, contains 2,300,000 limestone and granite stones weighing an average of two and a half tons each. Reliable estimates say it took 20 years to build this pyramid. If work was conducted 10 hours a day, seven days a week, how many stones were delivered each and every hour to the building site?

Problem #7

A rock quarry in the eastern desert not far from the site of the Great Pyramid furnished much of the blocks needed for its completion. If a team of 32 men could quarry one block every 67 minutes (on average), how many men would have been employed in the quarry? Use the answer from Problem #6 to assist you.(Round to nearest whole numbers.)

Problem #8

A delivery team of 20 men pulled the blocks from the quarry to the building site. If it took one team 3 hours and 20 minutes to complete one successful delivery and return back to the quarry, how many men were needed to deliver the daily amount of blocks needed to complete the pyramid in twenty years. (Again see Problem #6 and round to the nearest whole numbers when calculating.)

Problem #9

Hieroglyphic inscriptions were placed in tombs and temples by a team of skilled, learned workers who undertook the job in a precise sequence. A pattern layer placed drawings of the inscriptions on the walls after which a stone carver came in a cut a first draft. A second stone carver "proofread" the work of the first and made necessary corrections while a wall painter came in last to highlight the message in color.

In one tomb Getti, Abdul, Ewam and Nul were a team of four that wrote the inscriptions. Nul finished before Getti, but he wasn't first. Ewam worked after Abdul and was the most experienced of the four. Getti had nightmares of being locked in the tomb forever. Match each man to the job he performed.

Problem #10

Namil, a tax collector, was reviewing the various accounts on which he had collected over the past week. From each of four farmers he had collected 6 percent of his necessary total. From a potter, he had collected 10 percent; a fisherman, 9 percent; a stone mason, 15 percent; a brick mason, 12 percent; a baker, 10 percent. Finally, from a brewer he had collected the most—-180 basketfuls of grain. How many basketfuls of grain had he collected from each of the others? What was his total collection for the week?

Problem #11

Farouk, a baker in the ancient city of Thebes, was bent over his flat sandstone grindstone feverishly grinding wheat into flour with his granite roller. Pain from a cavity-filled tooth tormented him as he prepared his next order of bread. It made him think about how many of his customers had worn teeth similarly filled with cavities.

Just then, Fasete, a sea trader who specialized in fruit trade along the coast of the Mediterranean Sea, came in. His big smile lit up the dingy bakery as it did twice a year when he sailed up the Nile to pick up barge loads of dates and figs. To Farouk it seemed he had the perfect set of teeth, as if the gods were giving him a new set with each visit. Fasete made his purchase and was on his way while Farouk ended his moment of self-pity. He would visit the dentist to have his tooth removed that very day.

Why were the teeth of many Egyptians worn out and cavity-filled?

Problem #12

A potter in the ancient capital city of Memphis was getting ready to organize his shop one morning before opening. He had to arrange the following items on three shelves which hung directly above each other and on the floor beneath them: unpainted eating bowls, painted eating bowls, unpainted four-foot tall vases, painted four-foot vases, several bowls of red paint, large vases of mud, unpainted two-foot vases, painted two-foot vases, and several bowls of yellow paint.

Only one kind of item was on each of the top two shelves while the lowest shelf had two kinds of items on it. All working materials were placed in the middle on the floor arranged left to right by brightness of color. Shorter items were toward the top while unfinished items were toward the bottom. When on the same level, unfinished items were to the left of finished items. Draw a picture of the potter's display accurately placing each of the items.

Problem #13

A fisherman and a baker bartered with each other in order to attain goods they each needed. In this situation the fisherman gave fish, ducks and baskets of figs to the baker who exchanged baskets of beans, onions and barley. The fisherman gave one-and-a-half times as many items to the baker as the baker gave to him. There were 50 items in the total trade. Within the fisherman's trade goods, there were twice as many fish as ducks. In the baker's trade goods, there were twice as many baskets of barley as baskets of beans. Finally there were three times as many baskets of figs and beans as there were baskets of onions. How many of each item (fish, duck, figs, barley, onions and beans) were involved?

Problem #14

Ancient Egyptians mastered the science of mummification as bodies were prepared for the afterlife. Funerary services were priced for the elite, the middle class and for the very budget conscious middle class. One mummification business had three bodies to prepare in one week. 1769 feet of linen wrap was used in the three proceedings. The elite mummy received three times the wrap of the budget mummy while the mid-priced mummy had three-fifths as much wrap as the high-priced corpse. How much wrap was used on each mummy?

Chapter 2 Solutions

Solution for Problem #1

Students need to attack this problem in two stages. First, they need to diagram or illustrate the problem's parameters.

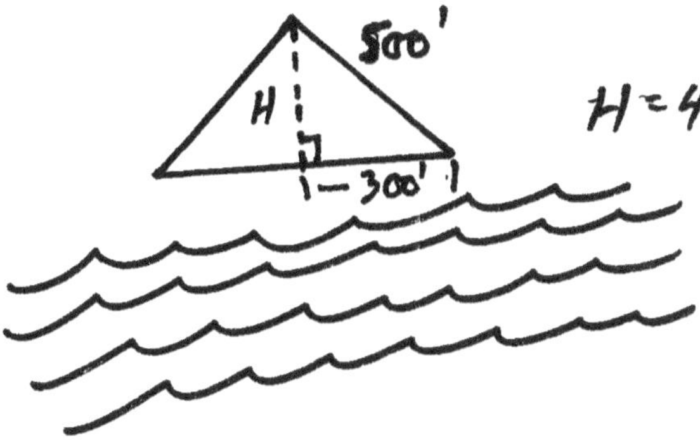

Using the measurements of the isosceles triangle, the height of the triangle may be deduced by figuring the Pythagorean Theorem (a2 + b2 = c2). In this case 300 squared + b2 = 500 squared. Since side b of the triangle in this equation happens to the height of the isosceles triangle, the height ends up being 400 feet.

Secondly, with the height calculated the pattern of water recession is the final clue necessary to obtain the answer. The first week's daily land exposure rates were given: 5,7,9,9,12,15,15. The pattern is +2,+2,0,+3,+3,0. To continue in this light would present the following figures for ten more days: 19,23,23,28,33,33,39,45,45,52. **After 17 total days, the entire 400 foot height of the triangle would be free of water.**

Interesting note: Every year in July the Nile would flood spreading a blanket of rich silt across the flood plain upon which farmers would produce the bounty to sustain Egypt. A year or succession of years without proper flooding would spell disaster for Egypt's people.

Solution for Problem #2

Students may attempt trial and error over a five day period to reach the 24,750 gallon total. Whether they guess or use basic algebraic equations, students must determine the base variable (the specific day in this problem around which most of the information centers).

The second day of Amut's labors happens to be the key. If we called day two *X*, here is how the other days relate to it:

Day One = 2X; Day Two = X; Day Three = 1.25X;
Day Four = 2.5X; Day Five = 1.5X

Having identified these relationships, students may insert various numbers at random trying to summon the 24,750 gallon total. Algebraically, the equation would be:

$$8.25X = 24750$$
$$X = 3000 \text{ gallons}$$

Therefore, on Day One Amut delivered 6000 gallons; Day Two, 3000 gallons; Day Three, 3,750 gallons; Day Four, 7500 gallons; Day Five, 4500 gallons

Interesting note: Along with maintaining irrigation canals, ancient Egyptian peasant farmers tilled the soil, harvested crops and worked on building projects for the pharaoh. The shadoof is still used in Egypt today to laboriously divert water into irrigation ditches.

Solution for Problem #3

Students will have to identify the rate of exchange to work this problem. A simple table may assist them in this.

	leopard skins	*spears*
previous rate of		
exchange	*100*	*150*
present rate of		
exchange	*100*	*300 (twice as much)*
expected price of		
present deal	*150*	***450***
	(50% more)	*(50% more)*

	Pairs of elephant tusk	*copper knives*
previous rate of		
exchange	*20*	*80*
present rate of		
exchange	*20*	*240 (three times more)*
price of present		
deal	*30*	***360***

Interesting note: The land of Punt was located along the "Horn of Africa" near present-day Somalia. Egyptians also traded copper tools and weapons for ebony, hunting dogs, baboons and incense.

Solution for Problem #4

Students could attempt trial and error with this problem with limited success. However, a more logical approach in the absence of advanced algebra would be to list possible weights of various numbers of each size jar and match up the three weights that total 308 pounds.

black powder		*green powder*		*animal fat*	
#/jars	Weight	#/jars	weight	#/jars	weight
5	15	5	25	5	100
6	18	6	30	6	120
7	21	7	35	7	140
8	24	8	40	8	160
9	27	9	45	9	180
10	30	10	50	**10**	**200**
11	33	11	55	11	220
12	36	**12**	**60**	12	240
13	39	13	65	13	260
14	42	14	70	14	280
15	45	15	75	15	300
16	**48**	16	80		
17	51	17	85		
18	54	18	90		
19	57	19	95		
20	60	20	100		

Marmez had 10 jars with twenty pounds of fat, 12 jars with five pounds of green mineral powder and 16 jars with three pounds of black mineral powder. (*Some other possibilities also exist.*)

Interesting note: The Egyptians ground mineral pigments into powder which was combined with animal fat to apply to the face. Black and green were the favorite cosmetic colors.

Solution for Problem #5

Students need to combine knowledge of basic area with the ability to identify the existing pattern within the problem to solve it. The large wall, measuring 17 feet by 35 feet has an area of 595 square feet. The completion pattern established in the first four days was 9 square feet, 10 square feet, 12 square feet, and back again to 9 square feet. So every three days, he completed a total of 31 square feet of wall.

595 square feet divided by 31 square feet gives 19 groups of 3 days with 6 square feet left over. **The wall painter will have worked 58 days** (19 X 3 with one extra day to complete the last 6 square feet) to finish painting the wall.

Interesting note: Cheops central tomb measured approximately 19 feet by 17 feet with a height of 35 feet. It is totally consumed by the 481 height of the entire pyramid. All that was left in the tomb when discovered by archaeologists was the red granite sarcophagus of Cheops (Khufu). Grave robbers had looted everything including the pharaoh's remains.

Solution to Problem #6

Students will need to do some protracted calculations for this item, so a calculator may be handy.

365 days X 20 years	= 7300 days
2,300,000 blocks/7300 days	= approx. 315 blocks a day
315 blocks/10 hours	= approx. **32 blocks an hour**

Interesting note: The entire weight of the Great Pyramid is estimated at 6,500,000 tons or 13,000,000,000 pounds set on a 13 acre base (or an area equal to seven city blocks).

Solution for Problem #7

One team of 32 men could quarry approximately 9 blocks a day (600 minutes/67 = 8.955.)

315 daily blocks/9	= 35 teams
35 teams X 32 men each	= **1120 men**

Solution for Problem #8

Again figure how many blocks one team of 20 men could deliver in a day:

10 hours/3 hours 20 minutes	= 3 blocks
315 daily blocks/3 per team	= 105 teams
105 teams X 20 men per team	= **2100 men**

Interesting note: Until recently, ancient estimates of 100,000 men working on the Great Pyramid for twenty years were considered gospel. However, excavations in the 90's have uncovered dwellings, bakeries and cemeteries employed by the workers. These workers were not slaves, rather artisans and peasants fulfilling what amounted to community service. Furthermore, archaeologists have simulated construction methods which have led them to believe the total of various work crews mentioned in problems 7 and 8 is a more accurate count of the labor actually employed. Total figures for all associated workers on the pyramid for Cheops probably was between 20,000 to 30,000 men.

Solution for Problem #9

Reason dictates that Getti's nightmares place him last in the sequence of four as the wall painter. Ewam's experience would place him as proofreader more than likely. Nul is before Getti and not first, so he would be the first carver while Abdul is first drawing the hieroglyphic pattern on the walls.

Interesting note: Most Egyptians could not read the "sacred writings" (what Greeks later named hieroglyphics). There was another script that utilized less illustrated sounds and symbols with which common people communicated.

Solution for Problem #10

The solution in this problem lies in identifying what percentage the brewer's tax of 180 baskets of grain equals. By adding the percentages of the other nine citizens, it happens to equal 80 percent. 180 baskets then equals 20 percent of the total amount collected. From there students may work backwards to identify the tax value for every other worker.

20/100	= 180/x
100(180)	= 20x
1800	= 20x
1800/20	= x
900	**= x (the total amount collected)**

Each farmer contributed .06(900)	**= 54 baskets each**
The potter chipped in with .10(900)	**= 90 baskets**
The fisherman gave .09(900)	**= 81 baskets**
The stone mason produced .15(900)	**= 135 baskets**
The brick mason contributed .12(900)	**= 108 baskets**
The baker gave .10(900)	**= 90 baskets**

Interesting note: Wheat and barley were the major grains with which taxes were paid. The food went to feed the pharaoh and nobility, but it was also judiciously stockpiled in storehouses to protect against possible future famine.

Solution for Problem #11

Students could be grouped to brainstorm possible causes for the dental problems of the ancient Egyptians—-too much sugar? poor dental hygiene? malnutrition? None of these was the case. Refined white sugar as we know it didn't exist; Egyptians used honey and date sugar while having a varied diet. They were even known to use various plant leaves to brush their teeth.

The secretive cause is actually within the problem. **Sand particles from the grindstone mixed in with the ground meal/flour and were consumed by the Egyptians gradually wearing out their teeth and causing cavities. In effect, they were eating small portions of sand with their bread!**

Interesting note: More often than not the local dentist was a person moonlighting from his chosen profession as a hair stylist.

Solution for Problem #12

All clues necessary to create the display exist within the problem. (Students may even use small objects to model the display.) The display would be as follows:

painted eating bowls

painted two-foot vases

unfinished eating bowls and two-foot vases*

unfshnd 4ft vases, yelw.pt,red pt, vases of mud, pnted 4ft vases

*These two items may be interchanged on either side of the shelf.

Solution for Problem #13

Students will need to combine a bit of algebra with trial and error to come up with this solution. For starters, it is critical that the amount of items each person brought to the trade is identified first. The fisherman had 1 1/2 times as many items:

1 1/2X + X	= 50
2 1/2X	= 50
X	= 50/2.5
X	= 20

The baker had 20 items in the trade while the fisherman had 30. From there students may play with the clues given to determine each man's trade goods. **The fisherman had 16 fish, 8 ducks and 6 baskets of figs. The baker traded 12 baskets of barley, 6 baskets of beans and 2 baskets of onions.**

Interesting note: While barter was the main means of economic activity (since there was no currency at the time), baskets of beans and onions often served the purpose as a form of common tender.

Solution for Problem #14

Students may use trial and error to work this problem, but it works best with a simple algebraic equation:

(X = amount of linen wrap used on budget mummy)	
3X + 1.8X(3/5 of 3) + X	= 1769 feet
5.8X	= 1769
X	= 1769/5.8
X	**= 305 feet for the budget mummy**
1.8X	**= 549 feet for the mid-priced mummy**
3X	**= 915 feet for the elite mummy**

Interesting note: Besides linen, natron, a salt-like substance abundant in the desert of Egypt, was used to dry out a corpse and prepare it for mummification. Note: The lengths of linen used in the problem are approximations only. Pharaohs were known to have had their bodies wrapped 20 to 80 times in the mummification process.

Chapter 3

The Ancient Near East

Problem #1

A Phoenician cargo ship sank approximately 1300 B.C. and was recently discovered by marine archaeologists. The hold of the ship (the lower level where cargo was stored) was pretty much intact. It was determined that from bow (the front of the ship) to stern (the back of the ship) there were nine separate sections holding five major types of merchandise—-stone anchors, pottery, copper ingots, tin ingots, and bluish glass. Copper occupied the most sections with pottery and anchors each taking up the second most number of sections. How many sections did each of these five cargos occupy?

Problem #2

(Use the information about the sunken Phoenician ship in Problem #1 to solve this problem.) The middle section of the cargo hold was filled with copper ingots while tin ingots made up the last section in the stern. All pottery sections were next to one copper section. The glass was sandwiched between sections of copper and pottery near the stern. All stone anchors were next to copper sections, the anchors were kept two sections apart from each other. Make a picture or diagram of the hold of the ship (looking as from above) and correctly label how the cargo was distributed from bow to stern.

Problem #3

To create hardened bronze a mixture of 90% copper and at least 10% tin was needed. Six tons of copper could help make 300 bronze helmets, 300 bronze corselets (front and back body armor), 3000 bronze short (18") swords and 3133 bronze spear points. If the corselet weighed five times as much as the sword, and the sword weighed half as much as the helmet but twice as much as the spear point, how much did each individual item weigh?

Problem #4

Joreb, a tailor in the Mediterranean coastal city of Tyre, had just finished a month of painstaking, tedious work. He had dyed 17 garments a royal purple by extracting 29,800 drops of purple dye from numerous shellfish known as murex. He had dyed head wraps, vests and robes. Head wraps took 1200 drops each, vests took 1000 drops each, and robes required 3000 drops of the precious dye. How many of each garment had he dyed?

Problem #5

A Phoenician trader of the ancient city of Byblos was set to hold a banquet for a couple dozen of his best customers. There was to be goat's meat, fish, barley bread, cheese, garlic, olives and figs. There was three times the amount of goat's meat as garlic. There was as great a quantity of olives as there was goat's meat and this was twice as much bread and figs that would be served. A combination of equal amounts of fish and cheese accounted for one-third of the total menu of 120 pounds of food. How much of each food was there?

Problem #6

Hazz was a Phoenician trader whose business was located on the harbor of the old city of Byblos. He had a problem. Workers had carelessly mixed imports from Egypt (papyrus, gold, turquoise, copper) with Phoenician exports for Egypt (cedar logs, glass, aromatic tree resin, pottery). His warehouse was a long rectangular building with four sets of side-by-side cargo stalls. There was an entrance on the dock side of the building closest to the export stalls and another on the opposite end by the four import stalls. Cedar logs and resin had to be moved to the end stalls. Glass was diagonally placed across from the cedar in front of the turquoise which was to the right of the copper. Gold was placed diagonally across from the copper. Pottery didn't have to be moved and papyrus was so light, moving it wasn't much to do. Explain where each cargo was to be placed in Hazz's warehouse.

Problem #7

During the period of time after the ancient Hebrews finished their Exodus from Egypt and before they were ruled by the monarchies of Saul, David and Solomon, they were a loose tribal confederation overseen by a series of twelve judges. The reign of four judges—-Othni-el, Ehud, Deborah, and Gideon—-took up two thirds of this time period. The peace secured by Ehud's rule was twice the time of the rule of the other three. The other eight judges ruled the remaining third of this time period. Abimilech for 4 years; Jephthah for 6% of the remaining third; Ibzan for 7%; Abdon for 8%; Elon for 10 years; Samson for 20%; Jair for 22% and Tola for 23% of the remaining third of time. How many years did each judge rule, and for how long a time period did the judges reign?

Problem #8

Moses was harried by all of the minor decisions he had to make regarding the Hebrew people he was delivering out of Egypt. His father-in-law gave him the following advice, "Find some capable,...men who hate bribes, appoint them as judges, one judge for each 1000 people; he in turn will have ten judges under him, each in charge of a hundred; and under each of them will be two judges, each responsible for the affairs of fifty people; and each of these will have five judges beneath him, each counseling ten persons." (Exodus 18: 21 *Living Bible*) If there were approximately 1,900,000 Hebrews in the wilderness with Moses, how many judges would have been appointed?

Problem #9

While wandering in the wilderness on their exodus from Egypt, the Hebrews built a tabernacle at which they would worship their God, Yahweh. They dedicated the altar with numerous sacrifices including 39 pounds of finished precious metals. There were two pound silver platters, one pound silver bowls and four ounce gold trays. They offered equal numbers of each item. How many platters, bowls and trays were used to dedicate the altar?

Problem #10

Solomon ordered that a grand palace be built for him and that a magnificent temple be built for the worship of Yahweh in the year 960 B.C. These construction projects required 80,000 stonecutters, 80,000 loggers, 140,000 common workers and 7,200 foremen. 20 stonecutters arrived on the first day and each day thereafter their numbers tripled from the day before until all had arrived. 36 loggers arrived on the first day and each day thereafter their numbers doubled from the day before until all had arrived. 120 common workers arrived on the first day and each day thereafter their numbers doubled from the day before until all had arrived. Finally, 10 foremen showed up on the first day and each day thereafter their numbers quadrupled from the previous day until all had arrived. How long was it before *everyone* that was needed for the projects had arrived?

Problem #11

King Solomon was an extremely wealthy monarch. The Bible says that "...silver and gold were as rocks on the street" during his reign. He sent traders to Egypt to buy the latest war technology—-chariots and horses to pull them. Chariots supposedly cost 600 shekels of silver while each horse would cost 150 shekels of silver. The Hebrew traders spent 1,065,000 silver shekels. They brought back 100 extra horses to go with the horse-drawn chariots. How many chariots and horses did they obtain?

Problem #12

Assyrian King Ashurbanipal gathered cuneiform tablets from his world-renown library in Nineveh. For personal study one evening he had his servants collect some tablets on Mesopotamian history which were 3/4 of an inch thick each. He also ordered astronomy tablets which were 1/2 inch thick each. Finally he also had brought to him some official documents that were on tablets 1 and 1/4 inch thick. Altogether, his stack of reading material was 18 inches high and consisted of 22 tablets. How many of each type of tablet did Ashurbanipal take?

Problem #13

Seven soldiers of the Assyrian army were chosen to impress the king with their various military skills. Two cavalry riders, three archers and two charioteers were selected. Three of these men were Assyrian while two were of Babylonian descent and two were Hittites. Their names were Senna, Karek, Ahiz, Ravi, Boz, Atalya, and Tola. Senna and Ahiz shared the same skill while Ravi, Karek, and Atalya all had different skills. Tola and Boz were of the same ethnic background while Karek, Ravi and Senna each came from a different background. Ravi suffered from saddle sores and did not perform well. Karek was depressed that only one other person knew his language. Which one was an Assyrian archer?

Problem #14

Rasu is the officer responsible for maintaining the strength of the Persian Empire's elite fighting force, the Immortals, at no less than 10,000 men. Whenever a casualty occurs, Rasu must be sure that the total number of men in the unit is no less than 10,000. Currently the Immortals have 11,800 men. In each of the last five months there have been the following number of casualties: 156, 180, 155, 190 and 160. At this casualty rate, how much time will pass before Rasu must add more men to the Immortals?

Problem #15

Beset by an opposing army, bandits and storms seven Persian couriers were delegated to take a message from Susa, near the Persian Gulf, to Sardis in western Asia Minor. The first six riders covered 21%, 16%, 17%, 15%, 13% and 10% of the total distance on the Royal Road. The last courier rode 128 miles. How long was the combined distance covered by these riders?

Chapter 3 Solutions

Solution to Problem #1

With a small bit of juggling the five cargos amongst nine sections (along with the allotted clues), students should rather easily determine that there were **3 sections of copper ingots, 2 sections of pottery, 2 sections of stone anchors, 1 section each of tin ingots and bluish glass.**

Interesting note: This cargo was part of the oldest shipwreck ever uncovered when it was found off the southwest coast of Turkey in the 1980's.

Solution to Problem #2

Using the various pieces of information presented in the problem, students should come up with the following diagram:

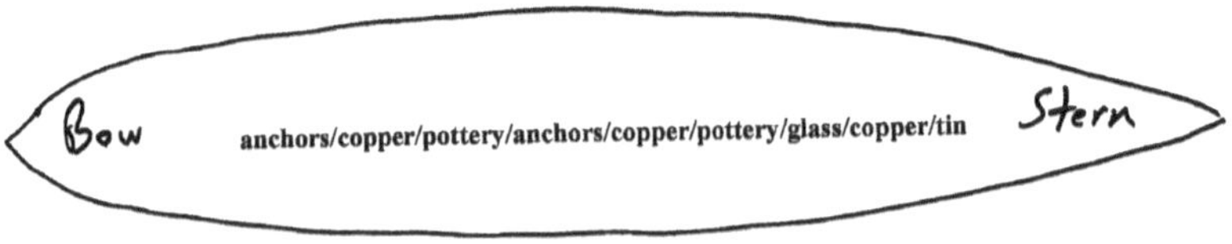

(Students may also use small objects to model the correct positions.)

Interesting note: Other cargo that was interspersed among the main items included ivory (hippo and elephant tusks), bronze swords and arrows, and a pleasant-smelling tree resin used in the Egyptian mummification process.

Solution for Problem #3

Students must first deduce the total weight of material that went into the making of the 6733 military items. 12,000 pounds (six tons) of copper were apparently needed, and this would have dictated the addition of 1333 pounds of tin to make the necessary total of 13,333 pounds of bronze. *(The actual figure will be 1333.333. Have students round to the nearest whole number.)*

From this point, students may make a simple table that lists all the military products and labels them with the proper mathematical relationships according to their weight. The corselets were five times the weight of the swords which were half the weight of the helmets and twice the weight of the spear points. Therefore, the spear points are the lightest item (X); the swords are second lightest (2X); the helmets (4X) and finally, the corselets (10X).

300 helmets	300 corselets	3000 swords	3000 spear points
4X	10X	2X	X

With a bit of trial error students will realize substituting 1 for X will result in the correct figures:

spear point	*1 pound x 3,133*	*= 3,133 pounds*
sword	*2 pounds X 3,000*	*= 6,000 pounds*
helmet	*4 pounds X 300*	*= 1,200 pounds*
corselet	*10 pounds X 300*	*= 3,000 pounds*
		Total = 13,133 pounds

Interesting note: The Bronze Age lasted from about 3800 B.C. when first bronze was smelted in ancient Iran till 1000 B.C. when the superior strength of iron displaced bronze's military importance.

Solution for Problem #4

In attacking this problem, students will serve themselves well by setting up a list of each type of garment and the various multiples of drops needed to dye numerous quantities of each.

Number of garments	*drops of dye vest*	*drops of dye head wrap*	*drops of dye robe*
1	1000	1200	3000
2	2000	2400	6000
3	3000	3600	9000
4	4000	**4800**	12000
5	5000	6000	15000
6	6000	7200	**18000**
7	**7000**	8400	21000
8	8000	9600	24000

The seventeen garments were comprised of 7 vests, 4 head wraps, and 6 robes.

Interesting note: The Phoenicians at Tyre specialized in purple dye due to their utilization of the murex shellfish. The meticulous nature of this process, taking only a few drops out of the animal at a time, made it very expensive. It was so extravagant that the purple color was synonymous for royalty thereafter because few members of society beyond the ruling class could afford it.

Solution for Problem #5

Students need to break the problem into two distinct entities and use basic algebra to solve one of the two portions. To begin, equal amounts of cheese and fish make up one-third of the 120 pound total.

1/3(120)	*= 40*
40/2	*= 20*

There was 20 pounds of cheese and 20 pounds of fish.

The remaining 80 pounds will be placed in an algebraic equation according to the clues given:

X =garlic;3X = goat's meat;3X = olives;1.5X = bread; 1.5X = figs

(As in any basic equation, all portions relate to one variable. In this case the smallest unknown is the amount of garlic. There was three times the amount of goat and olives as there was garlic. Yet the amount of goat and olives was only twice the amount of bread and figs. Therefore, there was 3/2 or 1.5 the amount of bread and figs as there was garlic.

3X + 3X + 1.5X + 1.5X + X	*= 80 pounds*
10X	*= 80*
X	*= 8*

There were 8 pounds of garlic, 24 pounds of goat's meat, 24 pounds of olives, 12 pounds of bread, and 12 pounds of figs.

Interesting note: The ancient city of Byblos (first known as Gebeil) is considered one of the first cities ever built dating back to 7000 B.C.

Solution for Problem #6

	back entrance	
papyrus		gold
copper		turquoise
pottery		glass
cedar logs		tree resin
	dockside entrance	

Interesting note: Byblos gave its name to the Bible because the papyrus from the city was used for Greek translations of the Hebrew prophets.

Solution for Problem #7

How many years were included in the final third of the judges' rule needs to be answered first. Abimilech's and Elon's reign account for a combined 14 years while the total percent of the remaining six judges total 86% of the final third time period. This means each of the percentages allotted for the remaining six actually equals the years they ruled since 14 years plus 86% would give 100 years. So **Abimilech ruled for 4 years; Jephtah for 6; Ibzan for 7; Abdon for 8; Elon for 10; Samson for 20; Jair for 22 and Tola for 23 years.**

Since 100 years was a third of the total reign by the judges, the two-thirds encompassed by the rule of Othni-el, Ehud, Deborah and Gideon would be 200 years for a total reign of **300 years.**

Using clues given the following equation can solve the specific number of years each of these four judges had control:

2X + X + X + X	= 200 (X=rule of Deborah, Gideon and Othni-el)
5X	= 200
X	= 40

The rule of Deborah, Gideon and Othni-el was 40 years each, while Ehud's was 80 years.

Interesting note: A thirteenth judge is mentioned in the Book of Judges in the Old Testament. The only thing mentioned about Shamgar, however, is that at one time he killed 600 Philistines with an ox goad!

Solution for Problem #8

For a thousand people there would have to be 131 judges (counselors) according to the breakdown given in Exodus 18.

1 per thousand + 1 per hundred (10) + 1 per fifty (20) + 1 per ten (100) = 131. Multiplying 131 by 1,900, the answer is **248,900 judges.**

Interesting note: The population figure given in the problem is an estimate based on the census taken in Numbers Chapter 1 by Moses of all males over twenty years of age able to fight in the army. That number totaled 603,550 for all the tribes of Israel excluding the priestly Levites who had another 22,000 males above the age of one month.

<u>*Solution for Problem #9*</u>

While some students may find success with trial and error on this problem, using a table to organize the data given will be a very efficient way to solve it.

NUMBER OF	*2LB.*	*1LB.*	*4OZ.*	*TOTAL WEIGHT*
ITEMS	*PLATTERS*	*BOWLS*	*TRAYS*	
1	*2lb.*	*1lb.*	*4oz.*	*3lb. 4oz.*
2	*4lb.*	*2lb.*	*8oz.*	*6lb. 8oz.*
3	*6lb.*	*3lb.*	*12oz.*	*9lb. 12oz.*
4	*8lb.*	*4lb.*	*1lb.*	*13lb.*
5	*10lb.*	*5lb.*	*1lb. 4oz.*	*16lb. 4oz.*
6	*12lb.*	*6lb.*	*1lb. 8oz.*	*19lb. 8oz.*
7	*14lb.*	*7lb.*	*1lb. 12oz.*	*22lb. 12oz.*
8	*16lb.*	*8lb.*	*2lb.*	*26lb.*
9	*18lb.*	*9lb.*	*2lb. 4oz.*	*29lb. 4oz*
10	*20lb.*	*10lb.*	*2lb. 8oz.*	*32lb. 8oz.*
11	*22lb.*	*11lb.*	*2lb. 12oz.*	*35lb. 12oz.*
12	*24lb.*	*12lb.*	*3lb.*	*39lb.*

There were 12 platters, 12 bowls and 12 trays dedicated at the altar.

Interesting note: 252 head of livestock (goats, rams, and bulls) were sacrificed as burnt offerings to Yahweh at the dedication of this tabernacle.

Solution for Problem #10

Each specific worker's pattern has to be worked out to see how many days it took to round up the entire lot.

1Worker/Day#	#1	#2	#3	#4	#5	#6	#7	#8	#9	#10	#11
stonecutter	20	60	180	540	1620	4860	14580	43740*			
logger	36	72	144	288	576	1152	2304	4608	9216	18432	36864**
common laborer	120	240	480	960	1920	3840	7680	15360	30720	61440	122880***
foremen	10	40	160	640	2560****						

***With 43740 stonecutters having arrived by the eighth day, all stonecutters would be in place by Day #9.**

****With 73692 loggers having arrived by the twelfth day, all loggers would be in place by Day #13.**

*****With 122,880 common laborers having arrived by the eleventh day, all laborers would be in place by Day #12.**

******With 3410 foremen having arrived by the fifth day, all foremen would be in place by Day #6.**

It would be 13 days before all workers were in place, ready to start the project.

Interesting note: According to 2 Chronicles Chapter 2 in the Bible, of the grand total of 307,200 workers that Solomon had gathered, exactly half (153,600) were conscripted foreigners living among the Twelve Tribes of Israel.

Solution for Problem #11

Students will need to establish an algebraic equation unless they want to use trial and error.

X = number of chariots bought	
600X + 150(X + 100)	= 1,065,000
750X + 15000	*= 1,065,000*
750X	*= 1,050,000*
X	*= **1400 chariots***
X + 100	*= **1500 horses***

Interesting note: 600 shekels would equal 15 pounds of silver while 150 shekels would be about 4 pounds of silver. Using $5 an ounce of silver as comparison, each chariot would have cost $1200 and each horse would've cost $320 at our current rate of exchange.

Solution for Problem #12

While students make take a chance on trial and error, an organized list of the three types of tablets is the most warranted method. Converting the fractional thickness to decimals may be a preferable option.

Type of Tablet	history	astronomy	documents
tablets/thickness	1 .75"	1 .5"	1 1.25"
	2 1.50"	2 1.0"	2 2.50"
	3 2.25"	3 1.5"	3 3.75"
	4 3.00"	4 2.0"	4 5.00"
	5 3.75"	5 2.5"	5 6.25"
	6 4.50"	6 3.0"	6 7.50"
	7 5.25"	7 3.5"	7 8.75"

There were 7 tablets on Mesopotamian history, 8 tablets on astronomy and 7 tablets dealing with official documents.

Interesting note: Ashurbanipal's library housed 22,000 cuneiform tablets written in Sumerian, Babylonian and Assyrian languages. They also included works on botany, math, letters and dictionaries.

Solution for Problem #13

While all problem solving relates to reasoning in one way or another, this problem lends itself almost solely to logic. Since Senna and Ahiz share the same skill, and Ravi, Karek, and Atalya each have a separate skill, Senna and Ahiz must be archers. Archery was the only skill represented by three soldiers.

Similarly, Tola's and Boz's common ethnic background must be Assyrian because three other soldiers mentioned each had a unique background and only one region (Assyria) had three men represented. Ravi, with saddle sores, would be in the cavalry and not be an archer. The fact that Karek has only one other person to converse with would indicate that he is not one of the Assyrians. **Therefore, Senna is the Assyrian who is an archer.**

Interesting note: In the seventh and eighth century B.C. the Assyrian military was the mightiest war machine in the Near East conquering lands from the Mediterranean to the Persian Gulf. Besides their fierceness, units of infantry, cavalry, archers and charioteers outfitted with iron weapons held superior technological advantage against most contemporary foes.

Solution for Problem #14

There is a distinct pattern that denotes the casualty count over the previous five months. +24,-25,+33,-30 are the spans between the monthly totals. As this pattern is extrapolated, it would continue as follows:

184, 159, 192, 162, 186, 161, 194, 164, 188, 163

Therefore, after ten months 1753 additional casualties would have been incurred forcing Rasu to add more men during the eleventh month.

Interesting note: The Immortals were the best fighters within the powerful Persian Empire of the fifth and sixth century B.C. They always led the remainder of the army into battle.

Solution for Problem #15

Students need to total the percentages of the distance covered by the first six riders to identify that the 128 miles covered by the last courier was 8% of the total distance. A simple proportion will allow the solution:

8/100	= 128/X
128(100)	= 8X
12800	= 8X
1600	= X

The Royal Road was 1600 miles long.

Interesting note: Before the Royal Road was created, the 1600 mile distance took three months to travel. Afterwards, with it guarded by the Persian Empire, a system of 100 relay stations allowed postal couriers to cover the entire course in 15 days!

Chapter 4

Ancient Greece

Problem #1

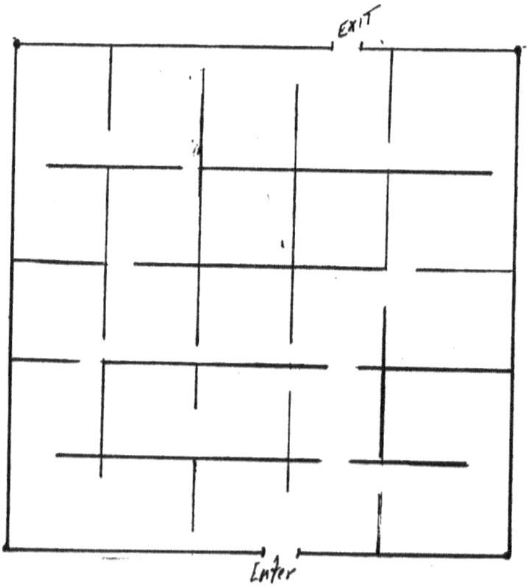

Problem #2

As of 708 B.C. the pentathlon was one of the events in the ancient Olympics. It consisted of five events—-running, wrestling, discus, javelin and long jump. The discus came before wrestling. The long jump was held after the discus, but it wasn't last. Running was held before the discus, but it wasn't first. What was the order of the five events?

Problem #3

Milo of Kroton was perhaps the most famous wrestler of the ancient Olympic era. He attempted to win his sixth consecutive Olympic wrestling title at the age of forty when he was finally defeated by a younger competitor. Since his titles came in succession, how old was Milo when he won his first Olympic wrestling competition?

Problem #4

Bull leaping was a pre-Olympic sport of ancient Crete. A man would stand on a bull, do a back flip off the bull and be caught by a female accomplice. In one religious festival five pairs of bull leapers competed. The best out jumped the worst by a distance twice as much. Second place was three meters better than third and four meters better than fourth. The total for all five competitors' jumps were fifty meters. How far did each competitor jump?

Problem #5

The ancient Olympic four-horse chariot race was held in Olympia in the hippodrome a wide, level, oval race course. The race entailed a little more than 18 laps of the oval. Each race started with 40 contestants. In one race the winner covered 14000 meters (14 kilometers; about 9 miles) while the tenth place finisher covered over 17000 meters (17 km.; about 10 1/2 miles). Each man completed the mandatory 18 laps and no more. Why was there such a major difference in how far they raced?

Problem #6

Dimos, a nobleman who grew olives on his large estate near the city of Corinth, was reviewing the sales accounts of the past year. He sold his crop in the form of olive oil in 20 gallon kegs. He had sold 16% of his previous year's oil to a Spartan merchant who put it in lamps for fuel. 20% of Dimos' sales went to an Asia Minor merchant. An Athenian cosmetic merchant who used the oil for soap bought 18% of his oil. Finally, a local Corinthian merchant bought 2070 kegs for cooking oil and for eating with bread. How many total gallons of olive oil did Dimos sell the previous year? Which merchant was his best customer?

Problem #7

Aristhenes celebrated his seventh birthday on the day he entered the military academy in Sparta. He was somewhat overweight, and his father knew that the academy would be the best place for his son to shed his baby fat and become the hardened soldier every Spartan father wanted his son to be.

The academy leader assured Aristhenes' father that in two days the boy would undergo his first physical inspection for body fat. In Aristhenes case, there would be plenty of running afterwards to decrease his fat content. These inspections would occur every ten days thereafter like clockwork. By the time Aristhenes enjoys his eighth birthday, how many times will his physique have been inspected for its fat content?

Problem #8

Spartans, those citizens who could trace their ancestry to the original inhabitants of the city, actually made up a distinct minority of that city-state's population. Helots, enslaved people who worked the region's farms, had about 12 times more people as the Spartan citizenry. Meanwhile, foreigners, who served as artisans and merchants, had a population 8 times greater than the Spartans. If the total populations of this ancient Greek city-state was 445,200, how many people were in each class?

Problem #9

Democracy in ancient Athens focused around males over the age of 20. Out of a total population of some 200,000, only about 50,000 were considered citizens as women, slaves and anyone whose ancestry could not be traced back to the original families of Athens would not be viewed as a citizen. Daily governmental control was exercised by the Council of Five Hundred. Each year, 500 citizens had their names drawn by random lot from a large vase, and they would run the government for that year. What was the probability that any particular citizen would be chosen for the Council of Five Hundred? If the average life expectancy of an Athenian male was 60 years of age, what was the probability that he would be chosen for the Council of Five Hundred once in his lifetime?

Problem #10

In a recent presidential election in the state of Ohio 65.8% of that state's registered voters cast ballots. At the time, there were 6,879,687 registered voters in Ohio. There were 8,247,986 voting-age adult citizens in Ohio. Use a calculator to determine what percentage of eligible adult citizens actually voted.

Problem #11

A captain of a Greek trireme looked over the roster of men on his ship. Out of every 40, 4 were officers and soldiers of upper class Athens; 18 had grown up along the sea and had tried their hand at fishing; 12 had been poor city dwellers; 6 had been farmers. The captain was keen to the fact that all told he had 90 men who had grown up along the sea. How many men of each background were there, and what was the total number of men on the ship excluding the captain?

Problem #12

The overseer of rowers on a Greek trireme warship was closely inspecting the rowing technique of nine rowers in one section of the ship. There were three rowers each on three separate levels (upper, middle and lower). Petri was above Balu and Antones but below Axos and Dimos. Cozen and Balu were in the middle of their levels. Stratos was above Cozen but not in the middle of his level. Leonidas was on the bottom two levels directly below Dimos who was not directly above Petri. Antones was below Marn. Correctly diagram the location of the nine rowers.

Problem #13

Alexander the Great perfected the infantry tactic known as the phalanx whereby eight rows of soldiers each armed with a long spear and shield would attack in unison at a set enemy position. Each row may have held 125 soldiers. Assume that the phalanx forces included men from Macedonia, Corinth, Athens and Thebes. Further assume that each 1000 man phalanx was divided into fifty men from one city-state beginning with the first fifty from Macedonia, the second fifty from Corinth, the third group of fifty from Athens and a fourth group of fifty from Thebes. Under this pattern what city-states would have been represented in the last row of the phalanx, and how many men from each city-state would there have been in the last row?

Problem #14

In 334 B.C. Alexander the Great was ready to invade the Persian Empire through Asia Minor (modern Turkey). Out of every 7000 man unit in his army, he had 1000 cavalrymen, 2400 archers, 1200 javelin throwers, 600 members of his personal Royal Guard, and 1800 spear-wielding men of the phalanx force. If the total number of men in the phalanx forces were 9000, how many men did he have altogether, and how many of each kind did he have?

Problem #15

Near the Greek colony of Syracuse, on the island of Sicily, a ship was loading wheat and barley for export to the Grecian homeland. There were 75 pound bags of grain and 200 pound barrels of grain. There were a combined total of 67 bags and barrels. Of these, there were three more bags and/or barrels of wheat than barley. The total weight of the grain was 8275 pounds. How many bags and barrels of each grain were there?

Chapter 4 Solutions

Solution to Problem #1

The maze presented here is to simulate the mythical Labyrinth on the island of Crete where the evil King Minos housed the horrendous half-man, half-bull ,Minotaur. The instructor should make two sorts of copies of the maze. One version should be the actual maze while the other should have the central maze blotted out with a white square of paper so that the copies only show the four exterior sides of the maze.

Students are paired off; one with the actual maze, one with only the four exterior sides showing. Using rulers, students with the copies of the actual maze will try to guide their partner through the correct corridors of the maze in a limited time period, say four minutes. (The students should be separated by an upright notebook or textbook.) Using light bond white paper or having a transparency of the completed maze will allow for easy checking when student pairs finish.

By not actually observing the maze and relying solely on the verbal instructions of their partner, students most closely simulate the problem-solving dilemma that awaited all those that entered Minos' maze (minus the Minotaur, of course).

Interesting note: There actually was a five-story castle on ancient Crete that had a maze-like entrance ways and interior passages.

Solution for Problem #2

The order of events would be as follows: **1. javelin 2. running 3. discus 4. long jump 5. wrestling**

Interesting note: Through ancient accounts, all that is known about the sequence of these events within the pentathlon is that wrestling was the final event.

Solution for Problem #3

At forty Milo was going for his sixth consecutive Olympic wrestling victory. Since the games were held every four years, **his first victory would have come at age 20.**

Interesting fact: Milo of Kroton was a man of legendary exploits. Supposedly his strength was so great that he at one time carried a cow on his shoulders!

Solution for Problem #4

Unless one has the aid of advanced algebra, the relatively low total distance allows this problem to be solved through simple trial and error. Students may begin with a figure of 20 meters and work downward in order to deduce the following answers:

1st place	14 meters
2nd place	12 meters
3rd place	9 meters
4th place	8 meters
5th place	7 meters

Interesting note: Back flipping off the bull seems rather dull when compared to how male contestants got on the bull. As the bull charged them, they grabbed its horns and used the bull's momentum to somersault onto its back!

Solution for Problem #5

The inside perimeter of the hippodrome was a stone or wooden partition that measured 769 meters. Meanwhile the outer perimeter measured twice that distance—-1538 meters. **The tenth place driver obviously started with the distinct disadvantage of an outside position**. With 40 drivers and 160 horses jockeying for inside position, numerous collisions were bound to occur. Note that the pole position in modern auto races, while not guaranteeing victory, does give an initial advantage to the driver who holds it.

Interesting note: The ancient Greeks also held Olympic chariot races involving two horses, two mules, two foals, and four foals.

Solution for Problem #6

Students need to account for all the information given in this problem first. All sales are according percentages of Dimos' total yearly take except the final dealing with the Corinthian merchant where a total number of kegs is given. The other three merchants equal 54% (16+20+18) of

Dimos' sales. Therefore, **the 2070 kegs sold the Corinthian must be 46% of his annual sales figure and therefore his best customer.**

His total sales would be calculated as follows:

46/100	*= 2070/X*
20700	*= 46X*
X	*= 4500 kegs*

4500 kegs times 20 gallons each equals **90000 total gallons of olive oil.**

Solution for Problem #7

While with a cursory look at this problem students might quickly deduce an answer of 36 times since there would be 363 days until Aristhenes next birthday, they need to plot out the 10 day pattern to arrive at the actual answer of **37 inspections.**

Interesting note: At age 20 Spartan boys utilized their previous fourteen years of training in becoming soldiers. They would be devoted to the military until the age of 60.

Solution for Problem #8

While students could attempt trial and error to eventually arrive at the answer, this problem begs for a basic algebraic equation:

(X = Spartan population)	
X + 12X + 8X	*= 445,200*
21X	*= 445,200*
21X/21	*= 445,200/21*
X	***= 21,200***
8X	***= 169,600 foreign merchants/artisans***
12X	***= 254,400 helots***

Interesting note: Around the eighth century B.C. Spartan rulers decided that severely oppressing their conquered masses was a better option than trying to include them within their society. This was the birth of their military state.

Solution for Problem #9

Students need to calculate basic probability. 500 names being drawn at random from a total of 50,000 would be a probability of **1 out of 100** (in its simplest form). At 60 years of age, an Athenian male would have been part of the Assembly (of citizens) for forty years. Therefore, his overall probability of having been chosen once for the Council of Five Hundred would have been 1/100 x 40 = **40 out of 100 (or 4 out of 10)**.

Interesting note: Major decisions, such as going to war, were made by the Assembly which included all male citizens. No man could serve on the Council of Five Hundred more than twice.

Solution for Problem #10

.658 X 6,879,687	= 4,526,834
4,526,834/8,247,986	= .5488 . .

Approximately 54.9% of eligible Ohio adults took advantage of their right to vote in the 1996 presidential election.

Interesting note: Compare this figure to the involvement of Greeks in Athenian democracy. Instructors may contact their local and state boards of election to cite similar statistics on a more localized level

<u>*Solution for Problem #11*</u>

Students may assist themselves with the construction of a table for this problem.

Total # = of men	*40*	*80*	*120*	*160*	***200***
Soldiers=	*4*	*8*	*12*	*16*	***20***
city dwellers=	*12*	*24*	*36*	*48*	***60***
Former = fishermen	*18*	*36*	*54*	*72*	***90***
Farmers=	*6*	*12*	*18*	*24*	***30***

There were 200 men altogether—-90 former fishermen, 20 soldiers, 60 city dwellers, and 30 farmers.

Interesting note: The 180 men that manned the three levels of oars on each side were generally poor citizens looking to make better wages. Officers and soldiers were usually from wealthier families.

Solution for Problem #12

Stratos	Axos	Dimos
Petri	Cozen	Marn
Antones	Balu	Leonidas

Interesting note: Athenians built triremes with silver from nearby mines. They had the largest navy in all of Greece.

Solution for Problem #13

Students need to plot out the pattern beginning with 50 Macedonians, 50 Corinthians and 25 Athenians in the first row followed by 25 Athenians, 50 Thebans, and 50 more Macedonians in the second row. **In the final row there will be 25 Corinthians, 50 Athenians and 50 Thebans.**

Interesting fact: The spears held by the phalanx were known as sarissa. They were reportedly anywhere from 14 to 20 feet long with iron tips. When properly trained, the phalanx force could defeat an oncoming enemy with ease since their sarise were so long!

Solution for Problem #14

For many students a table will be the most concrete method of envisioning a solution to the number of Alexander's armed forces.

	Total # of men				
Unit type and number	*7000*	*14000*	*21000*	*28000*	***35000***
cavalry	*1000*	*2000*	*3000*	*4000*	***5000***
archers	*2400*	*4800*	*7200*	*9600*	***10800***
javelin throwers	*1200*	*2400*	*3600*	*4800*	***6000***
Royal Guard	*600*	*1200*	*1800*	*2400*	***3000***
spear handlers of the Phalanx	*1800*	*3600*	*5400*	*7200*	***9000***

While the extreme right hand column of the table does give the correct answer, some students may be able to denote the five-fold increase of the phalanx force and transfer that concept to each unit in surmising the total number without the use of a table.

Interesting note: Amongst Alexander's objectives in conquering the Persian Empire was the fulfillment of his late father, Philip's, wish to defeat the empire that had invaded Greece almost two centuries earlier and to obtain needed wealth to pay his own army.

Solution for Problem #15

Organizing the multitude of information into a workable table is critical in solving this problem. One clue is that there is a combined total of bags and barrels of wheat that equals 35 while the barley totals 32 bags and barrels. (Wheat bags/barrels were three greater than those of barley.) The table starts at 10 bags or barrels.

#	*pounds/ wheat bags*	*pounds/ wheat barrels*	*pounds/ barley bags*	*pounds/ barley barrels*
10	*750*	*2000*	*750*	*2000*
11	*825*	*2200*	*825*	*2200*
12	*900*	*2400*	*900*	***2400***
13	*975*	*2600*	*975*	*2600*
14	*1050*	***2800***	*1050*	*2800*
15	*1125*	*3000*	*1125*	*3000*
16	*1200*	*3200*	*1200*	*3200*
17	*1275*	*3400*	*1275*	*3400*
18	*1350*	*3600*	*1350*	*3600*
19	*1425*	*3800*	*1425*	*3800*
20	*1500*	*4000*	***1500***	*4000*
21	***1575***	*4200*	*1575*	*4200*

By working with the various addends of 35 and 32 students will be able to coax the answer out of the table—-**There were 21 bags of wheat, 14 barrels of wheat, 20 bags of barley and 12 barrels of barley.**

Interesting note: Over time Greece became very dependent upon imported grain from colonies in the western Mediterranean such as Sicily. Grain production in Greece, while never very profitable, was virtually eliminated by colonial importation.

Chapter 5

Ancient Rome

Problem #1

About dawn several individuals, who were part of a Roman legion in southern Britain, were rising early determined to make a positive impression on the new commander of the legion. Each wanted to be ready to greet the commander at the site of a new bridge project along the river Thames. The surveyor arrived before the architect but after the centurion. The stone mason came after the stone cutter who was upset that he wasn't there on time to greet the commander. The engineer was also a part of this group. The stone mason wasn't last, and the centurion wasn't first. The commander was impressed by the three men who rose early enough to beat him to the bridge site. Name the order of arrival of all seven men to the bridge site that morning.

Problem #2

In 9 A.D. the Roman Empire suffered perhaps it most crushing and humiliating defeat at the hands of German tribesmen deep within forested central Germany. Half the dead represented three complete Roman legions or 3/28 of Rome's 140,000 man army. 1/6 of the dead were foreign auxiliary soldiers loyal to Rome. Finally, 1/3 of the dead were women and children attached to the supply train at the rear of the army. How many Roman soldiers, foreign auxiliary soldiers and women and children attached to the "Lost Legions" of Rome were killed that day? What was the disturbing total that Rome would find out about several months later?

Problem #3

At the time of Christ the Roman Empire's army had 140,000 Roman soldiers in it. This was 67% of the total forces it had; the others were foreign troops. By 400 A.D. the percentage of foreign soldiers in the army was three times as much as it was four hundred years before. If the number of men in the entire army was about the same, how many total troops were there, and of those, how many were Roman? **(Round all calculations to nearest whole number.)**

Problem #4

Roman legions were building a wall in northern Britain in 121 A.D. to isolate northern barbarian tribes. Supposedly for every mile of progress on the wall, they used 312,000 stone blocks, 100,000 gallons of water and 550,000 pounds of powdered limestone. (*The water and limestone were mixed to make mortar.)* During one building season of about eight months the soldiers had used 2,808,000 blocks. How much water and limestone had they used, and how long had the wall become in that season?

Problem #5

When Hadrian's Wall, located on the northern frontier of England, was finished, it initially included a small castle at the beginning of each mile of its total length. These castles housed 30 men each. Between the castles in one-third mile intervals were lookout towers with one man each. Hadrian's Wall was 80 miles long. To begin with, how many men were stationed on it?

Problem #6

(Use the information from Problem #5 to assist in solving this problem.) Eventually, sixteen forts were built along the wall, and a grand total of 20,000 men manned the wall. When the wall was at its maximum manpower, how many men were housed in each fort on average?

Problem #7

Four Roman soldiers on Hadrian's Wall went to a local tavern at the end of the day. They ate some servings of cheese at two silver coins a serving; some servings of bread at one silver coin each. Finally, they had some servings of beer at three silver coins each. Altogether, ten separate servings of food/drink were brought to the soldiers and the total bill was 20 silver coins. How many servings of each item was purchased by the four soldiers?

Problem #8

During the reign of Augustus Caesar (27 B.C.-14 A.D.), one eighth of Rome's male population was in the army. The overwhelming majority were common citizens, plebeians. The noble citizenry, the patricians, had little to do with the army since overseeing their land took up most of their time. At this time 90 per cent of Rome's citizens were plebeians and only 10 percent were patricians. The total number of men in the Roman legions at this time was 140,000 men. What was the total population of citizens in Rome at this time and how many were plebeians? How many were patricians?

Problem #9

Most Romans lived in apartments, and their rents were expensive. In fact, the closer the apartment was to street level, the more expensive it became. In one six-story building, apartments on the sixth floor went for 1 gold coin a month; those on the fifth floor cost 2 gold coins a month, and the fourth floor went for 4 gold coins a month. If the total rent for one apartment on each of six floors equaled 41 gold coins, what was the rate for renting apartments on each of the first three floors?

Problem #10

The Roman Empire paid tax collectors, called publicans, a fee to collect its taxes. One Athenian publican got a contract to collect taxes in a small neighborhood of ten families. Three families each paid him 1/13 of the total he collected. The other seven families paid the following amounts of silver coins: 12, 10, 10, 9, 8, 11, and 10. His contract was to have him collect 70 silver coins along with a 10 per cent service fee in addition for his efforts. Was he honest? If not, how many times higher was his actual collection fee?

Problem #11

Rome conducted extensive trade with the various portions of its empire and beyond. One ship sailed to North Africa with, among other things, large urns filled with olive oil and wine. The oil and wine were valued at 320 gold coins and were to be exchanged for bundles of bricks valued at 5 gold coins per bundle, slabs of marble (10 gold coins each) and slabs of granite (8 gold coins each). How much of each building material was headed back to Rome? Hint: There were equal amounts of brick bundles and marble and a lesser amount of granite.

Problem #12

During the *Pax Romana*, the Roman Empire enjoyed almost two hundred years of relative peace within and around its borders. Tiberius, Augustus, Claudius, Marcus Aurelius, Nero and Hadrian were six of the more powerful emperors of this era. Claudius ruled before Marcus Aurelius and Nero but after Tiberius. Hadrian came to power after Nero but wasn't the last of these six. Tiberius wasn't the first emperor. Give the correct sequence of these six emperors.

Problem #13

In the third and fourth century A.D. the Roman Empire began to weaken. Part of the reason for this decline were the inflating prices people had to pay for common necessities. For example the price of food needed for one person for a year might have been 216 silver coins. It had risen 20%, 12.5% and 28% in the three previous years. How much did a year's worth of food cost three years earlier?

Problem #14

It is the fifth day of the newly formed month of Julius during the rule of Julius Caesar. Biceptus Humongus, renowned Roman gladiator, has just defeated another gladiator in a duel to the death. He fights every third day, and if he wins his next three matches, he will win his freedom from his master whom he has made wealthy by his consistent victories over the past year. There is distinct pattern to the matches each day at the arena. A day of gladiators fighting bears, is followed by a day of fighting lions, a third day of men matched against men, and a fourth day of men pitted against numerous animals including elephants. What or who will Biceptus have to defeat in his last three matches to gain his freedom?

Chapter 5 Solutions

Solution for Problem #1

The order of arrival at the bridge site was as follows: **1. engineer; 2. centurion; 3. surveyor; 4. commander; 5. stone cutter; 6. stone mason; 7. architect**

Interesting note: Roman legions always included the various professionals and craftsmen mentioned in the problem. When not fighting, the legions were often referred to as "builders of the world" since they built forts, amphitheaters, canals, bridges, aqueducts, and thousands of miles of roads.

Solution for Problem #2

The key lies in identifying that half of the dead represented **15,000** Roman soldiers (3/28 X 140,000). That automatically deduces the total of **30,000** dead of which 1/6 would equal **5000** auxiliary soldiers and 1/3 being **10,000** women and children.

Interesting note: The Romans were done in by a German double agent. He had fought bravely in the Roman army for several years learning its strengths and weaknesses. He planned an ambush in heavily wooded terrain where the Romans could not properly use their normal battle formations.

Solution for Problem #3

Students must first identify the total number of men in the Roman army:

.67X	= 140,000
X	= 140,000/.67
X	= 208,955

Having determined the total, the original percentage of foreign troops needs to be identified which would be 33%. In 400 A.D. three times this percentage would be 99% leaving 1% Italian component to the army at that time. Therefore:

.01(208,955) = 2090

In 400 A.D. the Roman army consisted of approximately 208,955 total men of which about 2090 were actually Roman (Italian).

Interesting note: A major contributing factor to the decline of Rome, an over-dependence upon foreign soldiers left the empire with an army that had no real loyalty to Rome itself.

Solution for Problem #4

Students have a couple of options for this solution. Quicker students may see a direct nine fold increase between the 312,000 blocks used for one mile and the 2,808,000 incorporated for the entire building season. By multiplying the other building supplies' figures by nine their totals would be realized as would the nine mile distance that the wall had reached that year.

Some students may feel more comfortable employing a table that methodically plots out the increase as shown below:

Miles Built	*Blocks Used*	*Limestone Used*	*Water Used*
1	*312,000*	*550,000*	*100,000*
2	*624,000*	*1,100,000*	*200,000*
3	*936,000*	*1,650,000*	*300,000*
4	*1,248,000*	*2,200,000*	*400,000*
5	*1,560,000*	*2,750,000*	*500,000*
6	*1,872,000*	*3,300,000*	*600.000*
7	*2,184,000*	*3,850,000*	*700,000*
8	*2,496,000*	*4,400,000*	*800,000*
9	***2,808,000***	***4,950,000***	***900,000***

That season the wall utilized **2,808,000 stone blocks, 4,950,000 pounds of powdered limestone and 900,000 gallons of water while covering 9 miles.**

Interesting note: While the amounts of limestone and water in this problem are purely speculative, the number of stones is rather accurate considering a total of 25,000,000 went into the construction of Hadrian's Wall in northern England.

Solution for Problem #5

CASTLE————watchtower————watchtower————CASTLE

By illustrating the relationship of milepost castles and the two watchtowers between them, students should be able to tally the total number of troops. Eighty castles would subtotal 2400 men and 160 lookout towers would subtotal a like number of soldiers for a grand total of **2560 men on the wall.**

Interesting note: Hadrian's Wall was 20 feet high and 10 feet wide.

Solution for Problem #6

With one castle at the start of each mile, there would be 2400 men. The lookout towers at one-third mile intervals would take two men per mile for another 160 men. with these figures known, a simple algebraic equation could solve the problem's question:

2400 + 160 + 16X	*= 20,000 (X = number of men in one fort)*
2560 + 16X	*= 20,000*
16X	*= 20,000 - 2560*
16X	*= 17,440*
X	***= 1090***

There were (on average) 1090 soldiers in each fort.

Interesting note: The emperor Hadrian used the construction of the wall as a project to keep his men busy since he didn't wish them to expand northward and risk unwarranted losses in his eyes. Furthermore, he didn't want his officers idle concocting schemes to overthrow him.

<u>*Solution for Problem #7*</u>

The creation of an organized listing of the various menu items would be a more efficient way of deducing the answer than by routine trial and error.

Number of servings	*total cost of bread*	*total cost of cheese*	*total cost of beer*
1	*1 coin*	*2 coins*	*3 coins*
2	*2 coins*	***4 coins***	*6 coins*
3	*3 coins*	*6 coins*	*9 coins*
4	***4 coins***	*8 coins*	***12 coins***
5	*5 coins*	*10 coins*	*15 coins*

The soldiers had 2 servings of cheese (4 silver coins), 4 servings of bread (4 silver coins) and 4 servings of beer (12 silver coins).

Interesting note: As Hadrian's Wall expanded to include some 20,000 troops, civilians moved near the wall in order to conduct trade with the soldiers and so obtain Roman money. This was the official unit of exchange in Britain as throughout the entire empire.

<u>*Solution for Problem #8*</u>

Students need to start at the end of the problem with the 140,000 Romans in the legions. This represented 1/8 of the Roman male population (i.e. "citizenry"). **Therefore, the total male population would have equaled 1,120,000 men. Ninety percent were plebeians (1,008,000) while ten percent were patricians (112,000).**

Interesting note: Obviously most Roman legions came from humble backgrounds. Prior to Augustus, soldiers had been fiercely loyal to their commanding generals who shared the spoils of war with them. Augustus had coins minted with his likeness on them with which he paid the army and so gained its loyalty.

Solution for Problem #9

Students will need to recognize the pattern that exists amongst the rental rates on the various floors. Several patterns may be logical, but only one will fit the 41 gold coin total for all six floors. At first glance, 1,2,4, …would seem to be doubling the rate as the floors lowered. However, this will far surpass the 41 gold coin total for all six floors. Another pattern that exists here is that of adding one more to the difference of the preceding numbers—-1 (+1),2 (+2),4 (+3), **7 (+4), 11 (+5), 16**. This will indeed total 41 and is the correct answer.

Interesting note: For every 26 block long apartment building with six or more stories, there was but one privately-owned house in Rome. Furthermore, apartments were usually one room with about five or six people living in them.

Solution for Problem #10

Students need to piece all the parts of this mathematical puzzle together first. Working from the back of the problem will help. The seven families whose combined 70 silver coins are known also represent 10/13 of the total amount collected by the publican. That would put the entire tax gathered at 91 coins.

Including his 10 percent service charge, the publican should have collected 77 coins (7 as his fee). **He was obviously dishonest as he took a fee three times as high as it should have been.**

Interesting note: With no real accountability other than Rome's desire to gain wealth from her subservient provinces, unscrupulous publicans were quite common.

<u>*Solution for Problem #11*</u>

A listing for the possible amounts and equated values of the building materials would help students answer this problem. The list may start at a reasonably higher number such as ten.

	Value of building materials		
number of materials	*brick bundles*	*marble*	*granite*
10	*50*	*100*	**80**
11	*55*	*110*	*88*
12	*60*	*120*	*96*
13	*65*	*130*	*104*
14	*70*	*140*	*112*
15	*75*	*150*	*120*
	80	***160***	*125*

There were 16 bundles of bricks, 16 slabs of marble and 10 slabs of granite.

Interesting note: Roman trade extended into Northern Europe for amber as well as into China for silk.

<u>*Solution for Problem #12*</u>

The correct sequence of the emperors is as follows: **Augustus, Tiberius, Claudius, Nero, Hadrian, Marcus Aurelius**

Interesting note: While each of these emperors was known for particular vices and/or virtues, Marcus Aurelius was so popular that Romans believed him to have been a god sent to help them.

Solution for Problem #13

Students may find the trial and error method the most effective means of solving this problem. What number when increased by 20% will equal 216? **(180)** What number when increased by 12.5% will equal 180? **(160)** What number when increased by 28% will equal 160? **(125) The price of a year's worth of food three years earlier was 125 silver coins.**

Students may also work the problem algebraically as follows:

1.2X = 216	*1.125X = 180*	*1.28X = 160*
X = 216/1.2	*X = 180/1.125*	*X = 160/1.28*
X = 180	*X = 160*	*X = 125*

Interesting note: During this waning era of Roman influence, more gold was leaving Rome to pay for luxury items than was coming in from tribute and conquest. Therefore, money was losing value through inflation and people often resorted to bartering.

Solution for Problem #14

The month of Julius is our month of July. Students may list the days starting with the fifth and the corresponding match for each day to determine Biceptus' upcoming three contests.

5—men;6—many animals;7—bears;**8—lions;**9—men;10—many animals;**11—bears;**12—lions;13—men;**14—many animals**

Interesting note: Over the centuries, hundreds of thousands of animals were slaughtered at the daily gladiatorial contests. In fact, certain species of lion and elephant were exterminated for this purpose.

Chapter 6

The Ancient Maya

Problem #1

Itza and Imax met to do some weekly trading in their Maya village. All of the items to be traded had a value in cacao beans much as goods today have dollar values. Itza traded 8 fish, 3 baskets of seashells and a small string of pearls to Imax for 2 baskets of maize (corn), 3 baskets of tomatoes, 2 baskets of pumpkins and a dozen parrot feathers.

Use the table below to figure the value of each man's trade. Since a trading session had to end up with each side getting an exactly equal value of goods and/or beans, decide which man had to give the other a number of cacao beans to even up the trade and how many beans had to be given.

Item	Value in cacao beans
basket of maize	37
basket of pumpkins	18
basket of seashells	32
basket of tomatoes	18
one medium-sized fish	13
parrot feathers (one dozen)	77
pearls (a small string)	90
stingray spines (1)	10

Problem #2

Using the kernels of corn you obtained from your teacher, determine how many different outcomes you can obtain by tossing two kernels. How many outcomes would there be for three kernels? four kernels? five kernels?

What would be the probability of matching any particular outcome when tossing two kernels? three kernels? four kernels? five kernels?

Take two kernels and toss them thirty times while recording the results. Do the results match the probability that you determined earlier? If you took four kernels and tossed those 100 times, how many of those tosses could you predict would all be black-sided?

Problem #3

Similar to problem #2, if you tossed nine kernels 100 times, what number of your tosses could you predict to end up as all yellow-sided?

Problem #4

A Mayan farmer owned seven equally-sized fields. From west to east they were as follows: an unused field, a field of maize (corn), a bean field, a sweet potatoes field, a tomatoes field, a manioc (cassava) field, and another unused field. In order to enrich the soil and not wear it out, this farmer rotated his fields every two years by moving a specific crop two fields tot the east. Those crops in the farthest field east would rotate back to the westernmost fields. If he had just rotated his crops, in which fields would his crops be in seven years?

Problem #5

Archaeologists have determined that the major foods of the ancient Mayan population were corn, vegetables and some meat. They say that the Mayans ate ten times as much corn as meat and one-and-a-half times as much vegetables as meat. If these three foods made up 100 per cent of the Mayan diet, what percentage did each represent?

Problem #6

Use the answer from Problem #5 to solve this problem. If a Mayan family of four had consumed 1900 pounds of food in a year, how many pounds of corn, vegetables and meat did they eat as a family?

Problem #7

An ancient Maya trader, Mypex, was about to leave on a trading expedition along the Caribbean coast. His dugout canoe was loaded with wax, honey, salt, jade, cotton and feathers. It was an expensive cargo. The honey made up 10% of the value while the salt was 32%; the jade 20%; the cotton 17.5%; the feathers 13.5%. The wax alone was worth 840 cacao beans, and it was only 7% of the total value of his trade goods. What was the total value of Mypex's cargo in cacao beans? What was the value of each trade good (in cacao beans)?

Problem #8

Five Mayan merchants left the ancient city of Tikal on trading ventures. Moon Serpent left before Q Lux who left after Tac Lax. Quick Monkey left after Q Lux. Tac Lax left before Moon Serpent, but he wasn't the first to leave Tikal.

On the return to Tikal, Moon Serpent returned after Jaguar who came back before Tac Lax. Tac Lax wasn't last in returning. Quick Monkey returned before Jaguar, but he wasn't first.

Determine the order in which these five traders left Tikal and the proper sequence in which they returned.

Problem #9

Japac, a stoneworker, specializes in making spear points and wood cutting tools out of volcanic obsidian. He makes oval and triangular shaped spear points and rectangular and semi-circular shaped wood shaving tools. Four months ago Japac made 4 oval spear points, 2 triangular spear points, 7 semicircular wood shaving tools and 5 rectangular tools. Three months ago he made 6 oval points, 1 triangular point, 4 semicircular tools and 5 rectangular tools. Two months ago his work included 3 oval points, 3 triangular points, 8 semicircular tools and 6 rectangular tools. Last month he completed 5 oval points, 2 triangular points, 5 semicircular tools and 6 rectangular tools. How many more months will it take Japac to create 24 stone pieces in one month's time? How many of each will he create during that month?

Problem #10

Palex, a Mayan farmer, has brought his produce to market. He has breadnut fruit, bananas, papaya and manioc root to sell in woven bushel baskets. He has four times as many baskets of bananas than of papayas. His crops include twice as many baskets of breadnut fruit as bananas. The number of baskets of manioc root equals half the number of baskets of bananas. If Palex brought 90 baskets to market, how many baskets of each food did he bring?

Problem #11

A trader went on a long overland trading mission with a valuable cargo strapped to his back. He was carrying one pound chunks of jade worth 500 cacao beans each, five pound chunks of obsidian worth 150 cacao beans each and pearls worth 300 cacao beans each. If the pearls weighed less than an ounce each and his entire cargo was worth 10,500 cacao beans, how much did his load weigh?

Problem #12

The ancient Mayan city-state of Copan (in present-day Honduras) was at one time ruled by a succession of five related rulers. In no particular order they were: Yax Pac, 18 Rabbit, Smoke Monkey, Smoke Imix, and Smoke Shell. 18 Rabbit's great grandson was the last ruler of Copan. Smoke Shell's great grandfather was the initial ruler of this quintet. Yax Pac's grandfather was the middle ruler in the succession of five. Smoke Monkey was the grandson of Smoke Imix. Give the correct sequence of these five rulers starting with the earliest.

Problem #13

At the ancient Mayan city-state of Copan in the eighth century ruler Smoke Shell began numerous impressive building projects. One of these has been referred to as the Hieroglyphic Staircase because all along the surface of the stairs leading up to a temple at the top were glyphs that told about the marvelous things accomplished by the dynasty of his ruling ancestors. Each of these 72 stairs were 50 feet wide and one-and-a-half feet high. If the height of one story on a modern building is 12 feet, the height of the Hieroglyphic Staircase at Copan would have equaled how many stories on a modern building?

Problem #14

In the last years of the eighth century, Mayan city-states in what is now western Guatemala were in constant warfare with each other. Several city-states undertook extreme measures to build defensive positions. One such effort took place on a peninsula that jutted out into Lake Petexbatun. Punta de Chimino was a fort built by making an island out of the peninsula. Mayan citizens dug three moats across the neck of the peninsula in order to strengthen their position. Use the map below to answer the following questions:

1. How long was the longest moat (A)?

2. How wide was moat A?

3. How long was moat B?

4. From moat A to point C, how long was the man-made island?

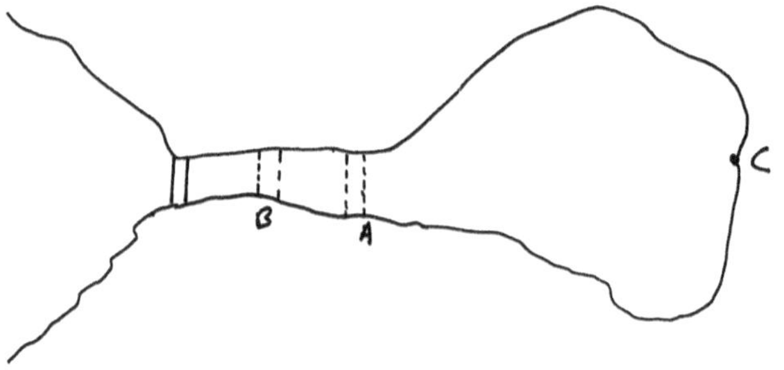

Scale 1/2 inch = 460 feet

Problem #15

The ancient Mayan people were able to thrive for centuries by clearing only small areas of forest for farming. In this manner they were able to grow the food necessary for the populations of their city-states. If the grid below represented some land in the Mayan domain, and the X'ed out numbers were plots where the forest would be removed in order to grow crops, identify six other numbered areas that would also be used for farming.

X 1 X 2 X 3   X 4____5___6____7

__8 X 9 ___10___11___12___13__14

_15 X 16___17___18___19___20__21

_22___23___24___25___26___27__28

_29___30___31___32___33___34__35

_36___37___38___39___40___41__42

_43___44___45___46___47___48__49

Problem #16

The ancient Mayans were able to create their own number system. It was a *vigesimal* system based on twenty (from the Latin word for twenty) unlike our base ten system. The following table shows the Mayan numbers 0-19:

0 was written as a shell-like figure:

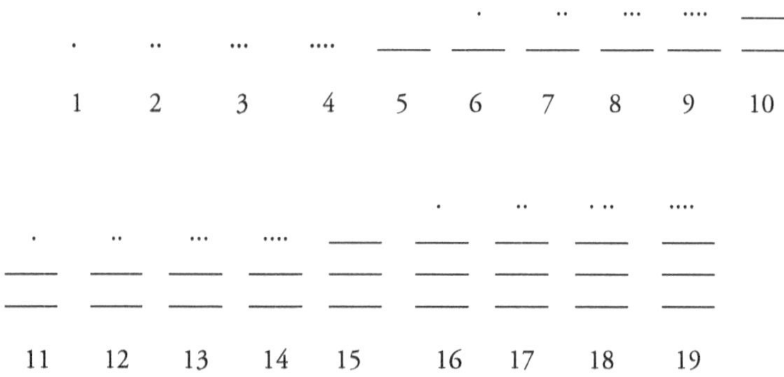

Mayans noted their numbers vertically with one's place value at the bottom, twenties value next followed by four hundreds value (20 sets of 20),

8,000's value (20 sets of 400), and so on. Addition was done horizontally. Observe the following addition problem:

```
  .     (8000)

  .                           . . . .
_____   (6 X 400= 2400)   +   ________   (14 X 400= 5600)

...
_____
_____   (13 X 20= 260)            .      (20)

  .
_____   (6)                    ______    (5)

        10,666            +        5625
```

The answer of 16291 would be written as follows:

```
.. (2 X 8000=16,000)
```

 (0 X 400 = 0)

```
....
_____
_____ (14 x 20 = 280)

_____
_____ (1 x 10 = 10)

   . (1)
```

Try to identify these Mayan numbers:

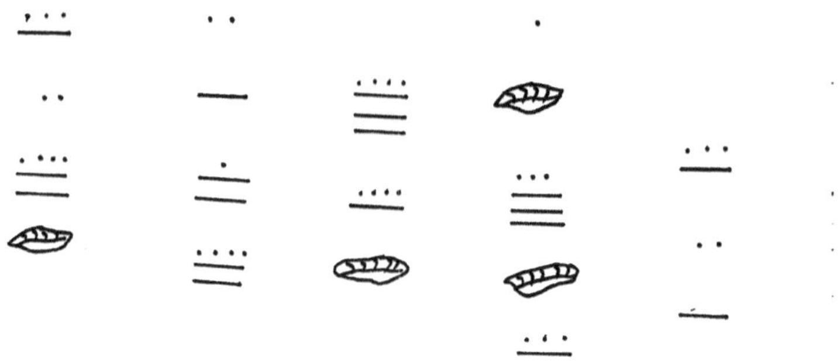

Chapter 6 Solutions

Solution for Problem #1

Each man's trade must first be converted into its value in cacao beans using a list.

Value of Itza's Goods	*Value of Imax's Goods*
8 fish X 13 beans = 104	*2 baskets of maize X 37 beans= 74*
3 baskets of seashells X	*3 baskets of tomatoes X 25 =*
32 = 96	*75*
1 small string of pearls =	*2 baskets of pumpkins X 18 =*
90	*36*
	1 dozen parrot feathers = 77

Since Itza forwarded items valued at 28 cacao beans greater (290-262), **Imax must give Itza 28 cacao beans.**

Interesting note:

Mayans used several items for currency besides cacao beans. Red beans, stone and shell beads, and feathers were also used for money. Normally, the cacao currency were just offered to equalize a bartered trade similar to the one in the problem.

Solution for Problem #2

Note: To proceed with this particular activity, the instructor needs to procure untreated kernels of yellow (or white) seed corn. Use black permanent marker or latex paint to darken one side of each kernel.

Similar to a coin, each kernel has two possible outcomes: Y (yellow) and B (black). Here are the possible outcomes for various numbers of kernels:

Two kernels (**3**): Y and B, Y and Y, or B and B
Three kernels (**4**): Y, Y, Y; B, B, B; Y, B, B; B, Y, Y
Four kernels (**5**): Y,Y,Y,Y; B,B,B,B; Y,Y,B,B; Y,B,B,B;
B,Y,Y,Y
Five kernels (**6**):Y,Y,Y,Y,Y; B,B,B,B,B; Y,Y,B,B,B; B,B,Y,Y,Y;
Y,B,B,B,B; B,Y,Y,Y,Y

The probability of tossing any particular outcome would be:

two kernels	1/3
three kernels	1/4
four kernels	1/5
five kernels	1/6

When tossing two kernels thirty times, student **answers will vary.** When tossing four kernels 100 times, the probability of getting four dark-sided kernels would be 1/5 or **about 20 times.**

Solution for Problem #3

In answering this problem, students need to have identified a pattern that developed in Problem #2. Each time a new two-sided kernel was added to a toss the possible outcomes increased by one.

Number of kernels	*1*	*2*	*3*	*4*	*5*	*6*	*7*	*8*	*9*
Number of outcomes	*2*	3	4	5	6	7	8	9	10
Probability of a specific outcome being tossed	*1/2*	1/3	1/4	1/5	1/6	1/7	1/8	1/9	1/10

With nine kernels the probability of tossing all yellow-sided kernels is 1 out of 10. In 100 tosses this would **predictably yield about 10 such tosses.**

Interesting Note:

A popular game of chance in Mayan marketplaces was the throwing of kernels of corn with one side darkened.

Solution for Problem #4

In undertaking this problem, students will first have to draw an accurate illustration or diagram depicting the farmer's fields as follows:

< *West* Untilled/Maize/Beans/Sweet/Tomatoes/Manioc/Untilled *east*>
Potatoes

Since the rotation had just been completed, there would be scheduled rotations two, four and six years hence. In seven years, the fields would represent the third rotation. Moving the crops two fields eastwardly three times would place them in the following setting:

< *West* **Maize/Beans/Sweet/Tomatoes/Manioc/Untilled/Untilled** *East*>
Potatoes

Interesting note:

Vegetables such as maize (corn), beans and pumpkins were often planted together. Corn would grow up in stalks while beans would cluster on lower growing plants. Pumpkins would extend from vine runners. This system kept soil intact and helped provide various nutrients for each type of plant.

Solution for Problem #5

Students may attempt basic algebra with this problem or go with trial and error.

Algebra would dictate the following equation:

x + 1.5x + 10x	= 100% (x = % of meat in diet)
12.5x	= 100%
x	= 8%

Therefore **meat represented 8% of the ancient Mayan diet, vegetables 12% and corn 80%.**

Those students utilizing trial and error could successfully work this problem by systematically replacing the variable X with assorted values (obviously less than 10 to begin with) that would bring about a total of 100%.

Interesting note:

The primary sources of domesticated meat for the ancient Mayans happened to be dog, duck, and turkey. Not until the arrival of the Spanish in the sixteenth century did pigs, cattle, and chicken come into their diet.

<u>*Solution to Problem #6*</u>

With the percentages garnered from Problem #5, students would do basic decimal multiplication to arrive at this problem's solution.

Corn = 80% of the diet (.8) Vegetables = 12% (.12) Meat = 8% (.08)	
.8 X 1900	= **1520 pounds of corn**
.12 X 1900	= **228 pounds of vegetables**
.08 X 1900	= **152 pounds of meat**

Interesting note:

Archaeologists have been able to determine the make-up of the ancient Mayan diet by sifting through their ancient garbage dumps, or middens.

<u>*Solution to Problem #7*</u>

Students need to realize that the solution rests with the fact the 840 beans worth of wax equals 7 per cent of the total cargo.

.07(t)	= 840
840/.07	= **12000 cacao beans**

Once 12000 beans has been realized as the total, students work backwards with the rest of the items by multiplying their decimal equivalents of their percentage values by 12000:

honey---.10 X 12000	= **1200 beans**
salt----.32 X 12000	= **3840 beans**
jade----.20 X 12000	= **2400 beans**
cotton--.175 X 12000	= **2100 beans**
feathers--.135 X 12000	= **1620 beans**

Interesting note: Some scholars believe that, at the height of Mayan power, some 4000 Mayan trade canoes were at sea at any given time.

Solution to Problem #8

Students will have to reason this problem through by notating sequence clues between pairs of traders before combining all the clues to produce an answer. Abbreviations are helpful
For the departure from Tikal, Moon Serpent (MS) left before Q Lux (QL); Tac Lax (TL) left before QL; QL left before Quick Monkey (QM); TL left before MS; Jaguar (J) is not specifically mentioned. Tac Lax was not first to leave.

MS BEFORE QL	TL BEFORE QL	QL BEFORE QM	TL BEFORE MS	J

Placing these relationships together we arrive at the following departure sequence:

	Jaguar	
	Tac Lax	
	Moon Serpent	
	Q Lux	
	Quick Monkey	

Working the arrival order in a similar manner, we see the following pairs:

J before M S	QM before J	J before TL	TL wasn't last
			QM wasn't first

The arrival sequence would be the following:

	Q Lux	
	Quick Monkey	
	Jaguar	
	Tac Lax	
	Moon Serpent	

Interesting note: Due to the numerous sea-based trade routes that the Mayans had, some archaeologists referred to them as the "Phoenicians of the New World".

<u>*Solution for Problem #9*</u>

Students will need to realize a pattern that exists in Japac's monthly production. Creating a table which depicts the production of each stone piece will assist them.

	oval points	triangular points	semicircular tools	rectangular tools	Total made
4 mth ago	4	2	7	5	18
3 mth ago	6	1	4	5	16
2 mth ago	3	3	8	6	20
1 mth ago	5	2	5	6	18

If students study their table, they will note each column has a distinct pattern. The monthly total column subtracts 2 then adds 4 regularly. Following it, one can conclude that 24 pieces in a month will be reached in **3 more months.**

Each individual item's column also has a noticeable pattern that would leave the following subtotals three months into the future: **1 oval point, 5 triangular points, 10 semicircular tools, 8 rectangular tools**

Interesting note: Some Mayan spear points have been found with carved facial profiles on them denoting the homage paid to the regional ruler. Archaeologists believe each of these foot-long royal lance points took hundreds of hours to create.

<u>*Solution to Problem #10*</u>

Students need to set up an algebraic equation unless they want to go the trial and error route. The base variable is the number of papaya baskets (p). Either way, they must realize the stated relationship between the various goods and the baskets of papayas. All the other commodities revolve around it. 4p = the number of baskets of bananas; 8p = the number of breadnut fruit baskets; 2p = the number of manioc root baskets

8p + 4p + 2p + p	= 90
15p	= 90
p	= 6

6 baskets of papayas; 12 baskets of manioc root; 24 baskets of bananas; 48 baskets of breadnut fruit

Interesting note: Mayans also grew and traded tobacco, cotton, rubber and vanilla.

<u>*Solution for Problem #11*</u>

Again, students may want to try trial and error. However, an organized list or table would be the best way to approach this problem. Total value and weight are the variables that need to be accounted.

VALUE IN CACAO BEANS			
Number	Jade (1 lb)	Pearls (< oz.)	Obsidian (5 lb.)
1	500	300	150
2	1000	600	300
3	1500	900	450
4	2000	1200	600
5	2500	1500	750
6	3000	1800	**900**
7	3500	2100	1050
8	4000	2400	1200
9	4500	2700	1350
10	5000	3000	1500
11	5500	3300	1650
12	**6000**	**3600**	1800

While this table could continue onward, it is at 12 where the first possibilities come into view. 6000 beans worth of jade (**12 pounds**), 3600 beans worth of pearls (**less than 12 ounces**) and 900 beans worth of obsidian (**30 pounds**) produce a little more than **42 pounds** of a load for the trader. **Please note: There are other possible answers for this problem. Accept any that fit the total value and where the student has demonstrated an understanding of the method.**

Interesting note: Jade jewelry signified high status in ancient Mayan society.

<u>*Solution for Problem #12*</u>

Students need to reason this situation through the creation of a lineage continuum similar to this one:

#5	#4	#3	#2	#1
Son	Father	Grandfather	Great Grandfather	Great Great Grandfather
Yax Pac	**Smoke Shell**	**Smoke Monkey**	**18 Rabbit**	**Smoke Imix**

Using the first clue given concerning 18 Rabbit, we can discern that he was the great grandfather of the last ruler (#5), so 18 Rabbit had to be ruler #2.

Since Smoke Shell's great grandfather was the initial ruler in this group, Smoke Shell must be #4.

Since Yax Pac's grandfather was #3 in this grouping, that made Yax Pac #5, the most recent ruler.

With Smoke Monkey being the grandson of Smoke Imix they are #3 and #1 respectively.

Interesting Note: These five men ruled Copan from 650 to 820 A.D. Only 18 Rabbit met a violent end, having been beheaded by a rival city-state's ruler.

<u>*Solution for Problem #13*</u>

Since each stair in the Hieroglyphic Staircase was one-and-a-half feet high, students multiply that figure by 72 stairs to obtain a total height (excluding the temple) of 108 feet. Divided by twelve feet, this would just about equal **an 8 story building**. (Perhaps the teacher can point out similarly sized buildings in his/her community by which students can better visualize the actual size of this staircase.)

Interesting note: There were a total of 1250 glyphs on the staircase making it the longest single written inscription in all of the ancient Americas. Only 30 of the stairs are left standing in their original order today.

<u>*Solution for Problem #14*</u>

Moat A is 1/2 inch long equaling **460 feet.**

Moat A is 1/8 of an inch wide. 1/8" = 1/4_of 1/2 or **approximately 115 feet.**

Moat B is 3/8 of an inch long. 3/8" = 3/4 of 1/2 or **345 feet.**

Moat A is three inches from point C or **2760 feet** (about a half mile)

Interesting note: The moats were dug thirty feet deep into bedrock. Some 50,000 cubic yards of the bedrock were removed and placed as a thirty-foot high wall along the island side of Moat A.

<u>*Solution for Problem #15*</u>

Students need to recognize that the X'ed out numbered squares—-1,2,3,4,9,16——are the four for squared numbers (1,4,9,16) and/or their square roots (1,2,3). Between 17 and 49, three more squared numbers and their roots will produce the six previously unknown sections—-**25 (5); 36 (6); 49 (7)**

Interesting note: One of the great achievements of the Mayan civilization was to maintain a dense population in a rain forest environment by clearing only limited areas for farming. Some archaeologists theorize that eventually even larger populations coupled with intensive warfare over territory stressed the Mayan environment to the point that over several centuries less farming was undertaken, the population moved or declined and the forest reclaimed the land.

Solution for Problem #16

1. 65,080	2. 18,234	3. 7,780	4.167,208	5. 3,245

Interesting note: In the ancient world only three civilizations were able to develop the concept of zero. They were the Mayans, Hindus and Babylonians.

Chapter 7

The Mound Builders

Problem #1

In a burial mound near the Mississippian culture center of Cahokia, (in southwestern Illinois) archaeologists found the burial mound of a chief or priest. This leader of high-standing was buried on top of a blanket made of 12,000 shell beads. In the first row of beads there were 2 black shells followed by 4 white shells, 3 gray beads and 1 pink shell. This pattern was repeated throughout the entire blanket. How many shell beads of each color were used?

Problem #2

Along the Mississippi Valley, mound building natives began growing corn and beans over twelve-hundred years ago. In the large land holdings of the Duck Clan, 1500 acres was devoted to corn and beans with some swampy areas between the two crops. There are three times as many acres of corn as there are acres of swamp and twice as many bean acres as swamp acreage. How many acres of corn, beans and swamp are there?

Problem #3

Wave Walker, master trader and bearer of all news in the Serpent River valley is about to go on another trading expedition. He will visit seven chiefs and their villages on a route he regularly takes. Included in this trip will be visits to **Badger**, whose village mines mica from a cliffside; **Roaring Sky**, whose coastal village has abundant sea shells; **Crippled Bear** who is situated near a large deposit of flint; **Bold Duck**, whose village has much greenstone which is used for ax heads; **Grey Eyes**, whose people grind acorns into fine meal; **Round Turtle**, whose village mines silver; and **Red Skull**.

As any master trader would, Wave Walker knows what his customers want. **Badger** needs flint for spear points; **Round Turtle** wants pretty purple cloth; **Red Skull** likes shell bead necklaces; **Bold Duck** prefers silver ornaments.

Wave Walker plans his trips so that the resources of one village provide the trade goods for the things needed in the next village. The last village on his route has provided the goods for the first village on this trip. He is especially excited to see his old friend **Crippled Bear** half way through this trip with trade offerings of his favorite food. List the order of the chiefs that Wave Walker will visit.

Problem #4

Monk's Mound, near East St. Louis, Illinois, is a one-hundred foot high mound that sits on sixteen acres of land. It was erected about one thousand years ago by Mississippian mound builders.

If these people were able to layer 8 inches of soil on the mound each year and the mound automatically lowered itself by 4 inches every other year (due to the soil settling), how many years did it take to raise this mound?

Problem #5

Along the islands of the present-day Georgia coast some very early mound builders were building fantastic seashell mounds. On one day five natives carried 540 baskets of shells in the construction of one of these mounds. Big Fish carried the most baskets. White Squirrel carried twice as many as Starfish Hunter. Beach Walker carried one-fourth the baskets that Big Fish carried, and Shadow Catcher barely lifted a finger in lugging but two baskets all day. How many baskets did each native carry?

Problem #6

A mound builder clan of fifty people went about building a mound. Each person used a basket that held eight pounds of dirt. Each person filled and emptied a basketful of dirt every five minutes. The clan worked ten hours a day for four days. How many tons of dirt did the clan use to make the mound?

Problem #7

The Newark Earthworks in Newark, Ohio represent one of the largest single areas of related prehistoric mounds in North America. Four large geometrically shaped mounds stand out in this grouping—-two circles of equal size, a square and an octagon two and a half times the size of the square. The octagon is the largest of the four mound sites. If the total enclosed area of these four mounds amounts to 130 acres, what is the area of each mound site?

Problem #8

At a tribal meeting held in the council lodge, one representative each of eight brotherhoods gathered and sat in a circular formation. The eight fellowships represented included Warriors, Healers, Star Watchers, Farmers, Stoneworkers, Potters, Weavers and Master Builders. The Warrior delegate sat directly opposite the Star Watcher. The Weaver sat opposite the Master Builder. The Master Builder sat to the left of the Star Watcher. The Stoneworker sat to the right of the Farmer who was closer to the Star Watcher than to the Warrior. The Healer wasn't seated opposite of the Farmer. The representative of the group that could keep track of the seasons sat closest to the lodge's opening. Diagram where each delegate sat in this circle.

Problem #9

Grave Creek Mound in West Virginia is the largest conical mound built by prehistoric man in America. It consists of 60,000 tons of earth. If we suppose that during the first decade of its construction, 1000 tons of earth were moved into place and in subsequent decades 2000, 4000, 5000 and 7000 tons were layered on top of it, how many years did it take to build Grave Creek Mound?

Problem #10

Recent archaeological studies indicate the Great Serpent Mound of southern Ohio may be much younger than previously believed. With the use of radiocarbon dating techniques archaeologists calculate the date of the Serpent Mound to be four years later than the brightest ever occurrence of Halley's Comet. The most recent sighting of this comet was in 1986. The brightest form of Halley's Comet was four score less than a millennium before 1986. According to recent archaeological theory, when was the Great Serpent Mound built?

Problem #11

Six Hopewell clan elders sat down to discuss concerns about the education of their clan's youth in the ways of their people. Two of the elders were in their forties; two were in their fifties, and two were in their sixties. Two were storytellers; two were shamans, or healers; two were chiefs. Walking Cat and Dark Night are the youngest. Straight Tree and Dark Night are clan chiefs. Bird That Sings is a shaman. Cold Moon is the oldest. Walking Cat has been mute since birth. Whispering Spring is the sixth member of the group. Who is the oldest storyteller in this meeting?

Chapter 7 Solutions

Solution for Problem #1

The repetitive pattern is made up of 10 beads. So:

1/10 of 12,000 beads are pink	= 1200
3/10 of 12,000 beads are gray	= 3600
4/10 of 12,000 beads are **white**	= 4800
2/10 of 12,000 beads are **black**	= 2400

Interesting note: Buried along with the leader were six male servants and fifty-three females. These people had been killed apparently to accompany their master into the next life.

Solution for Problem #2

Students may try simple algebra, $3X + 2X + X = 1500$, where X represents swamp acreage. In so doing, there are **250 acres of swamp, 500 acres of beans, and 750 acres of corn.**

Students may also attempt the trial and error method utilizing a relatively low rounded number to equal the swamp acreage to with which to begin, *i.e.* 100 which would give a value of 200 for bean acreage and 300 for corn acreage for a total of 600 acres. Continuing with 200 for the value of swamp acreage, students would net a total acreage of 1200. Using 300 for swamp acreage would reap a total of 1800 acres which would clue students toward the correct answer of **250 acres of swamp, 500 acres of beans and 750 acres of corn.**

Interesting fact: Archaeological evidence suggests that maize (corn) gradually worked its way north from southern Mexico into the Mississippi River valley.

Solution for Problem #3

Students need to create a table in which at the top all characters are listed with their specialty goods. At the left hand side a column listing all characters is placed. Where characters' needs intersect with another character's specialty product an "X" is marked.

Badger	Roaring Sky	Crippled Bear	Bold Duck	Gray Eyes	Round Turtle	Red Skull
mica	*shells*	*flint*	*green-stone*	*Acorn*	*silver*	
Badger		X				
Roaring Sky						
Crippled Bear				X		
Bold Duck					X	
Grey Eyes						
Round Turtle						
Red Skull X						

From this table and the fact that Crippled Bear is met halfway through the trading trip, the following order can be reasonably deduced:

Gray Eyes—-Crippled Bear—-Badger

Furthermore, Roaring Sky must precede Red Skull to fulfill his need for shells, Round Turtle must precede Bold Duck for similar reasons. However, Red Skull's specialty item was not mentioned. The only need not met by a known specialty item is Round Turtle's desire for purple cloth. Therefore, since Wave Walker travels this particular route routinely, it is logical to assume that Red Skull, at the end of his trip, provides the dye or cloth necessary for trade with Round Turtle at the start of the journey. The answer is as follows:

Round Turtle	Bold Duck	Gray Eyes	Crippled Bear	Badger	Roaring Sky	Red Skull

Interesting fact: The Hopewell Culture was not a homogeneous group of prehistoric natives. However, much as Europeans may be grouped together today based on a geographic similarity despite numerous cultural differences, the Hopewell were grouped together because of the penchant for mound construction and interrelated trade.

Solution for Problem #4

The answer is **200 years**. Every two years the actual gain in altitude of the mound would be 12 inches (16 inches of added soil with 4 inches deducted for settling). A gain of a foot every two years over 200 years would account for the 100 foot height of Monk's Mound.

Interesting fact: The sixteen acre base of Monk's Mound is about the size of sixteen football fields. This represents an area larger than the base of the Great Pyramid at Giza, Egypt.

Solution for Problem #5

Unless familiar with higher forms of algebra, students will more than likely use a trial and error method to secure the correct response. First, by eliminating **Shadow Catcher's 2 baskets**, students must total 538 baskets full of shells amongst Big Fish, White Squirrel, Starfish Hunter and Beach Walker. Since Big Fish carried the most baskets and Beach Walker's amount was one-fourth of Big Fish's total, students should begin with a relatively large rounded number for Big Fish's total (*100 might be a reasonable place at which to start)*

The correct total for **Big Fish** will be **200 baskets.** Therefore **Beach Walker** will have **50 baskets.**

With 288 baskets left to be accounted for, White Squirrel and Starfish Hunter will divide the remnant into three equal parts amongst themselves with **White Squirrel** having twice as much for **192 baskets** while **Starfish Hunter** had **96.**

Interesting fact: A four thousand-year-old circular shell mound on Sapelo Island, Georgia has a diameter of 300 feet and a height of seven feet throughout its circumference!

<u>*Solution for Problem #6*</u>

8lbs. X 12 (5 min.segments/hr.)	= 96 lbs./hr./person
96 X 10 hours/day	= 960 lbs./person/day
960 X 4 days	= 3840 lbs./person over the four days
3840 X 50 people	= 192,000 lbs. or **96 tons**

Interesting fact: The largest effigy, or animal-shaped, mound in the world is the Great Serpent Mound in southern Ohio. It measures over a quarter mile long. *See Problem #11.*

<u>*Solution for Problem #7*</u>

Once again, students are dealing with some educated trial and error. The octagon is the largest of the enclosures two-and-a-half times the size of the square. Remembering that the remaining circles are of equal size, it is best to begin working possible variables for the octagon. In so doing the solution works to be **octagon, 50 acres; square, 20 acres; circles 30 acres each.**

Interesting fact: Since the early 1900's, the octagonal mound and one of the circular mounds have been part of a prominent golf course in Newark. The country club has leased the grounds from the state historical society while providing continual maintenance for the grounds.

<u>*Solution for Problem #8*</u>

Students would be best served by diagramming the circular seating arrangement as follows using four diameters to denote specific seats:

		Stone Worker		
	Weaver		Farmer	
	Warrior		Star Watcher (entrance)	
	Potter		Master Builder	
		Healer		

Interesting Fact: The Hopewell Culture of mound builders (220 B.C.- 500 A.D.) of the Ohio River valley had such a food-rich environment that barely three hours a day was needed to obtain sustenance for each individual. This wealth allowed time for more specialized work such as the types alluded to in the problem.

Solution for Problem #9

Students will need to identify the pattern implicit within this problem. There is a +1000, +2000 pattern in the amount of dirt placed on top of the mound in subsequent decades—-1000 tons in the first decade *(+1000);* 2000 tons in the second decade (*+2000)*; 4000 tons in the third decade; (+*1000)* 5000 tons in the fourth decade, etc. A table may also be of some assistance:

Decade	Soil Added (tons)	Increase from Previous Decade	Total Amt.
1	1000		1000
2	2000	+1000	3000
3	4000	+2000	7000
4	5000	+1000	12000
5	7000	+2000	19000
6	8000	+1000	27000
7	10000	+2000	37000
8	11000	+1000	48000
9	**13000**	**+2000**	**61000**

It would have taken **90 years** for 60,000 tons of earth to have been heaped on to the formation known as Grave Creek Mound.

Interesting Fact: Grave Creek Mound is the largest conical mound built by any mound building culture. The Adena culture built this 69 foot high mound in stages between 250 B.C. and 150 B.C.

Solution for Problem #10

Students need to incorporate the reasoning of a lexicographer, or at least confer with Mr. Webster, in deducing the answer to this problem. The equation that needs to be addressed here is :

(1986 - (millennium - 4 score)) + 4	= *date of Serpent Mound*
(1986 - (1000 - 80)) + 4	= *d*
(1986 - 920) + 4	= *d*
1066 + 4	= **1070 (A.D.)**

Interesting fact: The Great Serpent Mound had not been exposed to significant archaeological study since the 1890's when in the early 1990's modern science produced the radiocarbon dates. Previously thought to have been built about 1000 B.C., the new evidence begs speculation that the mound may not be a serpent devouring an egg, rather the head of a heavenly serpent (comet) followed by its tail. Chinese astronomers are credited with recording the brightest occurrence of Halley's Comet in 1066.

Solution for Problem #11

Again students may want to construct a table in assisting them to reason the correct answer:

Name	Age	Position within clan
Walking Cat	40's	mute (not a storyteller)
Whispering		
Spring	?	?
Straight Tree	?	chief
Dark Night	40's	chief
Bird That Sings	?	shaman
Cold Moon	50's or 60's	?
	but not the oldest	

Straight Tree, Dark Night and Bird That Sings can be eliminated from consideration since they are not storytellers. Walking Cat, as a mute, may also be removed as a candidate; he being the second shaman. The two storytellers are Cold Moon, who is not the oldest and **Whispering Spring who is the oldest storyteller.**

Interesting Fact: It is believed that Hopewellian adults enjoyed a significant life span and that infant mortality was relatively low for the era in which they lived. Hopewell elders were living receptacles for the accrued political and natural history of their culture which they presented orally to the young.

About the Author

Max W. Fischer is a National Board Certified middle school teacher with nearly three decades of classroom experience in the fields of math and social studies. He has authored eight other resource books for teachers. His belief in the interdisciplinary approach to instruction was the motivation behind *The Math of Ancient History*.

0-595-23465-8

Ecuador and the United States Free Trade Agreement

JUAN CARLOS GACHET

Copyright © 2016 Juan Carlos Gachet

All rights reserved.

ISBN 1522802053
ISBN 9781522802051

DEDICATION

This book is dedicated to my family, in essence they have been an inspiration throughout all these years. My wife, Claudia Gachet; daughters, Monique and Lisette; son Jean Carlo, as well as my parents Juan E. Gachet and Bertha; brothers Augusto and Roberto Gachet; my grandmother, Laura Gachet, aunts, uncles, and cousins who supported me when I arrived in the United States.

CONTENTS

ACKNOWLEDGEMENTS

First of all, I would like to express my sincere appreciation to all the professors who guided me through college and the final stage of my postgraduate studies, their outstanding support, encouragement, and motivation. All of them were an invaluable source of inspiration during each step required to publish this book.

The journey through *Mejía* high school of Quito, Ecuador. Its Latin motto "*per aspera ad astra*" which means "through hardships to the stars" set an important milestone and inspiration for my academic endeavors that lead to the world of postsecondary education.

Chapter 1

HISTORICAL FACTS OF ECUADOR

The main focus of this book is to be aware of the effects of a Free Trade Agreement (FTC) between Ecuadorian sectors of interest and the United States with respect to the agricultural industry. The subject matter is geared to spell out the advantages, disadvantages, and the implications of free trade for the Ecuadorian people. It is not to convey at any time if Ecuador enters the Free Trade Agreement with the United States.

The objective is to inform the reader about the main economic issues at the micro and macroeconomic levels that would lead to an educated evaluation. The use of these tools would allow to come up with the right decision and have the necessary knowledge to inform others of the benefits generated by the Free Trade Agreement.

Free trade between different countries generates employment, attracts foreign and local investment, creates and boosts technological development, which in turn leads to industrial progress. Free trade mostly increases earnings in affected sectors and mainly in agriculture. Therefore, this literature will focus on how a fair Free Trade Agreement would benefit both markets.

An Ecuadorian economic deal with the U.S. has been in the development stages for a long time, and the politicians who get elected by popular vote to assume the presidency for

Palacio de Carondelet, Quito - Ecuador

Source: Wikipedia Turismo, 2015

a four year term never fulfill it. This is just a political campaign strategy to get the popular vote and keep the same *modus vivendi*.

NAFTA (The North American Free Trade Agreement) is the best example on how the Canadian, U.S. and Mexican markets benefit from each other. The Mexican economy benefited astronomically since the Free Trade Agreement between the three North American countries became reality. Mexico will become the 10th economic power by 2015 as a result of free trade.

Since this book focuses on the ongoing economic trade deal between the two American nations, it is imperative to take a close look to the basic historical facts of the South American country.

For about three centuries, Ecuador was under Spanish control after the capture and execution of Atahualpa, the last emperor of *Twantinsuyu*, the Inca Empire. Later on the northern part of the empire was known as the Territory of

Quito.

The French Geodesic Mission was in charge of locating and measuring the center of the world that divided the northern and southern hemispheres through the Equatorial Line. This event placed the region on the international spotlight and the government of Quito accepted the name Republic of Ecuador when the country separated from *Gran Colombia* in 1830.

Ecuador also refers to the Earth's equator that goes through to the very northern tip of Quito. The Spanish Secret News of America mentioned for the first time the name of the country, and during that period the territory of Ecuador was better known as the "Royal Audience of Quito."

The location of the country is on the north-west tip of the South American continent. Colombia is the northern neighbor and Peru borders Ecuador mainly to the east and south. The Pacific is to the west of the Ecuadorian seaboard, which places the country on a strategic location close to the Panama Canal and easy access to Europe and Africa.

The Galapagos Islands are part of Ecuador located in the Pacific about 600 miles from the coast. This archipelago is one of the main touristic attractions of South America. Charles Darwin conducted scientific research on the native species and developed his Theory of Evolution based on his observations and studies in the early part of the 19th century.

Ecuador covers a total area of 109,483 square miles and is the fourth-smallest in South America with an approximate population of 15,000,000.

The capital of Ecuador is Quito, located in the northern part of the country on a valley surrounded by snowcapped volcanoes at about 9,400 feet above sea level on the Andes Mountains, which contributes to spring like weather all year long.

Historians agree that the revolt for independence started in Quito with the massacre of the revolutionaries on 10 August 1809, which gave the city the name of "*Quito Luz de America*," Spanish for "Quito Light of America." Ecuador achieved independence from Spain after a long struggle for

Battle of Pichincha

Source: Ecuador History, 2014

freedom with the victory of *Simón Bolívar*'s most trusted warrior, *General Antonio José de Sucre* of *Cumaná*, Venezuela at the Battle of Pichincha on May 24, 1822.

In 1978, the city was declared "*Patrimonio Cultural de la Humanidad*," translated as World Cultural Heritage Site by UNESCO (United Nations Educational, Scientific and Cultural Organization) because of the unique and very well preserved colonial style architecture, religious art and tradition.

The new countries of Colombia, Ecuador, Venezuela, and Bolivia joined the *Gran Colombia*. Ecuador broke away from the Gran Colombia right after the Liberator *Simón Bolívar* resigned his presidency on April 27, 1830. Therefore, Ecuador became an independent country on May 13, 1830.

Ecuador is a republic where the President has the right of power as head of state and government, provided that political power is kept at that level and not beyond dictatorship, which was part of the political arena prior to 1979 in most South American countries.

All persons who are literate in the age group 18 to 65

have the right to vote for the political candidate of their choice every four years, and from a great deal of diverse political parties with many ideologies and many agendas.

The economy of Ecuador is very dependent on oil and the export of petroleum products since the dictatorship of *General Guillermo Rodríguez Lara*, who nationalized the industry under *CEPE* (*Corporación Estatal Petrolera Ecuatoriana*), the Spanish for Ecuadorian Oil State Corporation.

Other known industries are food processing on the west; tourism mainly in Quito, Cuenca and *Galápagos*; textiles on the coastal area; wood products from the Amazon, and chemicals. The oil industry represents over sixty percent of the country's export profits. The country experienced a great financial crisis in the late 1990s due to an earthquake that affected mainly the Amazon region and the oil pipelines, as well as the last war with Peru in 1995. However, Ecuador has recovered from its losses and it is developing very quickly.

Ecuador trades mostly with the U.S., Colombia, Peru, Venezuela, and Brazil. The European Community and Russia also have their trade posts in the South American nation.

Spanish is Ecuador's official language since the *Conquistadores* took over the Inca Empire and deposed its last Emperor, Atahualpa. An extraordinary ransom was paid to the Spaniards in exchange of Atahualpa's release. Roman Catholicism became the main religion and was imposed by the King of Spain, Carlos V, when Quito was a colony.

Some of the most breath taking and amazing places in the world are located in Ecuador. Cotopaxi, the highest snowcapped active volcano in the world is located a few hours south of Quito and attracts many tourists and Ecuadorians who love to climb mountains.

In order to understand what entails an economic deal, the first logical step is to look closer and see about its real meaning. The Free Trade Agreement is nothing more than an economic consensus between the interested countries to create a free trade area where daily business can be conducted without taking into accounting their borders expeditiously and without imposing any tariff at all.

The very well-known consensus that economic deals contribute to greater numbers in products sold abroad and overall output in a manageable period is attainable. Therefore, there is a great chance to implement an analysis of very well developed plans that lead to the development.

I certainly agree with this statement only because the European Union had to go through the motions of letting in countries with less developed economies that would not meet the standards such as Spain, Italy and Greece during the first years.

This was a significant economic headache, however, these countries with economies in trouble benefited from the membership. The injecting of needed capital to their markets contributed to the overall growth of the European Union to the point that the euro surpassed the market value of the U.S. dollar.

This book mainly focuses on the current situation and touches on historical facts about the developing issues of the Free Trade Agreement between Ecuador and the United States about the agricultural industry.

It incorporates the perspective from some scholars and economists in order to come up with a solution that would benefit both countries.

As of January 2004, the United States has signed free trade agreements (FTAs) with different countries from all over the world, such as Israel, Canada, Mexico, Jordan, Singapore, and Chile.

The U.S. is currently crafting bilateral FTAs with other 12 countries: Australia, Morocco, Central America, and the 5 members of the Southern African Customs Union, which some also count among the 33 partners in the negotiation of the Free Trade Area of the Americas (FTAA).

In addition, the U.S. Trade Representative (USTR) has notified the government of his intention to open talks with Bahrain, some South American countries, and Thailand in 2004. Current FTA partners account for more than one-third of total U.S. merchandise trade; upon completion of the new bilateral FTA negotiations and the FTAA, more than 40

percent of U.S. trade will be covered with free trade pacts (Schott, 2004).

In order to have a clear understanding and a big picture of Ecuadorian trade, the first necessary step is to take a look at its international trade from the macroeconomic perspective.

According to the 2012 Time Almanac, "Imports as of 2006: $12,114,000,000. Mineral fuels 21.1%; machinery and apparatus 20%; chemical products 15.3%; motor vehicles and parts 11.5%; iron and steel 6%. Major import sources as of 2008: U.S. 19%; Colombia 9.6%; Brazil 4.8%; Japan 3.6%; Mexico 3.5%.

Exports as of 2006: $12,728,000,000. Crude petroleum 54.5%; bananas and plantains 9.5%; fish 5.4%; shrimp 4.6%; refined petroleum products 3.9%; cut flowers 3.4%. Major export destinations as of 2008: U.S. 45.3%; Peru 9.2%; Chile 8.2%; Colombia 4.2%; Venezuela 3.8%.

This international trade macroeconomic view is more useful to the Ecuadorian economy than to the U.S. because Ecuador's exports account for 45.3% to the U.S., and its imported products from the United States account for 19%. The booming economy of the 1970s is attributed to the discovery of crude oil in the Amazon region, which revolutionized the Ecuadorian economy accounting for the majority of exports to other parts of the world.

A significant number of U.S. companies are conducting business in Ecuador, which is a good economic indicator because foreign investment stimulates the local economy by pumping in capital and creating jobs. This is beneficial for both markets no matter what the political views. It is simple math in economy; the "laws of supply and demand" do the job by generating profit for both countries. Therefore, it is a win-win situation.

American companies engaged in the Ecuadorian production industry that contribute to the local development are General Motors, Philip Morris, Mead Johnson, Avon known for cosmetics and lingerie; The Coca-Cola Co. famous for the soft drink market, and a few more.

U.S. fast food and restaurant franchises now operate in

Ecuador; the most famous are Pizza Hut, Kentucky Fried Chicken, Taco Bell, Burger King, McDonalds. The most recent ones established in the main Ecuadorian cities are Tony Roma's, Johnny Rockets, TGI Fridays, Chili's, Papa John's, Domino's Pizza, Subway, Quiznos, and Hooters.

According to U.S. Commercial Service, 2011. The U.S. banking sector is mainly conducted by Citibank and Bank of America. The transportation industry is represented by American Airlines, Continental and Delta. The U.S. has invested little by little in the textile and agricultural markets such as flowers, fruit and vegetables.

Twenty years ago, there were only a few firms operating down there, such as Texaco, Burger King and Kentucky Fried Chicken. Based on the number of U.S. firms operating nowadays in Ecuador, it is obvious that the South American country is practically Americanized.

The inflation rate was skyrocketing uncontrollably and in a sudden political, strategic move to control the Ecuadorian economy, the ultimate solution was to replace the sucre with the U.S dollar as the national currency at the beginning of the year 2000. Those turbulent times are already gone and the implementation of the U.S. dollar as the national monetary unit energized the economy in the long run.

The main problem with the free trade agreement is its stagnation. However, president Correa has decided to renegotiate the country's debt due to an earthquake in the Amazon region that destroyed the main oil pipelines, antagonized a free trade deal with the U.S., and yet decided to stop concessions for mining corporations.

Part of Correa's presidential campaign was to terminate the contract with the only U.S. military base in South America. That became a reality when the U.S. military base contract in Manta was not renewed due to an alleged surveillance scandal involving high-ranking officials.

There are different arguments that come to the negotiating table for a trade consensus between Ecuador and the United States that would benefit the Ecuadorian economy, or it would only place the Ecuadorian population in a position

of economic disadvantage.

According to Silkee in 2008, Ecuador's diplomatic dealings with the U. S. have been in good standing. However, political turmoil has jeopardized mutual relations. Although the U.S. has finalized free trade agreements with Ecuador's neighbors to the north and south, efforts for a mutual free trade agreement with Ecuador have been put off over a dispute with an oil company.

From a political and economic point of view the subject matter experts have advised the U.S. government not to contradict Correa, but to use diplomatic tactics to force him to maintain an open market and democratic policies, such as maintaining U.S. trade preferences for Ecuador.

During the administration of *General Guillermo Rodríguez Lara* (1972-1976), the oil industry was nationalized under *CEPE* (Spanish acronym) *for* Ecuadorian Oil State Corporation. More oil was discovered in the Amazon basin, this in turn revolutionized the Ecuadorian economy, and led to a transition from agricultural to oil producing.

Goodman, Gonzalez, 2008, stated that more than 50% of the economy of the South American country relies mainly on the oil industry and manufacturing is reserved for the local market. Local businesses and agricultural products are used for domestic consumerism and export.

The main commodities account for exports that generate a high influx of capital are oil, flowers, bananas, shrimp, and other goods. In the year 2010, crude and refined oil products generated 56% of total export income.

Ecuador is the world's leading banana exporter, as seen in the U.S. and European markets, generating about $2.1 billion and a major exporter of shrimp that brings in $830 million and cacao with a production of $402 million.

Exports of other products or better yet new to the market such as flowers account for $600 million, canned fish $600 million, and the automobile industry contributes with $375 million.

The Free Trade Agreement between Ecuador and the U.S. would not ultimately benefit the Ecuadorian economy in

the short term because the business industry over there is not very well developed yet.

In other words, Ecuador does not have economies of scale, and if, for example, Levis enters that market, it would help the consumer, but the local producer would be at disadvantage. The consumer will have the freedom of choice and get a better product at a higher price, however, the producer would be forced to go out of business because it could not produce in mass quantities, neither pays its employees enough salary, or support business operations.

The same would occur in the agribusiness, let's say the Ecuadorian government allows the dairy industry enter the local market at $1.00 per gallon, the local producer is selling the same product for $1.50. The consumer, of course, would choose the cheaper product forcing the local producer to lower the price. However, this local vendor would not be able to produce in scale, and ultimately be forced to go out of business.

As mentioned before, the Ecuadorian production is not an economy of scale yet, and it would take years to get to that level. Once that is accomplished the local market would be ready to engage a Free Trade Agreement with the United States. That is the main motive the different governments have not entered any agreement yet in order to protect the local market and mainly the local producer.

Other sectors of the Ecuadorian economy would benefit with an FTA, for example, the banking industry would be forced to compete, let's say with Bank of America. Why? Simply because Bank of America offers better services to the customer forcing the local bankers to compete by rethinking and reengineering their strategy and give the customer better deals with loans at better rates.

The main goal is to dig into the main facts and get the best answer for the Ecuadorian economy and the people who are contributing to its progress. Also, to study and get to the roots of the main issues why this process has come to a halt. There is much talk about this subject from the political point of view, as well as from a regular citizen, people who are in

favor and others who only oppose because of Yankee domination.

The objective is to perform the right analysis that impacts the Free Trade Agreement in regard to production, imports, exports, consumerism and incoming capital of the different economic agents of the Ecuadorian economy. This study analyses the results from various scenarios, with FTA and without FTA, for Ecuador using an applied model of general equilibrium, which includes data of the economic partners dealing with the FTA in South America, and the European Union.

Furthermore, this study takes into account the interdependencies between these trade partners, as well as the direct and indirect effects that an agreement with great and deep repercussions, such as the FTA, would bring upon the Ecuadorian economy.

Different scenarios were established in order to analyze the impact of the FTA in the Ecuadorian economy. On the first scenarios, Ecuador signs the FTA, and on the other scenarios Ecuador does not sign the FTA.

The scenario where Ecuador signs the FTA includes Colombia and Peru, which allow the U.S. free access to their markets. In exchange, the Andean countries consolidate the actual tariff preferences provided by the U.S., get various preferences by ATPDEA (Andean Trade Protection and Drug Eradication Act), and the U.S. eliminates subsidies in the production of the agricultural goods.

The scenarios in which Ecuador does not sign the FTA, but Colombia and Peru agree to do it would have these implications. Colombia and Peru allow the U.S. free access to their markets; in exchange these countries get preferences from ATPDEA.

Ecuador loses tariff preferential treatment from ATPDEA and it is imposed the same tariffs as Venezuela from the U.S. In other words, Ecuador gets more expensive tariffs than the ones imposed by the U.S. at the same price of the Venezuelan tariffs imposed by the United States.

On this other situation, Colombia and Peru allow the

U.S. free access to their markets. In exchange these two Andean countries get better preferences from ATPDEA, and the U.S. eliminates the subsidies in the production of agricultural goods. Ecuador lets go the tariff preferential treatment of ATPDEA and the country is obligated to pay tariffs equal to the ones imposed to Venezuela by the U.S.

The preliminary results will be if Ecuador signs the FTA as its Andean economic partners Colombia and Peru. If Ecuador does not obtain more leeway from the U.S. than the consolidation of ATPDEA, and assuming that there would not be any external or internal friction. This would implicate an immediate result of a commercial deficit, reduction in the income of the production factors, and the free fall of the welfare of the people in Ecuador.

If Ecuador signs the FTA and the Andean countries get better preferential tariffs and a greater quota of sugar. That translates as a 50% reduction of tariffs equivalent ad valorem by the U.S.

Ecuador would even experiment an immediate effect of the economic deficit, reduction of income of goods used in the production. That would lead to the free fall of the Ecuadorian welfare.

Again, if Ecuador signs the FTA and allowing the other Andean countries the tariff advantages already mentioned. As a result, the U.S. reduces its subsidies to the agricultural production. Ecuador would even end up with a negative impact on its commercial scale, a minor drop in production and demand, the income of the factors of production, and a minor fall in the welfare of the population.

Even without the FTA, Ecuador would experiment the same effects as with an FTA but in less proportion. The aggregated value of the majority of the sectors would show a fall, the income of the factors reflects that reduction in aggregate value, the aggregate demand falls, as well as the welfare of the community. In general, the economy does not show a better performance without an FTA, because of the fall of the gross domestic product.

However, if it is analyzed what would happen in the near

and long term, it is expected that with an FTA positive things would happen and the necessary mechanisms are in place to bring investment. Furthermore, it is expected that the FTA becomes the road for better technologies and innovation.

Colombia and Peru officially signed the FTA with the U.S. It was assumed that it would be difficult to protect some economic sectors threatened by competitive products imported from the United States. In particular, rice, cereals, and other products, because these would enter Ecuador from the bordering countries that signed the FTA with the U.S.

In an analysis of the different sectors, it is obvious that the sectors that would benefit with the FTA, in contrast of a reduction without the FTA, include the sectors that benefit with the tariffs provided by the ATPDEA, such as, these particular markets: fruit, vegetable, flowers, sugar and fishing.

According to the 2008 CRS Report for Congress, negotiations for a diplomatic free trade agreement with Ecuador have been at a halt until further notice over the dispute with the U.S. Oil company. In this case, the U.S. has to give in and reach an agreement with the Correa government or just wait with patience until his term comes to an end and the Ecuadorian people elect the next president.

The Ecuadorian foreign trade greatly depends on the U.S., for instance Ecuador imports from the U.S. are 19%, and exports 48% to the United States. Many investments come and are funded by U.S. businesses established in Ecuador.

The impact of a Free Trade Agreement according to Goodwin (2006), the Ricardian arguments on "Why Nations Often Resist Free Trade" mainly focus on efficiency, increased production and consumption, and why not the opportunities and benefits of a well-integrated global economy. Again, this is the case of the Correa Government in Ecuador. The president argues that an FTA agreement is not beneficial for the country for many reasons.

At this point, it is necessary to bring up an interview with President Rafael Correa on why "Ecuador is No Longer for

Battle of Yorktown

Source: Based on Virginia Memory.com, 2014

Sale" conducted by Journalist Greg Palast in 2008. Correa argued that the country is no longer on sale because of its sovereignty and nationalist government.
This is in essence true; some idiosyncrasies are used to isolation from the rest of the world. However, any economy needs foreign investment to stimulate the local markets in order to survive and get competitive.

This is a powerful interview that leads to the basic roots of the resistance from the Ecuadorian government to sign the FTA with the U.S. The main one refers to environmental and social damage in the Amazon region, as well as the opposition of U.S. military presence in Ecuadorian soil.

It is essential to talk about the main U.S. history facts to have a better understanding of the leading economy power in the world.

The Spanish, French and British colonized North America during the 16th and 17th centuries. Ponce de *León* explored Florida in 1513, Giovanni da Verrazzano under French command explored the North Atlantic in 1524, Hernando de Soto discovered the Mississippi River in 1542,

Sir Walter Raleigh founded Roanoke Colony in 1587.

The Continental Congress of the thirteen American colonies proclaimed the Declaration of Independence from the former British Empire on July 4, 1776.

The combined force of the Continental Army led by General George Washington and the French Army led by Comte de Rochambeau gave the final blow to the British and German forces commanded by Lord Cornwallis, who surrendered at the battle of Yorktown on 19 October, 1781. The Treaty of Paris (3 September, 1783) officially ended the American Revolutionary War.

Guayaquil, Ecuador

Source: Wikipedia Turismo, 2015

Chapter 2

FREE TRADE AGREEMENT IN THE AMERICAS

Not only does Ecuador straddle the Equator, it is virtually unique among the Latin American and Central American countries in general. It continues to straddle particularly in the line of its dealings with the U.S. for a common free trade agreement as part of a larger Free Trade Agreement of the whole continent. Indeed, despite ongoing negotiations, Ecuador stands out from its neighbors because it still does not have a common free trade agenda in place with the United States.

Over the past two decades, the trade agreements with different regions of the continent have been implemented with mixed results, but the Andean country of Ecuador remains excluded from these agreements and negotiations continue to this day. This chapter examines the background to these trade agreements and their current status with respect to Ecuador to identify potential barriers to progress and the likelihood for success.

The experience and overview on this economic subject are better portrayed by the adage that "a rising tide raises all boats." It is believed to especially apply to multilateral free trade agreements that increase commerce between countries

in ways that are mutually beneficial. Despite ongoing attempts to forge such an economic deal with Ecuador, both countries have failed to come up with a solution to the remaining obstacles to the former's inclusion in free trade deals with the U.S.

Nevertheless, the United States still represents Ecuador most vital export and import trading partner. Also, some observers suggest that Ecuador's accession to these free trade agreements are superfluous and the majority of the goods being traded between these two trading partners are already duty free.

Other analysts, though, suggest that by keeping Ecuador from membership in these multilateral trade agreements, the country is being denied equitable participation with neighboring Central American countries on a level economic playing field.

Although Ecuador enjoys other natural resources such as gold, timber and copper as well as a significant agricultural industry. The country's economy remains highly reliant on its known petroleum resources, which represent more than 50 percent of its exports and about 40 percent of its public sector gains in the last seasons (USDA at work in Ecuador, 2011).

During the early 1990s, Ecuador also experienced significant economic crisis that was caused by the banking crisis, causing GDP to shrink by 5.3% (Ecuador economy, 2012). As a result, the Ecuadorian government defaulted on external debts and poverty in the country increased substantially (Ecuador economy, 2012).

In response to the situation, the U.S. government authorized a number of structural reforms for Ecuador to help it recover from its economic downturn, including the use of the U.S. dollar for legal tender in Ecuador (Ecuador economy, 2012). Some economists describe the situation during this period thusly.

According to Matthews, Maniam & Balasundram (2006), Ecuador went through a complete overhaul of its monetary system, through world market macroeconomic shocks, which lastly obligated the country to relinquish its national currency,

the sucre, and adopt the U.S. dollar.

Following a series of structural reforms, Ecuador's economy improved and, due in large part to remittances from expatriates, higher oil prices and additional new exports (Ecuador economy, 2012). In fact, during the period 2002 to 2006, Ecuador's economy enjoyed a healthy 5.2 percent per year average increase, representing the most aggressive growth in a quarter century.

Following modest economic growth through 2007, the Ecuadorian economy zoomed to reach a 7.2% growth rate in 2008, attributable to increased petroleum prices worldwide as well as further investment in Ecuador's public sector.

Despite this sustained economy growth, things changed when President Rafael Correa assumed office in January 2007. Following his accession to office, the new president defaulted on Ecuador's sovereign debt in December 2008 in the approximate amount of $3.2 billion, an amount that accounted for about 30% of the country's public external debt.

Despite this economic setback, by May 2009, Ecuador was sufficiently solvent to redeem nearly all (91%) its defaulted debts (Ecuador economy, 2012). Subsequent actions by the new president's administration have created some cause for concern for the future.

The lingering aftereffects of the Great Recession of 2008, though, together with decreased remittances from expatriates and sharp reductions in global oil prices adversely affected the Ecuadorian economy and growth were just 0.4% in 2009. By 2010, however, economic growth had resumed healthier levels and achieved a 3.6% increase in 2010 and a 6.5% increase in 2011.

More recently, other indications suggest that the Ecuadorian economy be well poised for sustained growth in the future. For example, in 2011 alone, the Ecuadorian government:

1. Resolved a $2 billion loan deal with the China Development Bank.

2. Received $1 billion under a 2 year agreement sale of an oil covenant.
3. Conveyed $571 million in financing with the bank Exim of Chinas for a new hydroelectric project.
4. Proclaimed plans to secure more Chinese credits in 2012 (Ecuador economy, 2012).

Although the United States remains Ecuador's top export and import trading partner, since the Ecuadorian government defaulted in 2008, China has since then become Ecuador's largest overseas mutual lender which permits the Ecuadorian government wide latitude in social spending.

This latitude in social spending, of course, can also lead to politically motivated spending in ways that are intended to promote immediate chances for reelection rather than a long term economic development needs of the country. Moreover, it is difficult or even impossible to place a price tag on the country's leadership and ability to levy tariffs at will in response to changes in the marketplace in ways that contribute to their popularity.

Taken together, this powerful combination of political flexibility in the face of a dynamic social situation means that it is important to "follow the money" in order to determine who is benefiting from international trade and these issues are discussed further ahead.

The current status of free trade agreements in the Americas across the board with regional trading coalitions and mutual trade pacts have assumed a progressively significant role in the western hemisphere recently.

The U.S. continues to discuss an open trade arrangement with the Ecuadorian Government. At the time of writing, the editors report consumption, and Ecuador was also part of the AFTA negotiations but Peru managed to come to mutual terms on a unilateral deal with the great power of the north as the Ecuadorian status remained under negotiation.

There were some clear indications of a *quid pro quo* arrangement in the unilateral agreement forged between the U.S. and Colombia. For example, although the Ecuadorian

and Peruvian governments also had tax unrestricted admission to the U.S. marketplace for a mainstream of their goods pursuant to the AFTA, the agreement with Colombia and Peru, locks in and increases that tax-free admittance. In return, there was an immediate lifting of Colombian tariffs on almost 80% of U.S. manufactured products with an agreement to phase the remaining 18% out during the coming decade.

Furthermore, in 2003, the United States executive branch indicated that there would be subsequent negotiations to follow with respect to separate trade deal with the Republic Panama. The U.S. government ratified a consensus of assistance and accord for a United States-Uruguay Common Venture Accord at the end of 2006 (Faber, 2007).

Clearly, though, there are some "stumbling blocks" involved in these "building blocks." A Common Market Trade Area of the Americas, with Ecuador standing out like a sore thumb in an otherwise healthy hand by virtue of the complete absence of tariffs and duties that level the economic playing for free trade agreement members.

In reality, tariffs may be used in a great deal of diverse ways, depending on the desired outcomes and status of the trading parties. In some cases, tariffs are only designed to level the playing field while, in others, they are specifically intended for revenue-generation purchases and in others, they are intended as a sanction.

For instance, tariffs are taxes levied against products imported from abroad. International business is affected by tariffs and related trade restrictions; tariffs can be classified as either revenue or protective and a high tariff payment would make domestic products competitive. It is also known as duty. In addition, revenue tariffs are imposed to raise funds for any government, and protective tariffs, which are higher than revenue tariffs, are designed to increase the retail of imported goods.

As the only remaining Andean country that is without a formalized open trade treaty with the United States, these are essential issues since they relate to the ability of such agreements to help improve the standard of living for all

Ecuadorians rather than the select opulent elite.

In this regard, Venkataramany and Sivakumar (2008) report, "The countries that are presently part to the planned Free Trade Area of the Americas pact (FTAA) have exclusive requirements and worries, but nearly all are partaking to bring the remunerations of unrestricted trade down to the level of each associate of their community."

Therefore, the current status of the FTAA may not be at all surprising given the conflicting interests and dynamic political issues that are involved. The issue is just how long negotiators have been working to hammer out their differences.

In fact, the initial work with Ecuador and its status with respect to the FTAA directly shadowed the idea of the North American Free Trade Agreement during the mid-1990s and most observers at the time and in succeeding years have anticipated the formalization of this agreement by the established target date of January 1, 2005

This was the focus of a series of protests that occurred in Ecuador in response to the government's attempts to forge an AFTA agreement with the United States in 2006.

The social protest that emerged in response to Ecuador's determinations to forge an open trade treaty with the United States. In fact, unpopular among the majority of indigenous people, who believed the effort was an attempt to extend American hegemony in their country even more.

These are troubling issues in a purportedly post-colonial Latin America, but it is clear that in the 21st century where many multinational corporations' annual revenues exceed the budget of all Central American countries combined.

The manner in which the Ecuadorian government responds to these issues will likely dictate which course of action will be required to quash social dissent and to what extent military force will be used to achieve that goal. To date, reports from Ecuador sound more like something from the Middle East than in America's backyard.

The willingness of the Ecuadorian government to employ military force was also made clear during these

demonstrations. Not only did the government respond with military force, it suspended constitutional rights and implemented virtual martial law with broad arrest powers.

According to a report from *NotiSur* points out that, the police force and the army were mobilized to clear obstructed highways during the month of March, 2006. Then the administration declared a state of emergency in different parts of the country.

Ecuadorians reacted to the AFTA initiative with a particular young show of force. Approximately 1,500 natives and students marched peacefully through the capital, Quito, on March 22, protesting against a possible deal with the United States.

In a speech to councils of the Organization for American States (OAS) and the World Affairs Council in Washington, D.C. on January 16, 2002, President George Bush emphasized America's pledge to the establishment of a Free Trade Area of the Americas by 2005.

This was not the first time President Bush had made such a commitment to the Free Trade Agreement for the Americas initiative.

In fact, in a speech to the OAS, in 2001, the president reaffirmed the commitment of the United States to the promotion of the Free Trade Agreement for the Americas would make the American continent the largest free-trade area in the world, incorporating 34 countries and 800 million people. (Bush pushes Free Trade Area, 2002).

The target date, though, has come and gone and nothing substantive has changed in the interim except growing skepticism that any successful outcome can be achieved at all. The current status of the FTAA has remained virtually unchanged for the past 5 years: "Free Trade Agreement of the Americas talks crushed to a standstill in 2007 and remain pending" (Venkataramany & Sivakumar, 2008).

Indeed, even the Office of the U.S. Trade Representative's (USTR) Web site has been shut down with no indication of its future status and all of the references to Ecuador have been disabled (although they are still listed).

In reality, the concept of "free trade" sounds great for everyone at first blush. After all, more trade equals increased prosperity and who can argue with it being done "freely"? In this respect, Fandl (2007) reports that there are a number of desirable outcomes related with free trade agreements. By definition, free trade is intended to increase imports and exports between countries in ways that are mutually advantageous.

When the term "free trade" is used to describe agreements that are not necessarily fair, though, things become more abstruse. The economic results of free trade pacts on Andean countries following the money in international trade is complicated because of its multilayered processes.

In this respect, some observers suggest that this be the case with the FTAA, which is predicted to upsurge the economic prosperity of the participant countries in variable degrees, the rest of the world envisions it as a scheme of trade distraction. In an increasingly globalized marketplace, the Open Trade Agreement of the Americas concept may have become obsolete before it ever had a chance.

Furthermore, some analysts argue that the Free Trade Agreement of the Americas is superfluous since the vast majority of all goods are already exempt from government-imposed duties. For example, Kornis (2004) emphasizes that, as the United States begins talks with the South American counterparts for a United States-Andean free trade agreement, 86 percent of combined U.S. imports from these countries are already free of duty.

In the "follow-the-money" analysis, there must be some motivation on the part of Ecuador and the other Andean countries to hammer out any remaining differences they may have with the United States in general and the Department of Agriculture in particular with respect to completely free trade.

It turns out that much of the motivation for formalizing these free trade agreements relates to the need to establish a "first mover" status in international trade with the United States.

There have been subsequent efforts to negotiate the potential free trade agreement with Ecuador. The South American countries from the Andes region by this time enjoy duty-free action for the majority of their exports to the U.S. by virtue of the extended Andean Trade Preference Act (ATPA) formalized at the beginning of the millennium (Kornis, 2004).

Prior to its expiration, though, Colombia and the United States formalized a mutual trade treaty on February, 2006 after 2 years of intense negotiations.

The impetus for the bilateral agreement with Colombia to the exclusion of the Andean countries is due in part to the inability of the parties to resolve their remaining differences and it is unclear whether they will be able to do so in the foreseeable future.

The political nature of these negotiations clearly outweighs the underlying economic advantages that could be achieved through expanded free trade. As Fandl (2007), concludes, that new political events in Bolivia and Ecuador point out the improbability of obtaining settlements with those countries.

While Colombia managed to secure an open trade deal with the United States, the remaining obstacles to progress with respect to the Andean countries in general and in regard to Ecuador in particular relate to agricultural issues.

In this respect, Chauvin (2004) reports, "ATPA countries, which are significant markets for U.S. cereals and U.S. cotton are most subtle about admittance to their marketplaces for U.S. agrarian crops. Christian Espinosa, Ecuador's chief negotiator, had reportedly said, "The Andean countries will not yield ground on the agrarian front. We will protect our mechanisms to neutralize the effects of U.S. grants."

At present, pursuant to the Andean Community's so-called "price band" approach, tariffs in varying amount up to 100 percent are authorized to be applied to various agrarian goods that are being imported into Andean countries in order to stabilize prices in the event they drop below predetermined

levels (Chauvin, 2004).

In some cases, the Andean Trade Preference Act has been used to further American goals to reduce illicit drug trafficking between the Andean nations and the United States. According to Baker (2000), the ATPA promotes economic development in Bolivia, Colombia, Ecuador, and Peru. It was established under the George H.W. Bush administration from 1989 to 1993.

Its primary objective was to furnish these countries with an alternative to coca production by giving them more access to the U.S. market.

Similar approaches have been used by the United States in Colombia, for example, in its efforts to provide Colombian farmers with a viable crop alternative to coca. To this end, Burns reports, "Since the beginning of 2000, the United States has delivered different improvement support to more than 81,000 small agrarians, triggering in the farming of over 100,000 hectares of new crops and over 1,100 small infrastructure developments, such as schools, roads, and bridges that have been completed in communities participating in the program."

For the United States, these issues are more important politically than they are economically. For instance, U.S. trade with the ATPA nations remains fairly modest, and even combined.

In 2003, all ATPA countries accounted for just 1 percent of total U.S. transfers and even less 1 percent of total U.S. importations.

Despite the relatively modest percentages, though, the Andean countries of the ATPA were responsible for providing a number of strategically valuable products, including copper, as well as high-profit items such as asparagus and flowers (Kornis, 2004).

Moreover, the United States has experienced a collective shortage in its trade with ATPA countries since 1991.

By contrast, the export levels from the United States to ATPA countries have not increased substantially since 1999, at which point they had decreased fully 28 percent compared

to their value in 1998.

Furthermore, a veritable "perfect storm" of adverse economic conditions has hampered the ATPA countries' ability and desire to import products from the United States.

Much of the ATPA export trade with the United States is linked to oil and natural gas (about 33 percent); in addition, most of the machinery and equipment that is imported by ATPA nations from the United States is used in the petroleum and natural gas industry. In this area, it is clear that the U.S. enjoys a competitive advantage over these oil-exporting countries because they lack the refinery facilities needed to add value to them. For instance, Kornis reports, "Two principal U.S. export goods to the region are refined oil goods, fuel oils and lubricating oils."

In this environment, it is easy to see how the interests of the average Ecuadorian farmer could be overlooked. In an increasingly energy-hungry world, even the most spectacular cut flowers and tasty asparagus, two major Ecuadorian exports, will pale in comparison to the oil and gas that Ecuador has to offer.

Current trends indicate that these patterns will continue well into the foreseeable future as the search for viable alternative fuels continues. Unless and until that goal is achieved, U.S. trade with the ATPA nations will likely continue to be characterized by the strategic use of carrot and stick negotiations and unilateral agreements with Andean countries as the political situations in these countries plays out.

While it would be disingenuous to suggest that the U.S. Department of State be colluding with the Department of Agriculture to keep Ecuador a developing cousin to the south, it is reasonable to suggest that the United States continue to use cost-benefit analysis to determine how best to proceed with holdouts such as Ecuador.

In some cases, this may mean foregoing revenue-generating tariffs that could reduce the availability and adversely affect international pricing standards, issues that clearly worked to Ecuador's advantage.

Moreover, Ecuador already enjoys significant advantages in its existing relationship with the U.S. to the extent that it might not want to become tied to a more rigid framework that precluded its unilateral advantages in negotiating with the United States.

Indeed, Kornis adds, "The newly duty-free status of oil and offshoots from ATPA nations under ATPDEA was also liable for most of the reduction in the dutiable rate of U.S. imports and U.S. duty revenues from ATPA nations."

These outcomes are not all that surprising, though, since the ATPA is not necessarily designed to be a revenue-generating mechanism for the United States, but rather a framework in which political objectives can be achieved through various incentive programs. The ATPA also differs from other U.S. trade agreements because one of its main goals is to reduce trade at least illicit drug trade, among the ATPA nations and the United States.

The ubiquity of illicit drugs in these countries represents an ongoing and pervasive threat to the general safety of the great nation of the north, Sheppard maintains and suggests that this be one value-added activity that America could do without.

As already noted, since 1991, most of the goods exported from Ecuador enter the United States free of customs' duties by virtue of the Andean Trade Preference Act. As also noted, the ATPA has been used by the United States in its global war on terror and drugs in various ways with mixed results in the participating countries.

Concomitantly, the United States have not received any preferential treatment for its goods imported into the nations of the ATPA; the United States does receive another payoff in terms of international strategy and its need to establish a hegemonic arrangement with these countries.

The strategic value of the petroleum and natural gas products flowing to the United States and the bilateral relations with these countries are, therefore, likely far more important than their economic impact, but these levels are increasing. Some indication of these results can be discerned

by recent trade activity.

Currently, Ecuador exports totaled $22.29 billion in 2011, an increase from $17.49 billion in 2010. The country's major exports include petroleum, bananas, cut flowers, shrimp, cacao, coffee, wood, and fish (Ecuador economy, 2012) with its major export trading partners being the United States, Panama, Peru, Venezuela, Chile and Russia as shown in Table 2.1.

Table 2.1: Ecuador Major Export Partners

Country	**2009 % of Exports**
United States	**38.3%**
Panama	**10.1%**
Peru	**6.3%**
Venezuela	**5.3%**
Chile	**5%**
Russia	**4.6%**

Source: Based on tabular data, Ecuador economy, 2012

At present, Ecuador's imports included industrial materials, fuels and lubricants, nondurable consumer goods and totaled $23.58 billion in 2011, a considerable increase over its 2010 levels of $19.3 billion. Ecuador's major import partners are the United States, China, Colombia, Panama, Peru, Brazil and South Korea as set forth in Table 2.2.

Table 2.2: Ecuador Major Import Partners

Country	**2009 % of Imports**
United States	**27.9%**
China	**10.3%**
Colombia	**9.2%**
Panama	**4.6%**
Peru	**4.6%**
Brazil	**4.3%**
South Korea	**4.1%**

Source: Based on tabular data, Ecuador economy, 2012

As noted in the introductory chapter, though, increased trade with the U.S. has not necessarily translated into widespread acceptance of free trade agreements by the Ecuadorian people. In fact, this issue was the focus of a series of protests that occurred in Ecuador in response to the government's attempts to forge an AFTA agreement with the United States in 2006.

In fact, the social protest that emerged in response to Ecuador's struggles to forge an unrestricted trade agreement with the United States indicated many indigenous people believed the effort was an attempt to extend American hegemony in their country even more.

It would seem that U.S. government drug-related concerns are failing to take into account these localized and regional issues from an ethnocentric view. It is clear that the overarching theme that emerges from the relevant literature is the high-level concern for threats to national security pursuant to the Homeland Security Act that trump everything else.

The Andean countries represent a unique situation; however, there remains an overall thrust towards codifying the Free Trade Agreement of the Americas and this goal will likely remain to be the main focus of U.S.-Latin American dealings in the future (Gillgannon, 2004).

Nevertheless, there are some indications that there is far more involved in these ongoing negotiations that many Americans realize, or care, but which still have a potential effect on trade between the U.S. and Ecuador based on overriding issues involving national security.

Perhaps it is not so much a matter of the American public not caring about these issues. It may rather be an indication of just how potentially disruptive U.S. national security issues are today and most likely in the future with respect to trade with the South American countries of the Andean region in general and Ecuador in particular.

On the one hand, Gillgannon suggests that the U.S. public may not know because of a lack of information about these issues: "As usual, the American print and television media have underreported these trade talks and the American

public remains unaware and apathetic." Though, he posits a potential rationale in support of this paucity of timely coverage about the U.S.-Ecuadorian trade negotiations. Both of these foregoing reasons, though, do not take into account national security benefits that are convoluted pursuant to the Homeland Security Act.

In reality, the U.S. national security interests that are involved in Ecuador are widely perceived to be efforts to extend American hegemony in the Central American region even further and the image of the American is becoming widespread once again. In this regard, Gillgannon adds, "South Americans lean fearfully to trade settlements with Asia and the European Union to offset the pressures for trade supremacy and control of their national autonomy by the United States."

Although these negotiations have since been the focal point and the subject to the provisions of the Homeland Security Act. Some indication of how these efforts have been perceived in Ecuador can be discerned from Gillgannon's observation, "Defense Secretary Donald Rumsfeld met with the defense ministers of the Western Hemisphere in Quito, Ecuador, seeking hemispheric solidarity for the United States' war on terrorism."

Furthermore, there are some indications that the U.S. is not only willing to exert itself militarily if needed, it is downright anxious to do so in an effort to prosecute its global war on terrorism in its own hemisphere. These recommendations do not appear so benign, though, when the secretary of defense is taken into account, the other remarks are important.

These are important considerations when evaluating the future of U.S.-Ecuadorian trade relations. It is reasonable to suggest that if things are perceived to have gotten too far out of hand in any region of the Western Hemisphere, the U.S. will be ready to intervene based on the longstanding Monroe Doctrine and the more recent provisions of the Homeland Security Act. For many Americans, these issues are simply unknown because of a dearth of reporting in the U.S.

mainstream media, and it is reasonable to suggest that some Americans just do not know or care enough about these issues to have any impact on future negotiations between these two countries. As Gillgannon points out, "We have not read the signs of the times. We praise the abstractions of justice and peace, but we do not do the dirty work on the complicated details of the new world issues of trade and commerce."

Clearly, these are troubling issues in a purportedly post-colonial Latin America, but it is clear that in the 21st century where many multinational corporations' annual revenues. These exceed the budget of all Central American countries combined, the manner in which the Ecuadorian government responds to these issues will likely dictate which course of action will be required to quash social dissent and to what extent military force will be used to achieve that goal.

The recent condemnation of Syria by the new president of Egypt makes it clear that things are changing in the Middle East, and an Arab spring-like reaction might just occur in America's own backyard. To date, reports from Ecuador sound more like something from the Middle East than in America's backyard.

The willingness of the Ecuadorian Government to employ military force was also made clear during these demonstrations. Not only did the government respond with military force, it suspended the country's constitutional rights and implemented virtual martial law with broad arrest powers, a very unpopular move.

In this environment, trade between the U.S. and Ecuador may take a backseat to trying to resist American influence in the region while simultaneously seeking improved trade relations in the future.

The fact that Ecuadorians are taking to the streets at great personal risk and to demonstrate against so-called "free trade" that would mean more revenues suggests that nothing be really "free" when it comes to dealing with the United States and the Homeland Security Act, according to Clark (2012).

When Ecuadorian community members have encounters

with the legal system they must navigate not only court offices and legal procedures, but contradictory indigenous values.

Also, when they block roads as part of a national uprising against a free trade agreement, they must negotiate internally, as community members may also suffer from a civic strike, as well as with regional and national actors, both inside and outside the state.

Taken together, these events suggest that negotiations between the U.S. and Ecuador may require far longer and involve far more trade-offs than many observers might believe, and it is this problem that forms the focus of the study's methodology which is discussed further ahead.

Washington, DC., United States

Source: Wikipedia Turismo, 2015

Chapter 3

ANALYSIS OF POSSIBLE TRADE AGREEMENTS

This study used a mixed revision design that employed both qualitative and quantitative elements. The qualitative component consisted of an evaluation of the significant juried and academic literature regarding the Free Trade Agreement of the Americas and its implications for Ecuador, its Central American neighbors and the United States, with a focus on the most recent resources available.

This qualitative element is consistent with the guidance provided by numerous social researchers who stress the need to review the existing literature to determine what is known about a topic as well as any gaps that exist (Silverman, 2005).

The use of both qualitative and quantitative components is consistent with the guidance provided by Neuman (2003). He advises, "Both qualitative and quantitative research use several specific research techniques (e.g., survey, interview, and historical analysis), yet there is much overlap between the type of data and the style of research. Most quantitative-style researchers examine qualitative data and vice versa."

Similarities-Differences, Quantitative-Qualitative Analysis of Ecuador

Similarities

Analysis for both comprises inference - that is they both reach a decision based on proof.

Both involve a systematic process.

Both involve comparison, either internally or with related evidence from elsewhere.

Both strive to avoid errors, false conclusions and misleading inferences and seek valid description and explanations.

Differences

Quantitative analysis is highly standardized with little difference between projects; by contrast, qualitative analysis has many more possible approaches.

Quantitative analysis takes place at the end of data; qualitative analysis takes place during data collection.

Quantitative analysis tends to test hypotheses through the manipulation of numbers representing facts.

Qualitative analysis is less abstract and does not assume that real life can be measured by numbers.

Salience

From a qualitative perspective, the fact that the Office of the United States Trade Representative has disabled links to the relevant documents concerning U.S.-Ecuadorian negotiations is noteworthy because it reflects the international security issues that are involved (e.g., Homeland Security Act).

The historic levels of tariffs as evidence are identifiable, though, and these issues can be synthesized under the general equilibrium theory.

The research will proceed in a step-wise fashion.

Given the paucity of U.S. Government documents related to ongoing negotiations with Ecuadorian trade officials, it is

reasonable to suggest the situation remain dynamic and subject to overriding and classified security concerns.

The concept of equilibrium defines a complete system in which all of the parts behave in predictable ways. Thus, general equilibrium theory is a system with many of the characteristics of physical systems and, therefore, can be examined with many of the same research tools used for physical systems. Also, it provides the power of prediction, the ultimate goal of scientific investigation (Kirchhoff, 1999).

There are some distinct similarities between qualitative and quantitative approaches that can contribute in substantive ways to the research process if their differences are taken into account.

In fact, according to Jenson (2004), "It is at the level of methodology, defined as a theoretically informed plan of action in relation to the empirical field, that the distinction between qualitative and quantitative research becomes most apparent."

A quantitative and qualitative study can both employ the same methods, and both share the common goal of developing an accurate map of the underlying concepts onto empirical phenomena. In other words, "It is the specific linking of a necessarily selective empirical 'microcosm' with a theoretical 'macrocosm' which, above all, distinguishes quantitative and qualitative research as traditions and paradigms" (Jenson, 2004).

The quantitative component consisted of an analysis of the potential results from different scenarios, with free trade agreement and without free trade agreement using an applied model of general equilibrium, which includes data of the economic partners dealing with the FTA in Central and South America as well the United States and European Union.
Furthermore, this analysis takes into account the interdependencies between these trade partners, as well as the direct and indirect effects that an agreement with great and deep repercussions, such as the FTA, would bring upon the Ecuadorian economy in terms of the effect of increased trade on the national economy today as well as in the future as new

jobs are created and sustainable business ventures are established.

In a perfect world, increased trade between the United States and Ecuador would have a win-win outcome, but the exigencies of the drug trade flowing to the U.S. from the Andean countries suggest that the U.S. be willing to take an economic loss every time in favor of prosecuting its global war on terrorism, of which drug trafficking represents an important component.

Consequently, the Ecuadorian Government can either take America's offers, whatever they may be at present, or leave them, but the U.S. will likely continue to exert an enormous amount of influence in the area in ways that are not quantifiable.

The CIA has a long track record of clandestine operations in the Andean countries, and it is reasonable to suggest that they remain active today.

The Ecuadorian people, though, are not sitting idly by while these pressures are being brought to bear on them, and again, the fact that they are willing to risk arrest, injury or even death to protest these actions is based on the widespread recognition that increased trade will, in fact, be beneficial for their country.

For example, although trade between the United States and the countries of Latin America declined following the end of World War II through the 1980s, by 1990s witnessed impressive growth in international commerce in this region.

According to Hirst (2004), the rate of U.S. trade with Latin American countries increased significantly faster than the concomitant increase in U.S. global trade: "Latin America had become the fastest growing market for U.S. exports. The United States, at long last, did take notice of Latin America. Trade between some Latin American countries also boomed, especially within subregions such as the southern cone of South America, Venezuela and Colombia, the Central American countries."

The impact of this increased trade with the U.S. was manifested in several ways, including a number of free trade

agreements in the region that remain in place today. For instance, Hirst reports, "The establishment of formal freer-trade areas facilitated the growth of trade and other economic relations."

These free trade agreements included the following:

1. North American Free Trade Agreement (NAFTA): this agreement included Mexico, the United States, and Canada.
2. The *Mercado Comun del Cono Sur* (Southern Cone Common Market, more commonly known as MERCOSUR): this agreement is between Argentina, Brazil, Paraguay, and Uruguay.
3. The Andean Community: members include Bolivia, Colombia, Ecuador, Peru, and Venezuela.
4. The Central American Common Market (CACM).
5. The Caribbean Community (CARICOM).

The collective effect of these free trade agreements was pronounced and all of these countries, with the sole exception of Ecuador, have finalized their negotiations with the United States with respect to current free trade agreements.

The stalemate that exists today must be viewed with these events in mind since they have the potential to offset any potential gain that may be realized by either trading partner in the future.

Even though the United States remains Ecuador's largest import and export partner, Ecuador has far more at stake economically.

In this regard, Hirst emphasizes, "U.S. foreign direct and portfolio investment in large quantities flowed into Latin America and the Caribbean, financing the expansion of tradable economic activities. The speed of portfolio investment transactions, however, also exposed these and other countries to marked financial volatility and recurrent financial panics."

Of particular interest is Hirst's observation that

irrespective of the outcome of these eventualities for Ecuador, the stage had been set for what would follow post-September 11, 2001.

According to Hirst, "The transformation in hemispheric international economic relations-and specifically in U.S. economic relations with the rest of the hemisphere was already far-reaching as the twenty-first century began."

The Andean Trade Preference Act (ATPA) agreement started in December 1991 between the United States and four Andean countries, Bolivia, Colombia, Ecuador and Peru with the goal of supporting these Andean countries in their battle against drug production and trafficking by expanding their economics alternatives (Relyea, 2011).

The ATPDEA provides duty-free treatment for certain products previously excluded under the ATPA. The possible scenarios in which Ecuador does not sign the FTA, but Colombia and Peru agree to do it and alternatively, Colombia and Peru allow the U.S. free access to their markets.

Organizations that enter the global marketplace have several strategies available to them: exporting, licensing, joint ventures, direct ownership, importing, and countertrade. As logistics systems, improved consumption and production began to separate geographically.

Ecuadorian Free Trade Agreement Possible Scenarios

First Scenario

Colombia and Peru allow the U.S. free access to their markets.
Potential Implications

In exchange for this free access, these countries get preferences from ATPDEA. Ecuador loses tariff preferential treatment from ATPDEA and it is imposed the same tariffs as Venezuela from the U.S. In other words, Ecuador gets more expensive tariffs than the ones imposed by the U.S. at the same price of the Venezuelan tariffs imposed by the U.S.

Metrics to be Examined

1. Projected U.S. export levels to Colombia, Peru and Ecuador under this scenario.
2. The level of new tariff levels for Ecuador in line with Venezuela's.
3. The impact of more expensive U.S. tariffs for Ecuador.

Second Scenario

Colombia and Peru allow the U.S. free access to their markets.

Potential Implications

In exchange for this free access, these two Andean countries get better preferences from ATPDEA, and the U.S. eliminates the subsidies in the production of agricultural goods. Ecuador lets go the tariff preferential treatment of ATPDEA and the country is obligated to pay tariffs equal to the ones imposed to Venezuela by the U.S.

Metrics to be Examined

1. Projected U.S. export levels to Colombia, Peru and Ecuador under this scenario.
2. The level of new tariff levels for Ecuador in line with Venezuela's.
3. The impact of preferential treatment for Colombia and Peru under ATPDEA.

Rationale in Support of Research Design

According to Ackerman, Nadal, Benniti, Gallagher and Salas (2004), "The level of methodological sophistication that is involved in predicting the potential economic effects of a trade agreement depends on the scale of economic activity between the proposed trading partners."

The general equilibrium theory provides a useful

framework in which to develop the foregoing different "what-if" scenarios based on the propensity of economic systems to seek equilibrium.

For instance, according to Kirchhoff (1999), "Fundamentally, neoclassical economics proposes that a market economy take the form of equilibrium between supply and demand. The equilibrating mechanism is price. Although the market may not always be in equilibrium, the tendency of the economic system is to move toward equilibrium."

Morrison (1999) advises that, "In neoclassical general equilibrium theory, it is assumed that an auctioneer adjusts prices so as to achieve equality between demand and supply simultaneously in all markets."

The general equilibrium theory of international trade obviously concerns interactions of several markets; consequently, the extension of general equilibrium theory to the domestic economy can differentiate between theories of domestic and international trade (Marsh, 1991).
Therefore, the general equilibrium theory provides a useful framework in which to evaluate the quantifiable elements in each of the scenarios described above using the data collection techniques and database of study described ahead.

Data-Collection Techniques and Database of Study

Based on the framework set forth, the quantitative metrics corresponding to the two different scenarios were examined using archived economic performance data published by the United States and Ecuadorian governments, as well as international trade analysts and relevant nongovernmental organizations.

The qualitative data consisting of peer-reviewed and scholarly resources published within the past 10 years were collected from public and university libraries as well as reliable online research resources.

In addition, governmental and nongovernmental organizational resources were also consulted to develop the different scenarios using benchmarked trade data for the

countries of interest.

The qualitative and quantitative data were synthesized using the "reciprocal translation process" in which each study and resource help to inform the analysis of subsequent resources.

Limitations of the Research Design

Despite its ability to model economic markets, the general equilibrium theory is not without its detractors. As Marsh points out, "International-trade theory is no longer identical in scope with general-equilibrium theory. At one time, it is true, the theory of international trade was the only theory of general equilibrium, and it differed for this and other reasons from the partial-equilibrium theory which economists of that day applied to the domestic market."

The complexities involved in analyzing international trade are even more salient in the 21st century because of the increasingly globalized marketplace that is taking shape. In this regard, Marsh notes that, "Because it was a general-equilibrium theory, the theory of international trade, even at a very early stage, contained many insights of great importance to the whole field of economic analysis which were only tardily exploited in the theory of domestic trade."

One of the fundamental constraints involved in the application of general equilibrium theory to the situation with respect to U.S.-Ecuadorian free trade agreement negotiations is the quality of the variables that are used to evaluate its impact on each of the member states.

As Marsh points out, "Today theories of partial vs. general equilibrium have lost their (partly accidental) association with domestic versus international trade. General-equilibrium theory is applied to the domestic market as well, and the so-called general-equilibrium theory of international trade is an attempt to analyze international trade where each of the trading nations is itself a multi-market economy."

Other authorities suggest that the general equilibrium theory is simply not robust enough to capture all of the harsh

realities of doing business in the real world. For example, Kirchhoff (1999) reports, "Microeconomists find the assumptions of general equilibrium theory to be inconsistent with their measures of reality. Microeconomists reject the assumption of perfect markets where many buyers and many sellers meet to exchange goods and services."

These types of examples of markets are not typically found anywhere except in a few isolated situations where there are standardized products that are relatively insignificant to the entire economy (Kirchhoff, 1999).

In addition, there are some other, more straightforward methods available to quantify the outcomes in the two described scenarios. While a general equilibrium over the relevant period is a sufficient condition for the fixity of prices, it is not always necessary.

It is much easier merely to assume that prices are fixed over a period needed to calculate any aggregate quantity such as the national income or GDP.

Chapter 4

OUTCOMES AND TRADE DISCUSSION

This chapter will give a description of all the findings from this study, which includes the following sections: (a) the model; (b) results from the exploratory data analysis; (c) ordered modeling outcomes; (d) estimates and results of the group-based mixture modeling.

The Model

The design is considered to be a usual tool which is functional to the general symmetry model that has been following the custom of Whalley and Shoven (1986).

In these immobile functional common stability models, there are many agents that are in Ecuador and the United States. In Ecuador, most would reflect the United States, the Community of Andean Countries, and other places in the world.

Despite these limitations, the theory of general equilibrium is a useful tool to follow the money in ways that may shed new insights into the priorities of the respective parties to ongoing negotiations between Ecuador and the United States.

For notational determinations, the researcher will comply by what is considered the set of trade connections for each

market. The researcher will label the chief structures of the managers according to the case.

Data and Calibration

Given the researcher's statistics availability, he can regulate the parameters of the model for the simple reason that, in equilibrium, the official managers of the model imitate the same connections and additions in the real world in keeping with the Social Accounting Matrix.

A comprehensive clarification by using the practical general symmetry models and also the methods for calibration are depicted in Kehoe (1996). The limits can also be straight adjusted with the SAM.

The Addition comprises the standards of the consistent limits in the model markets also as the Communal Accounting Matrix of the nation of Ecuador and the United States. With the limits that are not being openly regulated from the statistics, the researcher explains how to allocate those standards.

Trade Partners' Income: The profits in regards to trade partners in the study have been taken out of the *Global Monetary Statistics* available by the Worldwide Financial Account.

Tariff Rates: The price amounts in the country of Ecuador and the United States are able to execute free trade of imports coming through their trade partners, which are removed indirectly from the tables of Input and Output. In such situation, the price proportions are successful cost charges instead of the minimal charges.

The researcher does make the extra supposition that Ecuador does impose a price degree of zero on every introduction that originates from the Andean Public ever since Ecuador is an associate of the bloc considered trading.

In order to determine the tariff charges that the transaction partners are imposing on does have imports that come from the countries of Ecuador or the United States. In this case, the current publications of the Trade Policy Evaluations by the

World Trade Society are used.

The cost charges are then executed by the United States and Ecuador, which are displayed in the Table. To regulate the rates of the tariffs that are forced by other parts of the world, the researcher does make the assumption for the country of Ecuador.

Elasticity of Substitution: The study showed that provided the static description of the model, the elasticity of replacement for imports and exports which cannot be standardized straight from the Input-Output tables. As an alternative, the researcher uses diverse sets of standards that are for these limitations.

Outcomes and Statistical Experiments

This part of the study displays the outcomes that are from the "benchmark imitation." For Ecuador authorizing an agreement of trade with the U.S., the researcher did examine the impact of trade freedom on different costs, entire local manufacture, trade capacity, and also welfare examination. Then, the researcher established an income as a social indicator that uses each of the purchaser real revenue index and the administration actual revenue index.

This equal difference can assess how much revenue the customer would need for foundation fees so it can get the utility related to the simulation.

Next, the benchmark imitation conducted as a reference, the researcher performed several numerical experimentations, which resulted as the insinuations on costs, production, job, and wellbeing. In this case, the researcher took into account a study that has been labeled as "incomplete" liberalization.

In fact, trade liberalizations are what occur in place that goes on over a changeover era. For instance, the countries that are involved started to focus on a prudently ordered agenda of a time where they slowly bring down the tariff rates.

The price taxes on main and factory-made goods are consistently set to 3% for Ecuador and to 2% for the United States. As a result, the researcher documented the impact of

this temporary step on fees, distributions, and well-being.

Second, in the benchmark statistical research, the elasticity of replacement for each exports and imports was expected to be similar across different sectors. The researcher examined how the benchmark results would alter when it came to the sectoral import elasticity taken from the estimated statistics displayed in the corresponding tables.

Third, the researcher has come to the realization that tariff incomes are a vital part of government income in Ecuador. This seems to be accurate in developing nations, where most of the taxes are serving as a protective extent and basis of funding from government.

Precisely, the researcher asked what is the valued added tax (VAT) to counterbalance the losing tariff income in order to maintain the state balance constant.

In the next experiment, the researcher observed that, by coming into Europe, Ecuador would have to accept a tariff schedule that is more protective than the former.

According to Ecuador's National Federation of Exporters, a trade deal with the EU could boost exports by some 300 million U. S. dollars over the next three years.

Numerical Experiments and Results

This is mainly vital for the circumstance of main goods, which Ecuador mostly brings in from the nations that are mostly external of the European Union.

Henceforth, the experiment takes a look at the theoretical example of Ecuador passing an agreement of free trade in regards to European Union implying that the South American country starts retaining its tariff schedule that involves other parts of the world. This study can deliver a valuable contrast on the diverse kinds of free trade.

Lastly, in this numerical experimentation, the researcher does accomplish a sensitivity analysis which was done in order to take a look at the relationship that is among dissimilar standards of the trade elasticity of replacement. Also, the nation's welfare that is derived from the trade of freedom.

In this case, the researcher was able to test the hypothesis that has been contended by Johnson (1954), which records the connection that is among the elasticity in regards of the replacement and the best tariff.

Consumption Good Prices

Table 4.4 displays the percentage alteration in the cost of consumption of products after the United States and Ecuador had both decided to authorize a free trade contract and with the country of Slovenia making a decision to join Europe. In regard to Ecuador, the largest failure in the price happened in the transport involving segments by a boundary of 1.55%. All of the other import segments (substances, cereals, textiles, and other products) documented drops in their cost. In this case, for Ecuador, the biggest one came down in the cost. This had taken place with products that involved the leather, food and the drink industries. All came down to around 1 percent.

The transportation area likewise is the one that displays a price weakening of 0.80%. On the other side of the coin, the products chronicle a price escalation of 0.44%. This escalation in the amount of main goods could be the outcome of trade alteration from connecting the European Union with the amount of diversion of trade that is displayed in the section.

For Ecuador, most sectors are from the import and export industries. This could bring to the detail that all of those sectoral influences are harder to differentiate for the United States than for Ecuador.

Domestic Production

Table 4.5 displays the percent alteration that represents the whole domestic production for Ecuador and the United States. Ecuador experimented the largest increase in domestic manufacture, it primarily took place in the banana sector with a growth of 16.03%. Other key export areas, such as shrimp, petroleum, and flowers have positive improvements that are in the local production. The sectoral shift in the local

production from trade liberalization is obvious in the case of Ecuador as all the export-concentrated segments undergo production that is higher at the expenditure of the rest of the economic segments. For the United States, a major recipient in terms of domestic construction is the textile industry, where production goes up by 35.41%, followed by other sectors with an increase of 22.83%.

Connecting manufacturing of the top transfer subdivision of the two countries, the United States profit more as the vehicle industry upsurges by more than 30%, and the petroleum subdivision in Ecuador goes by about 1%.

It went up from 24.62% to 32.59%, when it was placed on a disaggregate level, most of the exports to the United States had gone up in every area, with the biggest advance that had been displayed in particular products such as the banana sector that gave a growth of about 76%.

All four chief export areas: petroleum, flower sectors, banana, and shrimp display a sturdy increase of about 8% or more. Most of the exports that go to the rest of the world go all the way down by around 3%.

Imports that come from the United States in the transportation segment, that is the prevalent sector, positioned somewhere around 87%. The study observed that the other main import sectors for instance cereals, textiles and chemicals are displaying an upsurge that is somewhere around 30%.

Just as with the exports, even the imports which derive from all other parts of the world declined for all main import segments, excluding for the textiles and other agricultural sectors which displayed a small increase that went around 3% to 5%. In terms of a complete quantity, the imports from the rest of the world have gone down by 3%.

As for United States trade with the European Community, the entire imports and exports of goods go up by 47.71% and 33.53%, individually. As for the arrangement of goods trade by end, the portion of the United States export to the European Community still goes up from 76.42% to 86.61%, even though the share of import that derives from

the European Union has managed to increase from 69.49% to 78.13%.

On a sectoral level, the largest upsurge is in the exports to the European Union, which occurs in a place that is in the primary goods area, followed by food, drinks and the material's sectors. This reproduces the high trade barricades which are being set by the European Union in those segments that are before the United States that made the decision to join the customs' union.

Imports and exports of transport equipment that went to the European Union, which is the highest trade subdivision in United States, goes up by somewhere around 65% and 30%, correspondingly. Likewise, most of the imports that have been derived from the European Union have also gone up in the nutrition and drinks segment, followed by the textile segment.

Imports were the ones that had suffered the most, going down by about 12%, while the entire exports to the other parts of the world went all the way down to somewhere around 4%. In the case of Ecuador, the trade change here is easy to understand. In the case of the United States, it is noticeable particularly in the imports.

This is relevant on a disaggregated level for the imports that are coming from the rest of the world. The main merchandise declined by somewhere around 39% as an outcome, if the United States signed an agreement with the European Union.

Welfare

Finally, the researcher was able to inspect the influence of the trade liberalization that does happen on domestic welfare. For Ecuador, the administration's value income goes down by 33% after a deal with the United States. This income loss is reflected in the deterioration that is in the administration welfare, which went down about 3%. On the other hand, the customer's welfare gain is what provides a counter balance to the administration's welfare damage, and the general social welfare displays a minor upsurge of 0.22%.

In the United States, the tariff income appears to increase by somewhere around 6% as the nation starts to adopt the protectionist tariff rule in Europe. This explains the reason the government welfare has been going up in the United Sates, as opposite to the reduction that has been looked at in Ecuador.

Furthermore, the greatness of consumer welfare improvement is greater than Ecuador that gives the indication that the benefits that involve obtaining free admission to the European Community marketplace could maybe be better than the prices of trade development that is up under the implementation of European Union's protective tariff rule. The general common welfare also displays an upsurge of 1.67%.

Partial Liberalization

Consumption Good Prices. Table 4.8 displays the percent alteration in the value of consumption products. Ecuador and the United States jointly bring down the price of taxes to a static level of 2% on every manufactured good.

The United States and the European Union choose to lessen the tariff to an unchanging 2% rate, whereas recollecting its tariff charges in the direction of the rest of the world.

As for Ecuador, this incomplete liberalization can be understood as an in-between step before complete tariff elimination.

Trade Agreement vs. Custom Union Ecuador

In this part, the researcher was able to recognize an important glance of the theoretical situation of Slovenia authorizing a free trade arrangement with the European Union, which became a substitute of connecting the European Union as an entire associate.

This causes Ecuador and the European Union to take away their expenses on each other, though Ecuador summons

up its own tariff strategy with the rest of the world, instead of accommodating the tariff rule of the European Union.

For contrast purposes, the researcher takes the case of uniform import elasticity which is employed in the benchmark case and partner the outcome of two different liberalization rules. This assessment could transport a valuable understanding on the properties of dissimilar trade liberalization provisions.

Table 4.9 displays the percentage of change that is in the value of goods and also the whole local manufacture with different sectoral import replacement elasticity when Ecuador made the decision to sign over an agreement of free trade in connection with Europe.

One difference in the agreement of free trade experimentation happens in the primary sector, as the cost goes down under the agreement of free trade, associated to an increase on the benchmark situation.

The main decline in prices seems to occur in the leather, food and drinks, and transport subdivisions, by boundaries that are bigger than 2%.

For foodstuff and beverages and transportation subdivisions, the scope of price deterioration is 20.8% and 9.4% greater than the benchmark consequences, correspondingly. Concerning local manufacture, chief products division now shows a failure in production of 3.47%.

For other parts that are also experiencing improvements in production, the advances are greater under the free trade agreement than under the benchmark imitation.

Tables 4.10 and 4.11 display the percentage alteration in the volume of exports and imports to the European Union and the rest of the world.

With the entire trade potential with Europe, spreads of all primary and mass-produced goods did happen to go up by 48.59%, although imports from the European Union improved by 33.95%. Equated to the benchmark results, the figures are 8.3% greater for exports and 4.3% lesser for imports.

Concerning trade in merchandises with the entire world,

the exports have gone up by 9.57%, although all of the imports that do come from the rest of the world likewise went up by 8.23% ever since the free trade agreement does not cause trade alteration.

Observing at individual sectors, the ones that are considered to be the primary imports coming from the rest of the world goes all the way down by around 2%, associated to almost 37% weakening.

Furthermore, some sectors display a significant trade formation with both European Union and not to mention the rest of the entire world. For example, textile imports from the rest of the world upsurges by 33.37%, compared to 6.32% weakening under the benchmark situation.

This does not come at a price of lesser upsurge in imports coming from the European Union as the upsurge in textile imports deriving from the European Union.

Lastly, the researcher made sure to recognize the impact of signing a free trade agreement that comes with the European Union on national wellbeing. Likened to the society's union case, the customer prosperity does happen to go up much more under free trade pact as Ecuador does not grieve from trade and remains open to the European Union marketplace.

The consumer welfare does go up some more under the free trade settlement is about 30% greater than under the customs union standard event.

On the other hand, the growth in administration welfare is meaningfully lesser than under the customs' union event, reproduced in the administration price income losses.

The general social welfare moreover displays an upsurge of 1.68%, somewhat less than a customs' union setting. The results are summarized in Table 4.12A.

Suggestively, different from the benchmark results, as means for most of the import areas; which are 0.725 for Ecuador, and 0.714 for the United States. Regarding the values that are utilized in Rolleigh (2003), on the other hand, the regular elasticity is much higher, at around 0.95 for imports in Ecuador and 0.99 for the United States.

At disaggregated heights, the bigger the elasticity of substitution, the greater the influence on the value of consumption goods. Therefore, with parameters projected by Rolleigh (2003), the greatness of the price goes down in Ecuador, which is 4.4 times for the textile area and 23 percent bigger for the transport area than the standard outcomes.

In the United States, on the other hand, this is not always the case. In the food and beverage industry, with higher import springiness, the value goes down 75% more under the standard elasticity.

The transport segment, the amount of price deterioration is now lower, in spite of higher import elasticity. This is possibly because in the United States, most subdivisions are both export and import concentrated.

Before a transportation company can effectively market itself, it must first establish a strong brand and identity in order to generate interest for its services by utilizing a direct sales approach.

International Trade

These tables also show the present change in the magnitude of imports and exports that are for Ecuador and the United States using the elasticity given by Hummels while the other table shows the corresponding trade capacities by utilizing elasticity.

On the average, utilizing the estimates which are taken from Rolleigh (2003), the imports and exports of Ecuador with the United States go up by 59.61% and 85.49%, correspondingly. Conversely, using the estimates that are taken from Hummels (2001), the consistent numbers are somewhere around 39.44% and 51.11%, correspondingly.

Associated to the benchmark reproduction, the percent goes up in trade, which is a whole lot larger, particularly when approximations are brought from Rolleigh (2003).

Observing at the disaggregate level in the chief import areas, the import of transport equipment that derives from the U.S. goes up 135-173%, associated to 84% that is under

Pegaso Satellite

Source: Ecuador Economy, 2015

the standard situation.

For the textiles area, with the pliability which happens to be brought from Rolleigh (2003), the percent upsurge which is in regards to the imports that are coming from the U.S. is around 485%, which is the actual large upsurge when it is associated to 115% that is up under the benchmark event.

Other key import subdivisions for instance cereals and chemicals display a lesser increase in imports that derive from the United States.

Table 4.1: Sectorial for the Unites States and Ecuador

Ecuador	United States
Wild oats and nuts	Beverages and Foods
Breakfast cereal	Leather
Plants	Wood and Furniture
Fuel (Crude Oil)	Fabrics
Different Prime Goods	Transportation Equipment
Sea food	Other Productions
Fabrics	Services
Chemical Products	

Table 4.2: Rates of Tariff - Ecuador vs. United States (unit: %)

Sectors	Ecuador	U.S.
Bananas	.1	7.3
Cereals	14.5	1.8
Flowers	1.2	1.8
Oil	2	3.3
Other Primaries	9.2	9.1
Shrimp	20.2	2.4
Textiles	17.1	9.4
Chemicals	7.5	3.9
Transport	17	29
Other	7.8	4.9
Services	.1	2.3

Table 4.3: Benchmark Results

Sectors	**Hummels**	**Rolleigh**
Bananas	.60	.70
Cereals	.85	.75
Flowers	.56	.85
Petroleum	.83	.70
Other Primaries	.81	.75
Shrimp	.77	.90
Textiles	.82	.95
Chemicals	.86	.65
Transport	.89	.85
Other	.83	.80
Services	.84	.75

Table 4.4: Outcome of Trade Liberalization on Consumption Good Cost

Sector	**Price**	**Sector**	**Price**
Bananas	.55	Primary	.57
Cereals	-.51	Food-	-1.08
Flowers	.58	Textiles	-40
Petroleum		Leather	-1.45
Other	.30	Wood	.28
Shrimp	.45	Transport	-.96
Textiles	-.17	Other	.02
Chemicals	-.25	Services	.89
Transport	-1.45		
Other	-.56		
Services	.56		

Table 4.5: Effect of Trade Liberalization on Domestic Manufacture

Sector	Quantity	Sector	Quantit
Bananas	15	Primary	2
Cereals	-9.3	Food-	-3
Flowers	2.85	Textiles	30.5
Petroleum	1.2	Leather	-2.78
Other	.78	Wood	-5.14
Shrimp	5.45	Transport	20.73
Textiles	-.9	Other	4.05
Chemicals	-2.5	Services	-2.3
Transport	-5.99		
Other	-2.68		
Services	-.69		

Table 4.6: Result of Trade Liberalization on Chief Imports (unit: %)

Segment	U.S.	ROW	Segment	EU	ROW
Cereals	75.45	-6.52	Primary	36.05	-38.9
Other	70.85	3.91	Food	70.26	-11.7
Textiles	109.06	1.91	Textiles	65.97	-5.89
Chemicals	37.25	-0.44	Leather	30.48	3.56
Transport	85.2	-11.37	Wood	12.40	-7.90
			Transport	39.50	-5.00
Other	5.89	-0.42	Other	25.45	-9.80

Table 4.7: Result of Trade Liberalization on Welfare (unit: %) Institution

Institution	Ecuador	United States
Consumer Welfare	1.00	2.45
Government Welfare	-5.34	2.56
Social Welfare	3.23	3.00

Table 4.8: Tariff Charges

Sector	Price	Subdivision	Price
Bananas	0.39	Primary	-0.22
Cereals	-0.47	Food	-1.08
Flowers	0.24	Textiles	-0.34
Petroleum		Leather	-0.78
Other	0.20	Wood	0.34
Shrimp	0.37	Transport	-0.54
Textiles	-0.20	Other	0.02
Chemicals	-0.11	Services	0.55

Table 4.9: Effect of Free Trade Agreement-Price-Production, Slovenia (unit: %)

Customs Union		FTA	CU	FTA
Primary	0.23	-0.25	1.23	-2.37
Food	-1.04	-1.21	-2.00	-2.56
Textiles	-0.23	-0.21	29.41	28.63
Leather	-1.11	-1.23	-3.59	-3.06
Wood	0.30	0.24	-6.18	-5.25
Transport	-0.87	-2.04	19.83	20.21
Other	0.02	-0.05	4.04	3.66
Services	0.46	0.88	-1.17	-1.31

Table 4.10: Effect of Free Trade Agreement on Exports – Ecuador

	Customs	FTA	Customs	FTA
Primary	223.00	277.34	-17.21	7.56
Food	221.21	203.91	4.48	19.64
Textiles	109.89	115.45	-6.34	13.14
Leather	21.34	18.45	6.34	18.68
Wood	-2.23	-2.49	-11.56	0.89
Transport	78.34	67.21	1.78	18.23
Other	36.45	34.78	-378	7.22

Table 4.11: Effect of Free Trade Agreement on Price and Production – Ecuador

	Customs	**FTA**	**Customs**	**FTA**
Primary	217.56	258.86	-18.14	7.89
Food	202.13	205.71	3.67	21.45
Textiles	123.60	143.50	-5.34	15.23
Leather	20.25	17.54	6.22	17.81
Wood	-1.76	-2.32	-11.78	0.34
Transport	66.92	77.45	2.89	16.56
Other	34.56	38.89	-5.78	7.03

Table 4.12: Effect of Free Trade Agreement on Imports – Slovenia

Primary	36.07	29.45	-36.56	-0.80
Food	68.45	67.4	-12.21	-3.74
Textiles	67.34	56.7	-7.56	34.78
Leather	23.56	26.5	3.47	-0.67
Wood	16.25	17.3	-4.67	-1.34
Transport	45.30	41.3	54.45	16.44
Other	45.34	45.23	-9.70	7.61

Table 4.12A: Effect of FTA on Welfare – Ecuador

	Customs	**FTA**
Customer Welfare	1.4	1.8
Administration	2.9	0.9
Community	1.7	1.6

Cuenca, Ecuador

Source: Wikipedia Turismo, 2015

Chapter 5

POLITICAL ISSUES

In the first scenario, both Colombia and Peru allow the U.S. free access to their markets. In return for this free access, these Andean countries receive preferences from ATPDEA. Concomitantly, Ecuador foregoes tariff preferential treatment from ATPDEA and it is imposed the same tariffs as Venezuela from the U.S. meaning that Ecuador is assessed more expensive tariffs than the ones imposed by the U.S. at the same price of the Venezuelan tariffs imposed by the U.S.

The projected U.S. export levels to Colombia, Peru and Ecuador under this scenario are discussed further with respect to the level of new tariff levels for Ecuador in line with Venezuela's and the impact of more expensive U.S. tariffs for Ecuador.

The outcomes of this scenario may be miniscule in comparison to the larger geopolitical considerations that are swirling around these negotiations today, but the likelihood of this situation remains uncertain given the stalled nature of the discussions between the U.S. and Ecuador. What is certainly known is that the United States is Ecuador's main trading partner, accounting for more than a third of the country's exports.

In the second scenario, Colombia and Peru allow the U.S. free access to their markets. In return for this free access, both Andean countries receive improved preferential

treatment from ATPDEA, and the U.S. eliminates the subsidies in the production of agricultural goods.

In addition, this scenario comprehended Ecuador foregoing the tariff preferential treatment of ATPDEA wherein it is required to pay tariffs equal to the ones imposed to Venezuela by the U.S. The projected U.S. export levels to Colombia, Peru and Ecuador under this scenario are discussed further.

This situation also required a level of new tariff levels for Ecuador to be consistent with Venezuela's and an examination of the impact of preferential treatment for Colombia and Peru under ATPDEA.

The research for this book showed that while Ecuador has a number of natural resources including, most especially, oil and gas. It also has gold, timber and copper as well as a significant agricultural sector.

At present, Ecuador's economy continues to depend heavily on its proven oil and gas resources which account for more than half of the nation's exports and around 40 percent of its community sector revenues.

During fiscal year 2000, Ecuador's economy suffered a downturn as a result of the banking crisis that caused economic performance to decline, causing the country to default on its external debts.

Various structural reforms and efforts to increase international trade have been implemented since that time in an effort to help Ecuador regain its financial vigor with mixed but positive results.

These outcomes are consistent with economic theories that consider free trade as a viable way to promote economic development in emerging nations and there are a number of examples from the historical record that confirm that this approach can work (Ackerman, 2004).

For instance, according to Ackerman, "The more the market forces are allowed to rule and the more of an open economy to free trade and competition, the more efficient and flourishing the economy will be."

Neoclassical economic theories that subscribe to these

fundamental tenets are not new, but rather can be traced to the classical economists of the early 19th century who assumed that factors of production remain mobile between industries within a given country but not between countries and that all production technologies are available within all countries in ways that provide the country with what is termed "static comparative advantage" (Ackerman, 2004).

Static comparative advantage is contrasted with dynamic comparative advantage which varies from time to time and over time.

The dynamic comparative advantage that Ecuador possesses with respect to its current levels of goods and services relates in large part to its strategic geographic location and the potential for the country to be used as a source and conduit for illicit drug manufacture and distribution and how these potential eventualities affect foreign policy decisions with the United States and its neighboring countries.

Critics of America's efforts to promote increased free trade with the Andean nations, including the Ecuadorian people themselves, argue that free trade agreements with the United States are ill-disguised efforts to exert further American hegemony in the hemisphere with the Andean countries being a minor linchpin in the process.

These criticisms are congruent with the observation by Ackerman and his colleagues that, "There are good reasons in theory to doubt the simple prescriptions of free trade, and ample historical evidence that successful development has almost never occurred in a free trade environment." Consequently, any dynamic comparative advantage that might accrue to free trade agreements between the U.S. and Ecuador are far more complex than a casual analysis might indicate.
Recent heated and sometimes violent demonstrations by Ecuadorians in response to their proposed membership in the Free Trade Agreement of the Americas suggests that a free trade be viewed from this perspective by many average Ecuadorian citizens who want to pursue an Ecuador solution for Ecuador's problems.

In fact, the overwhelming majority of Ecuadorians

approved a new Constitution that provided the country's leftist President Rafael Correa with even stronger controls on the nation's economy as he pursues a socialist approach comparable to Venezuela's president, Hugo Chavez (Brodzinsky, 2008).

According to the country's president, "Today Ecuador decided to found a new country" and the new country was founded with nearly seven out of ten Ecuadorians voting for Correa and the new charter. With this mandate under his belt, Correa felt justified in boasting that the old power structures have been defeated.

Following Venezuela's political lead to the left, Correa was adding an enormous bargaining chip to his foreign policy negotiations with the United States because the last thing U.S. policymakers want is a growing threat to their security from the south, but this is just how Ecuador might be perceived by the U.S. intelligence community, particularly in light of recent events in the Middle East and the groundswell of anti-American demonstrations taking place across the Muslim world.

For instance, according to Brodzinsky, "With the passage of the new Constitution, Ecuador became the first country after Venezuela in the region to institutionalize its leftward shift. This is a lurch to the left on the part of Correa."

Some of the main major issues that have emerged dearly from recent events from the region include the fact that although Correa is pursuing political initiatives that resemble those taking place in Venezuela, the Ecuador approach is not a "one-size-fits-all" solution but is rather specifically tailored to focus on Ecuador's problems. In this regard, Ecuador's recent shift to the left does not make Correa a Venezuelan "wanna-be" who follows Chavez's lead, but rather a like-minded political leader who recognizes the limitations of overreliance on U.S. relations (Brodzinsky, 2008).

Nevertheless, even with these differences, there are also a number of similarities between the Ecuadorian Constitution and the Venezuelan charter approved in 1999.

By and large, Ecuador's new Constitution is not a

Marxist agenda but is rather geared towards a more popular and equitable distribution of the nation's wealth, at least in theory. Among other guarantees, the new Constitution warrants all Ecuadorians the right to have access to water, universal healthcare, pensions, and free state-run education all the way through the university level; in addition, the Constitution permits the Ecuadorian government to confiscate land that is fallow and to then distribute it to the poor.

The Constitution, like Venezuela's, also codifies sovereignty in political and economic terms and allows presidents to serve two consecutive terms, meaning Correa himself could remain in office until 2017 backed by a recent constitutional amendment.

This eventuality also means that many Ecuadorian citizens may regard this continuity in office as guaranteed political stability in an uncertain world as highly desirable, but it remains clear that Correa serves at the pleasure of the people and Correa must walk a fine line and take care of his actions in order to avoid an Arab spring-style uprising in his own country.

This was foreshadowed by recent violent demonstrations in the streets of the country's capital and the responses from the Ecuadorian government that indicates social unrest on the level being seen in the Middle East will not be tolerated. Moreover, the revised status of Ecuador's central bank pursuant to the new charter was one of the issues considered in a referendum in Venezuela that was rejected by voters there in 2007.

In response to his new mandate, President Correa stated that the widespread approval of the Constitution "goes in line with the changes that all of Latin America is seeing," clearly referring to Venezuela, but Brodzinsky also emphasizes that Correa frequently makes it clear that he is not a poor cousin version of Chavez and stresses "His own 'Citizens' Revolution' responds to the demands of his countrymen, not Venezuela's regional agenda."

The mandate established by Ecuador's leftist-leaning

Constitution is regarded as being helpful to Chavez's regional project, but it is not caused by it. Both Chavez and Correa are responding to long popular demands for change in their own countries.

Besides the autonomy of the country's central bank issues, there are some other distinctive differences between Correa and Chavez, particularly with respect to foreign policy issues.

For instance, Brodzinsky (2008) reports that Ecuador has not joined the Bolivarian Alternative for the Americas, the option developed by Venezuela in response to the Free Trade Agreement of the Americas, and Ecuador has continued to remain without formal diplomatic relations with Columbia despite Venezuela having done so.

Moreover, Correa's negotiations with the United States have historically been less confrontational than those with Venezuela. This "kinder and gentler" version of Venezuela's hard-line approach to foreign policy negotiations with the United States may be part of the reason negotiations have not been concluded, with Ecuador holding out for optimal trade provisions with the United States and others.

Furthermore, Correa has some additional bargaining chips at his disposal that Venezuela does not enjoy. For instance, following his election in 2006, Correa reported that the U.S.'s lease on a strategic Air Force base in Manta, Ecuador would not be renewed because the new Constitution specifically prohibits any foreign military bases in Ecuador; however, Correa has avoided confrontation with the U.S. in ways that further differentiate him from Chavez.

As noted, with the potential for Correa to remain in power until 2017, Ecuador is enjoying a period of relatively political stability that is truly uncharacteristic given its troubled past with eight presidents in the last decade.

This sense of stability has been a powerful political drawing card, and current signs indicate that Correa will continue to enjoy enough popular support in the future to remain in power without having to resort to the brinksmanship tactics used by Venezuela.

In fact, the current Ecuadorian president has made it clear concerning which approach he will use in developing Ecuador's infrastructure. Certainly, oil is not Ecuador's only export resource that is being affected by ongoing free trade negotiations with the United States.

According to a report from Llana (2006), fully 25 percent of all of the cut flowers sold in the United States originate from Ecuador, especially the Cayambe Valley region where a $350 million industry has emerged.

These types of local industries would appear to stand to gain the most from streamlined trade with the United States since that market represents the bulk of their current trade.

The demonstrations, though, were not limited to Quito and many, if not most, of Ecuadorians believe that the provisions of the Free Trade Agreement of the Americas are somehow contrary to their best interests irrespective of any other considerations based on growing animosity towards all things American in the region, fueled by the leftist and anti-American vituperations emanating from Chavez. Nevertheless, it would seem that some Ecuadorians, at least, would stand to benefit from further concessions that facilitated trade with the U.S.

"However, Canada and the European markets are always open for a new crop, and other products. This prediction was based on recent trends in the U.S. government that indicated time was running out on the interim agreements that affected this industry" (Llana, 2006).

Unfortunately, virtually all reports concerning free trade negotiations are likewise couched in terms of "pending," "ongoing," "cautious optimism," and so forth, but there apparently has been little substantive progress in concluding these negotiations.

This lack of progress is perhaps a reflection of the sensitive nature of the negotiations given the backdrop in which they are taking place. The hard-line stance adopted by Correa in his negotiations with the U.S. may have some unexpected and unintended adverse consequences for some Ecuadorian industries.

In response to these events, regional representative for Expoflores, a local flower growers' association in Ecuador, emphasized that the Ecuadorian flower industry was enormously concerned and that ongoing negotiations with the United States over trade-related issues will likely have significant repercussions in this country and the other Andean nations as well, but these others have already concluded their free trade negotiations with the U.S. while Ecuador remains the sole holdout.

Clearly, though, just as the U.S. tries to stay solvent by approving ever-higher budget deficits, these efforts do not represent long-term solutions to whatever constraints remain to concluding negotiations between the U.S. and Ecuador in ways that are mutually advantageous.

Ecuador's strategic location and increasing involvement with the drug trade, though, have complicated these negotiations in ways that are not necessarily being reported in the mainstream press, but which can be discerned, at least in a general fashion, from the few relatively recent reports on these negotiations that do exist.

The impact of stalled negotiations and their eventual outcomes are having a profound effect on tens of thousands of Ecuadorian pursuing their livelihoods in sectors outside oil and gas.

While Chavez and Correa joust with the U.S. over optimal trading conditions, these sectors, are being overlooked and the adverse effect on trade might be far graver than observers predict in the wider geopolitically charged environment in which they are taking place.

For instance, in an interview with Mario Castro, the head legal counsel at the regional government of Cayambe, Llana (2006) found that little of the country's oil wealth has trickled down to this part of the country where citizens rely heavily on the floral industry for their income.

In this political environment, it is reasonable to suggest that these stakeholders seek to maximize their own interests to the extent possible within the parameters established by the new Constitution.

If Correa tries to leverage his unique position in the Andean community as the sole holdout with the potential for years of more of the same, he gains significant face in the Latin American community and a sense of pride among his fellow Ecuadorians where strong leaders are viewed approvingly.

For example, Castro emphasized the importance of Ecuadorian communities' investment in the flower industry and argues that any trade agreement that places this investment at risk does so at the peril of the indigenous populations rather than the policymakers sitting in Quito.

Indeed, almost half of the cultivatable land in the Cayambe Valley is used for flower production at present and the surrounding communities have built an infrastructure based on this industry (Llana, 2006). In fact, the population of these communities has grown by 30,000 in recent years and the so-called "ghost-town" effect would quickly materialize if the flower industry were significantly disrupted.

Clearly, these stakeholders would be willing to go just so far in supporting any trade negotiations with the United States that would work to their disadvantage. Notwithstanding the increasing importance of Ecuador's flower industry, trade liberalization with the United States continues to be a charged issue, even among those who stand to gain from a successful conclusion to the negotiations.

As noted in 2005, "Thousands of citizens blocked roads and burned tires in the northern Andes, fueled by rumors that the government was on the verge of sealing a free-trade agreement with Washington" (Llana, 2006).

The key question, then, relates to what parts of the negotiations that are the most salient in preventing popular approval of free trade agreements with the United States, and these must be sufficiently powerful since they directly impact the livelihoods of the stakeholders involved.

This stalemate was the result of a law passed in Ecuador that increased tariffs on private firms during periods of oil price increases; in addition, other events that have had larger implications for these negotiations with the U.S. include the

fact that Ecuador's state oil company seized an oil field operated by U.S. owned Occidental Petroleum Corp in response to an alleged breach of contract.

Given Correa's 70 percent vote in the most recent election, it is reasonable to conclude that he enjoys a mandate to follow through on this commitment, but not everyone is of a like mind when it comes to how the negotiations should proceed.

Not surprisingly, Ecuadorians are "divided on the issue" and some indication of the equality of this divisiveness can be discerned from the results of a survey conducted by Quito-based Cedatos-Gallup that found nearly half (42%) of those polled indicated they were in favor of a free-trade pact, but slightly more (44%) were Ecuadorians opposed the free trade agreement (Llana, 2006).

There are some vested interests involved in any negotiated outcome for free trade with the United States and following the money can help determine which side of the free trade agreement stakeholders might come down on given current trade levels and the potential effects of streamlined trade with the United States in the future.

With thousands of jobs at stake in Ecuador's flower industry, any moves by U.S. corporations to alter their existing trade arrangements are widely regarded with alarm.

Ironically, government subsidies of small farmers in Ecuador would be required to the extent that free trade agreements reduced their competitiveness, suggesting that any new trade generating by such agreements might not be so free after all.

In this regard, Llana notes that Chiriboga also argues, "Free trade is the only way to obligate the government to support farmers who can't compete in a globalized world. Currently only 8 percent of Ecuadorian farmers receive subsidies, mostly from NGOs (Non-Government Organizations).

Free trade is regarded as a way to adjust institutions. "We have to do something or we'll continue to export people" (quoted in Llana, 2006). In an interview with President

Correa, a reporter for the Christian Science Monitor, A. P. Lowenthal (2011) discussed the implications of the free trade agreement and a number of salient issues emerged of interest to this study.

According to President Correa, "What is needed is to develop new and equitable rules of the game and truly democratic processes of decision-making. That takes time" (Lowenthal, 2011).

With this renewed emphasis on forging fair and equitable trade agreements with the interests of the Ecuadorian people foremost in mind, Correa is walking a very fine line between too much and too little regulation. Clearly, the Ecuadorian president's goals with respect to free trade with the United States include a number of domestic agenda items, as well.

When asked about his views concerning U.S. policies toward Latin America and toward Ecuador under President Obama and whether there are ways for Ecuador and the United States to cooperate on shared concerns and interests, the U.S.-educated Correa responded, "I admire US society, and I had a very positive experience in the United States during the four years I studied for the Ph.D. at the University of Illinois. But US foreign policy has historically been antagonistic to progressive change in Latin America, and has been marked by attitudes of domination and arrogance" (Lowenthal, 2011).

The Ecuadorian president also apparently studied the Monroe Doctrine during his college experience, and his country's experiences with heavy-handed American corporations have convinced Correa that a tough negotiating stance against the United States is the only viable course for his country.

The changing winds of hope are, in fact, blowing through the halls of government in Ecuador. Following the assumption of office by the country's new congress on January 5, 2007, President Correa was sworn in on January 15, 2007 and just two months later, the Ecuadorian Supreme Electoral Tribunal discharged 57 congressmen based on violations of the country's campaign laws (Background note:

Ecuador, 2012).

Thereafter, the new Congress became stalemated in their decision-making process and the entire assembly was essentially replaced by referendum at the end of September 2007 (Background note: Ecuador, 2012).

In an effort to help forge popular support for his policies, the president has spent some of the country's money on social programs and further reforms.

While the country's priorities for increased social spending are being addressed, the Ecuadorian business environment has deteriorated to some extent, especially since these policies continue to undergo change.

This dynamic political situation may help explain why free trade agreement negotiations are taking so long and why they have been deadlocked over unnamed security-related issues that may fall under the provisions of the Homeland Security Act.

The Ecuadorian economy is comprised of two primary sectors: petroleum production (primarily for the domestic market and commerce), and agricultural production (for domestic consumption and export); the main exports are petroleum, bananas, shrimp, flowers, and other primary agricultural products.

There have been some other reforms in Ecuador in recent times that have the great potential to affect free trade agreement negotiations with the United States. For example, in 2000, Ecuador adopted the dollar as its official national currency in the wake of the crisis in its banks and a recession in 1999. This initiative improved Ecuador's economic stability and performance through 2006 at which point economic growth was once again constrained by falling oil prices.

By late 2008, though, the Great Recession adversely affected Ecuador's economy. In addition, the Ecuador government defaulted on its 2012 and 2030 global bonds in December 2008 with an approximate value of U.S. $3.2 billion; however, by June 2009, the Ecuadorian government had repurchased almost all 91% of these global bonds at a discounted rate.

Cóndor Assembled in Ecuador

Source: Ecuador Economy, 2012

Finally, there are a number of unknowns involved in the two "what-if" scenarios. The 2012 budget of $26 billion for Ecuador is based on a projected average price per barrel of Ecuadorian crude oil of $79.80

President Rafael Correa

Source: Based on News.harvard.edu/gazette, 2014

President Barack Obama

Source: Based on Newsone.com, 2014

Chapter 6

FREE TRADE BEST SOLUTION

The research for this book on free trade was consistent in showing that Ecuador has sought negotiated solutions to international problems over the years, and these trends can be reasonably expected to continue in the future as resolutions to the free trade agreement negotiations with the United States continue to be hammered out.

Currently, Ecuador is a member of the United Nations and most of its specialized agencies, the Organization of American States, as well as numerous regional organizations, including the Rio Group, the Latin American Energy Organization, and the Union of South American Nations (UNASUR).

Ecuador served the rotational presidency position in this organization, and the Community of Andean Nations during 2009-2010. In addition, Ecuador is a member of the Community of Latin American and Caribbean States.

During the Correa administration, Ecuador has succeeded in diversifying its economy and improving its relations with countries in Latin America, Europe, and Asia.

Conversely, the Correa administration has also drawn what appears to be a line in the geopolitical sand that suggests that the country no longer accept trade conditions it deems

unsatisfactorily to the country's long-term best interests, but the volatility of the petroleum industry and the nascent state of diversification of the Ecuadorian economy, the likelihood of either scenario playing out in the real world remains uncertain.

What can reasonably be concluded is that Ecuador may not be another Venezuela, but it is shaping up to be a regional contender that can exert influence far above its economic clout because of its strategic geographic location and potential for a staging ground for terrorist organizations in America's own backyard.

In the final analysis, a Free Trade Agreement of the Americas that includes all regional countries, including Ecuador, may not ever be realized while President Correa is in office, and he is expected to be president for another 5 years.

This means that, notwithstanding the political stability, this projected tenure provides particularly in view of Ecuador's turbulent political past, scenario one and two outcomes will positively affect some stakeholders more than others.

With the country's citizens equally divided about the desirability of the provisions of the Free Trade Agreement of the Americas, it is reasonable to conclude that irrespective of which scenario is pursued.

Correa will please some constituents while dissatisfying others. The key question then will become which of these stakeholder groups has the most at stake and how best to balance their needs with the other sectors in the Ecuadorian economy.

The historical record suggests that big multinational oil corporations prevail, but the line in the geopolitical sand drawn by Correa may help preclude this eventuality and promote trade practices that benefit historically marginalized sectors.

The U.S. elections set for November 2012, reelected Obama and Biden with 51% of the vote. The Ecuadorian people decided the political outcome of the country by casting their vote in February 17, 2013. As a result, Correa was

reelected because the country has experienced a great deal of progress under his leadership. Even though, he opposes such deal with the United States, the majority of the population supports his political agenda.

The book focused on the agroeconomic perspective in order to determine the implication of an economic deal of such magnitude, advantages, disadvantages, and the best decision for the Ecuadorian community and the U.S.

The Free Trade Agreement between Ecuador and the U.S. would not ultimately benefit the Ecuadorian economy in the short term because the industry is not very well developed yet. It would eventually take at least 20 years to get the right infrastructure to push Ecuador in that direction.

The elements of the marketing mix play an important role in the development of Free Trade Agreement talks between the major economic players.

Price is the main determinant of whether a country imports or exports a product. World price is the value prevailing in world markets and is the agreed value at which we can sell or buy products.

Domestic price is the local price without trade. Ignoring transportation costs, if the world's price is greater than the domestic price before trade, therefore, the good is ready for export. If the world price is less than the domestic price before trade, then we would import the good.

Trade allows anybody to buy goods at a cheaper price from international markets and sell those at a valuable higher price than if we were restricted to the domestic price. For goods that we import, domestic prices fall and consumers benefit while producers are hurt as evidenced by a reduction in purchases.

Tariffs by definition are taxes levied on businesses for imported goods. Tariffs traditionally raise the domestic price above the world price by the amount of the tariff. The increase in the domestic price will lead to a decrease in domestic quantity demanded, and an increase in domestic quantity supplied.

Quotas are restrictions in the maximum amount of

goods that may be imported, and have a similar effect as tariffs. These quotas restrict the amount available to domestic consumers and push up the price resulting in a negative effect similar to that of a tariff.

The main difference is the distribution of the surplus. A tariff raises revenue for the government, whereas import quota creates surplus.

The North America Free Trade Agreement has increased trade and investment among Canada, Mexico, and the United States by gradually eliminating tariffs, import quotas, and barriers to foreign ownership.

Canada's economy is much more dependent on international trade than the United States. Canadians are taxed more and paid less than their U.S neighbors. Canadian corporations have invested heavily abroad, especially in the United States. Canadian Pacific has almost a third of its railroad tracks in the United States and is one of the largest owners of luxury operated hotels in America.

Before NAFTA was the *maquiladora* sector. Established by the government, maquiladoras are foreign factories that can import parts and materials into Mexico duty free, as long as they are used to make products for export.

Mexico is the United States' second largest trading partner, with two-way trade between the countries, as a result, Mexico will be the 10th largest economy by 2015.

The United States is still the world's largest economy with a GDP of $15.7 trillion dollars and the largest management for foreign investment. The United States also leads in knowledge creation, financial services, and information technology, positions that will be important in the years ahead.

Today, American multinationals operate around the world, with many enjoying dominant positions in their respective industries followed by China, Germany, South Korea and Japan with corresponding GDPs in excess of $1 trillion.

The different regions would specialize in those commodities that could be produced most efficiently, excess

production could be shipped economically to other producing or consuming areas, and needed goods not produced locally were imported.

This market principle, when applied to world markets, helps to explain the high level of international trade that takes place today. Efficient logistics systems allow world businesses to take advantage of the fact that lands, and the people who live there are not equally productive, and logistics is the very essence of trade. It contributes to a higher economic standard of living.

When a logistics organization operates globally, the best business is usually one in which the planning and control functions are centralized and the operation's functions are decentralized.

Systems have to be in place for rapid product introduction, focused market needs, quick-response delivery, expanded services, innovative channels, working capital, inventory, credit, investment in building and equipment.

Ecuadorians are friendly people, who welcome foreigners from all over the world because they love to learn their language and culture, and even better if there is capital investment to pump up the local economy.

There are approximately 10,000 American expatriates living in Ecuador and about 700,000 Ecuadorians residing in the United States. Therefore, a Free Trade Agreement would benefit both countries.

The U.S. is a technological powerhouse that exports automobiles, machinery and other manufactured products to Ecuador. However, China is the new manufacturing house for the world, and India the world's customer service back office. Ecuadorians have a special fondness for "Made in USA" products, and are willing to pay anything to get such items because of the good quality

On the other hand, Ecuador benefits from its duty- free exports, such as, crude oil, shrimp, agricultural products and raw materials to the U.S., and welcomes American investment led by manufacturing and wholesale sectors.

The Free Trade Agreement between Ecuador and the

United States would give the common Ecuadorian citizen a better chance to purchase, let's say, a vehicle that sells in Detroit for $20,000; instead of a $40,000 ticket in the Ecuadorian market. If the automobile ticket price were the same as in Detroit, the Ecuadorians would increase the demand for more cars, which in turn would benefit both economies.

The whistle-blowers, about U.S. government intrusion in its citizens private affairs, Assange (Australian founder of WikiLeaks) and Snowden (U.S. National Security Agency Analyst) cases, both seeking political asylum with the Ecuadorian Government, are recent events that have slowed down and compromised the trade agreement with the United States.

A quick history throwback that hopefully won't repeat during this process of negotiations. Although the independence movement in South America achieved its primary goal, none of the leaders enjoyed the benefits of victory.

The French invasion of Spain gave the American colonies the opportunity to revolt. The final blow for Spain took place at the battle of Ayacucho when *Antonio José de Sucre* was leading the patriot army defeated the Royalist army.

When the Liberators, *Bolívar* and *San Martín*, met in Guayaquil, the historic encounter was not a success and subsequently *San Martín* withdrew from all his posts. *Bolívar*, then returned to hold together his creation of Gran Colombia, but it turned out to be an impossible task when his health started to fail.

Gran Colombia fell apart with the secession of Ecuador and Colombia due to political differences. Both, *Bolívar* and *San Martín* died in exile.

It is sad to realize the fathers of South American freedom ended up in the desolation after sacrificing everything in the name of one powerful nation.

Bolívar's dream was to join all Spanish America under one flag to keep up with the foreign powers. However, his dream of Pan Americanism faded away with him. In 1830, on

his death bed, Bolívar wrote: "America is ungovernable. Those who have served the revolution have ploughed the sea."

Former Venezuelan President *Hugo Chávez* passed away in March 2013, after 14 years in power. *Nicolás Maduro*, assumed the presidential post, and the country is currently going through political and economic chaos, as well as, a fierce revolt by the opposition

President Rafael Correa visited the United States to speak at Harvard, Yale University and the Massachusetts Institute of technology during the second week of April, 2014. Correa talked about his political agenda and accomplishments during his tenure as president. He also secured the student exchange program between Ecuador and the best universities of the United States.

Between 2007 and 2013, Correa said, 1.13 million Ecuadoreans were lifted from poverty, and the incidence of extreme poverty fell from 16.9 percent to 8.6 percent, while growth averaged 4.2 percent per year. The country also claims the lowest unemployment rate in the region, 4.1 percent, and according to a U.N. development report it is one of three countries with the greatest upward mobility (Harvard Gazette, 2014).

16 April 2016, a magnitud 7.8 eartquake on the northern ecuadorean coast left a death toll of more than 600 people. The next presidential election will be held on February 19, 2017.

Now, it is just a matter of both governments to discuss the terms of the Free Trade Agreement that has been on the negotiation table for so long. The trade talks are currently on hold due to political differences designed to benefit the few who control the Ecuadorian economy.

Teleférico, Quito - Ecuador

Source: Wikipedia Turismo, 2015

New York, United States

Source: Wikipedia Turismo, 2015

MAPS

Map 1.1: Republic of Ecuador

Source: Based on Ecuador Economy, 2012

Map 1.2: The United States of America

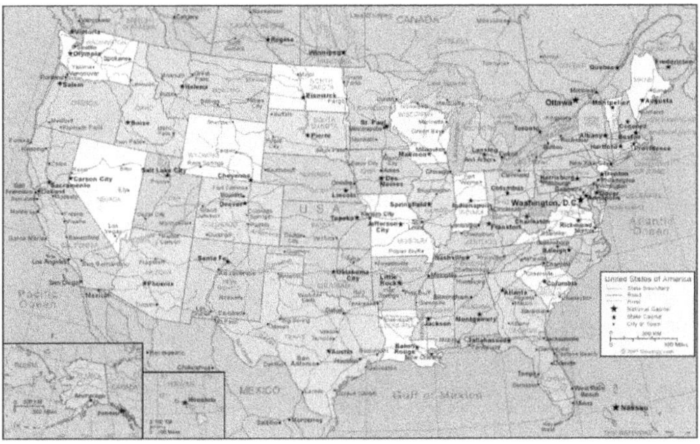

Source: Based on Geology.com, 2014

SELECTED BIBLIOGRAPHY

Ackerman, I., Nadal, F., Benetti, A., Gallagher, C., Salas, C. (2004). *The flawed foundations of general equilibrium theory: Critical essays on economic theory*. New York: Routledge.

Baker, J. C. (2003). *Financing international trade.* Westport, CT: Praeger.

Bush pushes Free Trade Area. (2002, February 11). *The New American*, 18(3), 1.

Chauvin, L. O. (2004, July 30). Farm tariffs, intellectual property rights seen as main obstacles to U.S.-Andean FTA. *International Trade Daily, Bureau of National Affairs.*

Ecuador Economy. (2012). *U.S. government: CIA world factbook.* Retrieved from https://www.cia.gov/library/publications.

Ecuador: Militant opposition to Andean Free Trade Agreement blockades several regions, moves to capital. (2006, March 24). *NotiSur.*

Faber, E. M. (2007, Winter). Pregnancy discrimination in Latin America: The exclusion of "employment discrimination" from the definition of "Labor Laws" in the Central American Free Trade Agreement. *Columbia Journal of Gender and Law*, 16(1), 37-44.

Fandl, K. J. (2007). Bilateral agreements and fair trade practices: A policy analysis of the Colombia-U.S. Free Trade

Agreement (2006). *Yale Human Rights and Development Law Journal*, 10, 111-115.

Gillgannon, M. (2004, December 17). Globalization proceeds without practical ethical guides. *National Catholic Reporter*, 41(8), 19-21.

Goodman, A., Gonzalez, J. (2008). *Democracy Now. Ecuadorian President Rafael Correa on the Lawsuit Against Chevron, Eradicating Foreign Debt and Why He Says 'Ecuador is No Longer for Sale."* Retrieved May 6, 2012 from www.democracy.org.

Goodwin, N. (November, 2006). *Global Development and Environmental Institutions*. Retrieved June 8, 2012 from http://www.eoearth.org/article/Global_economy.

Hirst, M. (2004). *The United States and Brazil: A long road of unmet expectations*. New York: Routledge.

Hummels, D. (2001), "Toward a Geography of Trade Costs", *GTAP Working Papers* 1162, Center for Global Trade Analysis, Purdue University.

Jensen, K. B. (2004). *A handbook of media and communication research: Qualitative and quantitative methodologies*. London: Routledge

Kirchhoff, B. A. (1999). *Entrepreneurship and dynamic capitalism: The economics of business firm formation and growth*. Westport, CT: Praeger Publishers.

Kornis, M. (2004, November-December). The expanded Andean Trade Preferences Act and a U.S. free trade agreement. *International Economic Review*, 19-22.

Maniam., B. (2005). The Global Perspective of Free Trade Area of the Americas. *The Business Review, Cambridge, 21*(16), 12-23.

Matthews, D. A. & Maniam, B. (2006, September 1). Reassessing the case of Ecuador's Dollarization. *Journal of Economics and Economic Research*, 7(3), 22-26.

Neuman, W. L. (2003). *Social research methods: Qualitative and quantitative approaches, 5th ed.* New York: Allyn & Bacon.

O'Neill, R. (April, 2014). *HARVARD Gazette.* Retrieved April 14, 2014 from http://news.harvard.edu/gazette.

Palast, G. (February, 2008). *Democracy Now.* Retrieved May 6, 2012 from www.democracy.org.

Relyea, C. W., Liliana, S., Fish, K. E. & Shankar, G. S. (2011, January 1). Laissez faire and international trade: A critique of the proposed United States - Colombia Free Trade Agreement. *Journal of International Business Research*, 10(1), 37-39.

Rolleigh, M. (2003), "Plant Heterogeneity and Applied General Equilibrium Models of Trade: Lessons from the Canada-US Free Trade Agreement", Mimeo, University of Minnesota.

Schott, J. (2004). *Assessing US FTA Policy. Institute for International Economics.* Retrieved from: www.iie.com.

Seelke, C.R. (2008). *CRS Report for Congress. Ecuador: Political and Economic Situation and U.S. Relations.* Retrieved from: www.fas.org/sgp/crs/RS21687.

USDA at work in Ecuador. (2011). *U.S. Department of Agriculture.* Retrieved from: www.fas.usda.gov/icd/tim/FINAL.

Wikipedia Turismo. (2015). www.wikipedia

ABOUT THE AUTHOR

"SOLDIER BALANCES CAREER, EDUCATION TO SUCCEED"

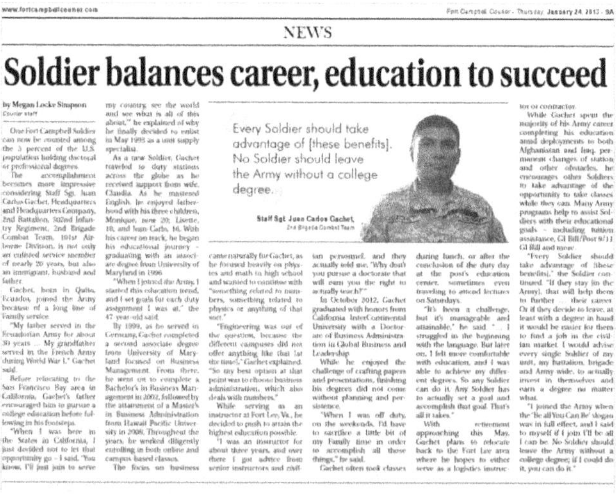

www.fortcampbellcourier.com — Fort Campbell Courier · Thursday, January 24, 2013 · 9A

NEWS

Soldier balances career, education to succeed

by Megan Locke Simpson
Courier staff

One Fort Campbell Soldier can now be counted among the 3 percent of the U.S. population holding doctoral or professional degrees.

The accomplishment becomes more impressive considering Staff Sgt. Juan Carlos Gachet, Headquarters and Headquarters Company, 2nd Battalion, 502nd Infantry Regiment, 2nd Brigade Combat Team, 101st Airborne Division, is not only an enlisted service member of nearly 20 years, but also an immigrant, husband and father.

Gachet, born in Quito, Ecuador, joined the Army because of a long line of Family service.

"My father served in the Ecuadorian Army for about 30 years … My grandfather served in the French Army during World War I," Gachet said.

Before relocating to the San Francisco Bay area in California, Gachet's father encouraged him to pursue a college education before following in his footsteps.

"When I was here in the States in California, I just decided not to let that opportunity go – I said, 'You know, I'll just join to serve my country, see the world and see what is all of this about,'" he explained of why he finally decided to enlist in May 1993 as a unit supply specialist.

As a new Soldier, Gachet traveled to duty stations across the globe as he received support from wife, Claudia. As he mastered English, he enjoyed fatherhood with his three children, Monique, now 20; Lisette, 18, and Juan Carlo, 16. With his career on track, he began his educational journey – graduating with an associate degree from University of Maryland in 1996.

"When I joined the Army, I started this education trend, and I set goals for each duty assignment I was at," the 47-year-old said.

By 1999, as he served in Germany, Gachet completed a second associate degree from University of Maryland focused on Business Management. From there, he went on to complete a Bachelor's in Business Management in 2002, followed by the attainment of a Master's in Business Administration from Hawaii Pacific University in 2006. Throughout the years, he worked diligently enrolling in both online and campus based classes.

The focus on business came naturally for Gachet, as he focused heavily on physics and math in high school and wanted to continue with "something related to numbers, something related to physics or anything of that sort."

"Engineering was out of the question, because the different campuses did not offer anything like that [at the time]," Gachet explained. "So my best option at that point was to choose business administration, which also deals with numbers."

While serving as an instructor at Fort Lee, Va., he decided to push to attain the highest education possible.

"I was an instructor for about three years, and over there I got advice from senior instructors and civilian personnel, and they actually told me, 'Why don't you pursue a doctorate that will earn you the right to actually teach?'"

In October 2012, Gachet graduated with honors from California InterContinental University with a Doctorate of Business Administration in Global Business and Leadership.

While he enjoyed the challenge of crafting papers and presentations, finishing his degrees did not come without planning and persistence.

"When I was off duty, on the weekends, I'd have to sacrifice a little bit of my Family time in order to accomplish all those things," he said.

Gachet often took classes during lunch, or after the conclusion of the duty day at the post's education center, sometimes even traveling to attend lectures on Saturdays.

"It's been a challenge, but it's manageable and attainable," he said. "… I struggled in the beginning with the language. But later on, I felt more comfortable with education, and I was able to achieve my different degrees. So any Soldier can do it. Any Soldier has to actually set a goal and accomplish that goal. That's all it takes."

With retirement approaching this May, Gachet plans to relocate back to the Fort Lee area where he hopes to either serve as a logistics instructor or contractor.

While Gachet spent the majority of his Army career completing his education amid deployments to both Afghanistan and Iraq, permanent changes of station and other obstacles, he encourages other Soldiers to take advantage of the opportunity to take classes while they can. Many Army programs help to assist Soldiers with their educational goals – including tuition assistance, GI Bill/Post 9/11 GI Bill and more.

"Every Soldier should take advantage of [these benefits]," the Soldier continued. "If they stay [in the Army], that will help them to further … their career. Or if they decide to leave, at least with a degree in hand it would be easier for them to find a job in the civilian market. I would advise every single Soldier of my unit, my battalion, brigade and Army wide, to actually invest in themselves and earn a degree no matter what.

"I joined the Army when the 'Be all You Can Be' slogan was in full effect, and I said to myself if I join I'll be all I can be. No Soldier should leave the Army without a college degree; if I could do it, you can do it."

"Every Soldier should take advantage of [these benefits]. No Soldier should leave the Army without a college degree."

Staff Sgt. Juan Carlos Gachet,
2nd Brigade Combat Team

Source: Fort Campbell Courier, 2013

One Fort Campbell Soldier can now be counted among the 3 percent of the U.S. population holding doctoral or professional degrees.

The accomplishment becomes more impressive considering Staff Sgt. Juan Carlos Gachet, 101st Airborne Division, is not only an enlisted service member of nearly 20 years, but also an immigrant, husband and father.

Gachet, born in Quito, Ecuador, joined the Army because of a long line of Family service.

"My father served in the Ecuadorian Army for about 30 years … My grandfather served in the French Army during World War I," Gachet said.

Before relocating to the San Francisco Bay area in California, Gachet's father encouraged him to pursue a college education before following in his footsteps.

"When I was here in the States in California, I just decided not to let that opportunity go – I said, 'You know, I'll just join to serve my country, see the world and see what is all of this about,'" he explained of why he finally decided to enlist in May 1993 as a unit supply specialist.

As a new Soldier, Gachet traveled to duty stations across the globe as he received support from wife, Claudia. As he mastered English, he enjoyed fatherhood with his three children, Monique, now 20; Lisette, 18; and Jean-Carlo, 16. With his career on track, he began his educational journey – graduating with an associate degree from University of Maryland in 1996.

"When I joined the Army, I started this education trend, and I set goals for each duty assignment I was at," the 47-year-old said.

By 1999, as he served in Germany, Gachet completed a second associate degree from University of Maryland focused on Business Management. From there, he went on to complete a Bachelor's in Business Management in 2002, followed by the attainment of a Master's in Business Administration from Hawaii Pacific University in 2008. Throughout the years, he worked diligently enrolling in both online and campus-based classes.

The focus on business came naturally for Gachet, as he focused heavily on physics and math in high school and wanted to continue with "something related to numbers, something related to physics or anything of that sort."

"Engineering was out of the question, because the different campuses did not offer anything like that," Gachet explained. "So my best option at that point was to choose business

administration, which also deals with numbers."

While serving as an instructor at Fort Lee, Va., he decided to push to attain the highest education possible.
"I was an instructor for about three years, and over there I got advice from senior instructors and civilian personnel, and they actually told me, 'Why don't you pursue a doctorate that will earn you the right to actually teach?'"

In October 2012, Gachet graduated with honors from California InterContinental University with a Doctorate of Business Administration in Global Business and Leadership.

While he enjoyed the challenge of crafting papers and presentations, finishing his degrees did not come without planning and persistence.
"When I was off duty, on the weekends, I'd have to sacrifice a little bit of my Family time in order to accomplish all those things," he said.

Gachet often took classes during lunch, or after the conclusion of the duty day at the post's education center, sometimes even traveling to attend lectures on Saturdays.
"It's been a challenge, but it's manageable and attainable," he said. "… I struggled in the beginning with the language. But later on, I felt more comfortable with education, and I was able to achieve my different degrees. So any Soldier can do it. Any Soldier has to actually set a goal and accomplish that goal. That's all it takes."

While Gachet spent the majority of his Army career completing his education amid deployments to both Afghanistan and Iraq, permanent changes of station and other obstacles, he encourages other Soldiers to take advantage of the opportunity to take classes while they can. Many Army programs help to assist Soldiers with their educational goals – including tuition assistance, GI Bill/Post 9/11 GI Bill and more.

"Every Soldier should take advantage of [these benefits]," the Soldier continued. "If they stay [in the Army], that will help them to further … their career. Or if they decide to leave, at least with a degree in hand it would be easier for them to find a job in the civilian market. I would advise every

single Soldier of my unit, my battalion, brigade and Army-wide, to actually invest in themselves and earn a degree no matter what.

"I joined the Army when the 'Be all You Can Be' slogan was in full effect, and I said to myself if I join I'll be all I can be. No Soldier should leave the Army without a college degree; if I could do it, you can do it." (Megan Locke Simpson, 2013).

Military Tradition

Juan Carlos Gachet Castro, born in Quito, Ecuador, joined the Army because of a long line of family service and adventure. Father, Juan Edmundo Gachet Valencia, served in the Ecuadorian Army for about 30 years.

During World War I, France called to its ranks all the firstborn of French citizens and therefore Augusto Gachet travelled to Europe and was part of this global conflagration. At that time Philippe Pétain was in command of his assignment.

Andrés Augusto Gachet Baca

World War I Veteran. Tour of Duty: 1914 – 1919.
"Distinguished citizen born in Ecuador, registered at the French embassy, completed his military service in France delivering his quota for the freedom of the world, happily married to Laura M. Valencia, distinguished Quiteña wife, had

five distinguished sons and five beautiful and spiritual young ladies, many grandchildren and great grandchildren and thousands of noble and unforgettable memories. Sadness and nostalgia for not having him with me." (Gabriel Gachet.)

Juan Edmundo Gachet Valencia

Army Tour of Duty: 1956-1981.
Beloved father and distinguished citizen born in Quito – Ecuador, served in the Mechanized Infantry as a tanker, Transportation Corps as driver, and the Quartermaster Corps as logistician. Happily married to Bertha Castro de Gachet, they have three sons: Juan Carlos, Augusto, Roberto, and seven grandkids. His quota for the freedom of the country is greatly appreciated.

Juan Carlos Gachet Castro

Army Tour of Duty: 1993-2013.
Originally from Quito, Ecuador. He completed Basic Training in Fort Jackson, South Carolina, and Advanced Individual Training in Fort Lee, Virginia. Upon graduation, he was assigned the Military Occupational Specialty 92Y (Logistics and Supply Management). His assignments included Italy, Germany, Texas, Hawaii, Virginia, and Kentucky. He deployed once to Iraq and Afghanistan twice, served with the Department of Veterans Affairs as Facility Revenue Technician, and continues to serve with the Defense Commissary Agency as Contract Specialist.

Chester Native Serves in Afghanistan

KABUL, Afghanistan (Feb. 21, 2011) – A Chester, Va. Native is currently serving as the supply sergeant for the U.S. National Support Element located on Camp Kabul International Airport.

U.S. Army Staff Sgt. Juan Gachet is responsible for over $500,000 of equipment and materials, directly supporting U.S. service members entering Afghanistan through Kabul. In addition to his regular duties, he also assists in the administrative processing of incoming and outgoing troops, providing advice for Soldiers being deployed for the first-time to the region.

"It is my duty to serve my country," said Gachet. "I'm here to support the American troops stationed at KAIA, who go out the wire on a daily basis to keep a safe environment for the civilian population."

When asked about why Gachet joined the U.S. Army, he said it was "family tradition."

"It runs in my blood," added Gachet. I just wanted to keep the family tradition going, which started with my grandfather during World War One."

Gachet believes he needs to help and support Afghans in building their country.

"My motivation to provide a safe and secure Afghanistan stems from my family," explained Gachet. "It's difficult being away however with their support and knowing I am making a difference, I am able to continue and commit to my mission 100 percent." (Story by U.S. Navy Ensign Peter Lee, U.S. Forces, Afghanistan.)

Kabul, Afghanistan – U.S. Army Staff Sgt. Juan Gachet (right) participates in a half-marathon on Camp Kabul International Airport during a Morale, Welfare and Recreation event.

Kabul, Afghanistan – Wearing over 50 lbs in body armor and protective gear, U.S. Army Staff Sgt. Juan Gachet (right) marches 22 km in full "battle rattle."

Instituto Nacional Mejía

The journey through Instituto Nacional Mejía of Quito, Ecuador. Its Latin motto "per aspera ad astra" which means "through hardships to the stars" set an important milestone and inspiration for my academic endeavors that lead to the world of postsecondary education.

I bet everybody misses high school to a certain degree. Six years, the best of my life, were spent at Mejía, where the transition from boyhood to adolescence took place. Three years were devoted to junior high, learning the traditional subjects, and the last three were spent taking Math and Physics as my majors. Mejía is the largest secondary level educational establishment. Ecuadorian political leaders and celebrities experienced and shared their learning activities.

Instituto Nacional Mejía. 21 August, 2015

We graduated 32 years ago, there was no ceremony since only three of us made it to the official lists, but we graduated even a couple of months later.

We participated in the protests against social injustice of that time, therefore, the government suspended classes indefinitely and this event was one of the ingredients for the official induction ceremony cancellation, but the last day of classes of 1983, we made the pledge at the monument to José Mejía Lequerica, our patron, to meet in the future after a few decades.

That oath was realized in December 2014, 31 years after the gathering on the steps of Patrón Mejía and there comes the feeling of the symbolic act of commencement that came true on 21 August 2015.

California Intercontinental University, Class of 2012

"**O**nly three percent of Americans hold doctoral or professional degrees, which is what makes one Fort Campbell soldier, Staff Sgt. Juan Carlos Gachet, among the United States elite. Gachet, a soldier of 20 years, is also a husband, father and immigrant. The 47-year-old's accomplishments are not only honorable; they're proof that education is within reach for all military personnel.

Gachet began pursuing his higher education in the 1990s, graduating from the University of Maryland with an associate degree. These goals had a quick snowball effect, as Gachet completed a second associate degree in 1999, this time in Business Management, then continued his education in the field. He received his Bachelor's in Business Management in 2002 and a Master's of Business Administration in 2008.

He attributes his success to his own determination and the support of his family, but also the financial assistance provided by the Army. Gachet's feelings are exactly why CalUniversity supports our troops and encourages our military service men and women to use their VA and GI Bill education benefits at our school." (Angelo Lioudakis, 2014).

Attendees at the Commencement Ceremony

Dr. Finian Tan, President
Rax Ribadia, Director Finance
Dr. Troy Roland, Chief Academic Officer
Dr. Fathiah Inserto, Dean of Academics
Randy Saban, Director of Admissions
Chris Ho, Manager – Student Affairs
Verna Zak, Staff Accountant
Andrea Clarke, Manager – Student Services
Thai Le, IT & Designs Project
Eupeterson Lewis, Military Education Liaison
Terry Ruiz, Senior Admissions Advisor
Marc Mu, Senior Admissions Advisor
Michael Winn, Admissions Advisor
Ray Soriano, Admissions Advisor
Dr. Jennie Wong, EdD Instructor
Dr. Kamalnayan Inamdar, EdD Instructor
Dr. Kenneth Phillips, EdD Instructor
Benjamin Perez, MSc Instructor

Graduates

Dr. Scott Vowels, DBA-EBM
Dr. Juan Gachet, DBA-GBL
Dr. Kim Erwin, DBA-GBL
Dr. Marshall Cowell, DBA-GBL
Dr. Augustus Ynares, DBA-HCML
Dr. Blanche Shenosky, DBA-HCML
Dr. Julie Ann Carson, DBA-HCML
Dr. Monserrate Jurgensen, DBA-HCML
Dr. Sulma Gandhi, DBA-HCML
Dr. Tracy McClelland, DBA-HCML

Dr. Andrea Clarke, DBA-HCML
Dr. Hartinder Singh Johal, DBA-ISERM
Dr. Thomson Lukose, DBA-ISERM
Dr. Wajdi Aghnatios, DBA-ISERM
Sr. Fabio Sotelo, BBA-IME

Saturday, 6 October, 2012

Dear Juan:
Let me congratulate you on an excellent presentation last Friday morning. It was very interesting and informative. You passed your Oral Defense in an exemplary manner.
Best regards.

Dr. Dennis Toutant
EdD Instructor

It was such a pleasure meeting you and seeing your family at the commencement. As I mentioned before, I loved your speech!! As you said, we as parents should set the standards high for our kids but we need to be the example. That was just wonderful.
And once again, congratulations on completing your degree with us at Calu, Dr. Juan Gachet!!
Thank you.

Maritza Sandoval
Student Services Advisor

Family

Parents, Juan E. Gachet and Bertha; brothers, Augusto and Roberto Gachet; wife, Claudia Gachet; daughters, Monique and Lisette; son, Jean Carlo.

My parents Juan E. Gachet and Bertha Castro have lovingly devoted much of their life to their children; Juan Carlos, Andrés Augusto and Luis Roberto, who in turn corresponded to that sacrifice with their academic

achievements. Augusto is Professor of Physical Culture and Roberto is an Economist. Our parents have essentially been an inspiration throughout the years.

My wife, Claudia Gachet, Teacher of Early Childhood Education, has given everything for our children, Monique, Lisette and Jean Carlo.

They have responded to that sacrifice and Monique

graduated with a Bachelor of Science in Accounting, Lisette got a Bachelor of Science in Information Systems, and Jean Carlo accomplished his first college degree, an Associate of Science in Business Administration in May 2016.

Ecuador y los Estados Unidos Tratado de Libre Comercio

JUAN CARLOS GACHET

Copyright © 2016 Juan Carlos Gachet

All rights reserved.

DEDICACIÓN

Este libro es dedicado a mi familia, en esencia ellos han sido una inspiración a lo largo de todos estos años. Mi esposa, Claudia Gachet; hijas, Monique y Lisette; hijo Jean Carlo, así como mis padres Juan E. Gachet y Bertha; hermanos Augusto y Roberto Gachet; mi abuela, Laura Gachet, tías, tíos, y primos que me brindaron todo su afecto cuando llegué a los Estados Unidos.

CONTENIDO

RECONOCIMIENTO

En primer lugar, me gustaría expresar un agradecimiento sincero a todos los profesores que me guiaron por el colegio, la universidad y la etapa final de mis estudios de postgrado, su apoyo excepcional, y motivación. Todos ellos fueron una fuente inestimable de inspiración durante cada paso requerido para publicar este libro.

La trayectoria por el colegio Mejía de Quito, Ecuador. Su lema latino "*per aspera ad astra*" el que significa "por los ásperos caminos hacia la cumbre" puso un precedente importante y la inspiración para los logros académicos que me llevaron al mundo de la educación universitaria.

CAPÍTULO 1

HECHOS HISTÓRICOS DE ECUADOR

El enfoque principal de este libro es más que nada de concientizar los efectos de un Tratado de Libre Comercio (TLC) entre los sectores de interés ecuatorianos y de los Estados Unidos con respecto a la industria agrícola. El tema principal está diseñado para detalladamente explicar las ventajas, desventajas, y las implicaciones del libre comercio para la gente ecuatoriana. No es para convencer si Ecuador entra en el Tratado de Libre Comercio con los Estados Unidos.

El objetivo es informar al lector sobre las cuestiones económicas principales a niveles micro y macroeconómicos que conducirían a una evaluación culta. El uso de estas herramientas permitiría llegar a la decisión correcta y tener el conocimiento necesario para informar a otros de los beneficios generados por el Tratado de Libre Comercio.

El libre comercio entre países diferentes genera empleo, atrae la inversión extranjera y local, crea e incrementa el desarrollo tecnológico, que por su parte conduce al progreso industrial. El libre comercio sobre todo aumenta ganancias en sectores afectados, principalmente en la agricultura. Por lo tanto, el contenido se centrará en cómo un Tratado de Libre Comercio justo beneficiaría a ambos mercados.

Captura de Atahualpa

Fuente: Historia del Ecuador, 2014

Un acuerdo económico entre Ecuador y los EE.UU. ha estado en etapas de desarrollo por un largo tiempo, y los políticos que son elegidos por voto popular para asumir la presidencia por un período de cuatro años nunca lo cumplen. Esto es sólo una estrategia política de campaña para obtener el voto popular y mantener el mismo modus vivendi.

NAFTA (*The North American Free Trade Agreement*), como es conocido por sus siglas en inglés es nada más ni nada menos el Acuerdo de Libre Comercio de América del Norte, el cual es el mejor ejemplo de cómo los mercados de Canadá, EE.UU. y México se benefician mutuamente. La economía mexicana se benefició astronómicamente desde que el Tratado de Libre Comercio entre los tres países de América del Norte se convirtió en realidad. México se ha ubicado en treceava potencia económica en 2015 como resultado del libre comercio.

Dado que este libro se centra en el acuerdo comercial económico en curso entre las dos naciones americanas, es imperativo poner más atención a los hechos históricos fundamentales del país sudamericano.

Por aproximadamente tres siglos, Ecuador estaba bajo

control español después de la captura y ejecución de Atahualpa, el último emperador del Twantinsuyu, el Imperio Inca. Más tarde la parte norte del imperio era conocida como Territorio de Quito.

La Misión Geodésica Francesa fue la encargada de localizar y medir el centro del mundo que divide los hemisferios norte y sur a través de la Línea Ecuatorial. Este evento pone a la región en el centro de atención internacional y el gobierno de Quito aceptó el nombre de República de Ecuador, cuando el país se separó de la Gran Colombia en 1830.

Ecuador también se refiere al ecuador del globo terráqueo que va hasta la punta norte de Quito. Las Noticias Secretas de España en América mencionan por primera vez el nombre del país, y durante ese período el territorio de Ecuador fue mejor conocido como la "Real Audiencia de Quito".

La ubicación del país está en el extremo noroeste del continente sudamericano. Colombia es el vecino del norte y el Perú limita con Ecuador, principalmente en el este y el sur. El Pacífico se encuentra al oeste de la costa ecuatoriana, lo que coloca al país en una posición estratégica cerca del Canal de Panamá y el fácil acceso a Europa y África.

Las Islas Galápagos son parte de Ecuador ubicadas en el Pacífico a unos 600 kilómetros de la costa. Este archipiélago es uno de los principales atractivos turísticos de América del Sur. Charles Darwin llevó a cabo la investigación científica sobre las especies nativas y desarrolló su teoría de la evolución basada en sus observaciones y estudios en la primera parte del siglo 19.

Ecuador tiene una superficie total de 109,483 millas cuadradas y es el cuarto país más pequeño de América del Sur, con una población aproximada de 15 millones.

La capital del Ecuador es Quito, ubicada en la parte norte del país, en un valle rodeado de volcanes nevados a unos 2,850 metros sobre el nivel del mar en la Cordillera de los Andes, lo que contribuye a un clima primaveral durante todo el año.

Batalla de Pichincha

Fuente: Historia del Ecuador, 2014

Los historiadores están de acuerdo en que la rebelión por la independencia comenzó en Quito con la masacre de los revolucionarios el 10 de agosto de 1809, que dio a la ciudad el nombre de "Quito Luz de América". Ecuador logró su independencia de España después de una larga lucha por la libertad con la victoria del guerrero más confiable de Simón Bolívar, el general Antonio José de Sucre de Cumaná, Venezuela en la Batalla de Pichincha el 24 de mayo de 1822.

En 1978, la ciudad fue declarada "Patrimonio Cultural de la Humanidad" por la UNESCO (*United Nations Educational, Scientific and Cultural Organization*), por sus siglas en inglés, y conocida como (Naciones Unidas para la Educación, la Ciencia y la Cultura), debido a la arquitectura de estilo colonial único y muy bien conservado, el arte religioso y la tradición.

Los nuevos países de Colombia, Ecuador, Venezuela y Bolivia se unieron a la Gran Colombia. Ecuador se separó de la Gran Colombia después de la renuncia a la presidencia del Libertador Simón Bolívar el 27 de abril de 1830. Por lo tanto, Ecuador se convirtió en un país independiente el 13 de mayo de 1830.

ECUADOR Y LOS ESTADOS UNIDOS TLC

Ecuador es una república donde el Presidente tiene el derecho de poder como jefe de Estado y de gobierno, a condición de que el poder político se mantiene en ese nivel y no más allá de la dictadura, que formaba parte de la escena política antes de 1979 en la mayoría de los países de América del Sur.

Todas las personas que saben leer y escribir entre los 18 y 65 años tienen derecho a votar por el candidato político de su preferencia cada cuatro años, y de un buen número de diversos partidos políticos con muchas ideologías y un sinnúmero de agendas.

La economía de Ecuador es muy dependiente del petróleo y la exportación de productos derivados del petróleo desde la dictadura del general Guillermo Rodríguez Lara, quien nacionalizó la industria bajo CEPE (Corporación Estatal Petrolera Ecuatoriana).

Otras industrias conocidas son las de elaboración de alimentos en el oeste; el turismo, principalmente en Quito, Cuenca y Galápagos; textiles en la zona costera; productos de madera de la Amazonia, y productos químicos. La industria del petróleo representa más del sesenta por ciento de las ganancias de exportación del país. El país experimentó una gran crisis financiera a finales de 1990 debido a un terremoto que afectó principalmente a la región del Amazonas y los oleoductos, así como la última guerra con Perú en 1995. Sin embargo, Ecuador se ha recuperado de sus pérdidas y está desarrollándose muy rápidamente.

Ecuador negocia principalmente con los EE.UU., Colombia, Perú, Venezuela y Brasil. La Comunidad Europea y Rusia también tienen sus puestos de comercio en la nación sudamericana.

El español es el idioma oficial del Ecuador desde que los conquistadores se apoderaron del Imperio Inca y depusieron a su último emperador, Atahualpa. Se pagó un rescate extraordinario a los españoles a cambio de la liberación de Atahualpa. El rey de España, Carlos V impuso la fe católica romana apostólica, la cual se convirtió en la principal religión, cuando Quito era una colonia.

Algunos de los lugares más impresionantes e increíbles en el mundo se encuentran en Ecuador. Cotopaxi, el volcán activo más alto del mundo se encuentra a pocas horas al sur de Quito y atrae a muchos turistas y ecuatorianos que les gusta escalar montañas.

Con el fin de entender lo que implica un acuerdo económico, el primer paso lógico es mirar más de cerca y sacar una conclusión de su verdadero significado. El Tratado de Libre Comercio no es más que un consenso económico entre los países interesados para crear una zona de libre comercio, donde se puede llevar a cabo los negocios cotidianos sin tener que preocuparse de las fronteras con rapidez y agilizar las transacciones económicas sin imponer tarifas en absoluto.

El muy conocido consenso que los acuerdos económicos contribuyen a un mayor número de productos vendidos en el extranjero y la producción total en un período establecido es alcanzable. Por lo tanto, hay una gran oportunidad para poner en práctica un análisis de los planes estratégicos que conducen al desarrollo.

Estoy totalmente de acuerdo con esta conjetura ya que la Unión Europea tuvo que pasar por todos los inconvenientes de permitir el ingreso de los países con economías menos desarrolladas que no cumplían con los requerimientos, tales como España, Italia y Grecia durante los primeros años.

Esto causo dolores de cabeza significantes a nivel económico, sin embargo, estos países con economías problemáticas se beneficiaron por ser miembros. La inyección de capital necesario para sus mercados contribuyó al crecimiento global de la Unión Europea hasta el punto de que el euro superó el valor de mercado del dólar de EE.UU.

Este libro se centra principalmente en la situación actual y aborda hechos históricos sobre los temas que están desarrollándose con respecto al Tratado de Libre Comercio entre Ecuador y los Estados Unidos a nivel de la industria agrícola y otros sectores económicos.

Incorpora la perspectiva de algunos estudiosos y economistas con el fin de llegar a una solución que beneficie a

ambos países.

A partir de enero de 2004, Estados Unidos ha firmado tratados de libre comercio (TLC) con diferentes países de todo el mundo, como Israel, Canadá, México, Jordania, Singapur y Chile.

Los EE.UU. está actualmente elaborando acuerdos de libre comercio bilaterales con otros 12 países y regiones: Australia, Marruecos, América Central, y los 5 miembros de la Unión Aduanera del África Meridional, que algunos también cuentan entre los 33 socios en la negociación del Área de Libre Comercio de las Américas (ALCA).

Además, el Representante Comercial de EE.UU. (USTR) por sus siglas en inglés, ha notificado al gobierno de su intención de iniciar conversaciones con Bahréin, algunos países de América del Sur y Tailandia en el año 2004. Los socios actuales del TLC representan más de un tercio del comercio total de mercancías de EE.UU.; tras la finalización de las nuevas negociaciones bilaterales de libre comercio y el ALCA, más del 40 por ciento del comercio de EE.UU. será cubierto con los pactos de libre comercio (Schott, 2004).

Con el fin de entender más claramente así como una visión global del comercio ecuatoriano, el primer paso necesario es echar un vistazo a su comercio internacional desde la perspectiva macroeconómica.

De acuerdo con el *2012 Time Almanac* "Las importaciones a partir del 2006: $12,114,000,000. Combustibles minerales 21.1%, maquinaria y aparatos 20%, productos químicos 15.3%, vehículos motorizados y repuestos 11.5%, hierro y acero 6%. De fuentes de importación importantes a partir del 2008: EE.UU. 19%, Colombia 9.6%, Brasil 4.8%, Japón 3.6%, México 3.5%.

Las exportaciones a partir del 2006: $12,728,000,000. Petróleo crudo 54.5%, bananos y plátanos 9.5%, pescados 5.4%, camarones 4.6%, productos derivados de la refinación del petróleo 3.9%, flores 3.4%. Los principales destinos de las exportaciones a partir del 2008: Estados Unidos 45.3%, Perú 9.2%, Chile 8.2%, Colombia 4.2%, Venezuela 3.8%."

Este punto de vista macroeconómico del comercio

internacional es de más beneficio para la economía ecuatoriana que para los EE.UU. ya que las exportaciones de Ecuador a los EE.UU. representan el 45.3%, y sus productos importados de los Estados Unidos representan el 19%. El auge de la economía de la década de 1970 se atribuye al descubrimiento de petróleo crudo en la región amazónica, que revolucionó la economía ecuatoriana y al mismo tiempo representa la mayor parte de las exportaciones a otras partes del mundo.

Un número significante de empresas de Estados Unidos están llevando a cabo negocios en el Ecuador, lo cual es un buen indicador económico, ya que la inversión extranjera estimula la economía local mediante el incremento de capital y la creación de puestos de trabajo. Esto es beneficial tanto para ambos mercados sin importar las opiniones políticas. Es simple matemática de economía; las "leyes de la oferta y la demanda" hacen el trabajo mediante la generación de ganancias para ambos países. Por lo tanto, es una situación de triunfo económico.

Empresas estadounidenses dedicadas a la industria de la producción ecuatoriana que contribuyen al desarrollo local son General Motors, Philip Morris, Mead Johnson, Avon conocido por los cosméticos y lencería; Coca-Cola Co. famoso por el mercado de refrescos, y unos cuantos más.

Las franquicias estadounidenses de comida rápida y restaurantes actualmente operan en Ecuador; los más famosos son el Pizza Hut, Kentucky Fried Chicken, Taco Bell, Burger King, McDonalds. Las más recientes establecidas en las principales ciudades ecuatorianas son TGI Fridays Tony Roma, Johnny Rockets, Chili, Papa John, Domino Pizza, Subway, Quiznos y Hooters.

Según el Servicio Comercial de Estados Unidos del año 2011. El sector bancario de Estados Unidos se lleva a cabo principalmente por el Citibank y Bank of America. La industria del transporte está representada por American Airlines, Continental y Delta. Los EE.UU. han invertido poco a poco en el mercado textil y agrícola, como flores, frutas y verduras.

Hace veinte años, sólo había unas pocas empresas que operan allí, como Texaco, Burger King y Kentucky Fried Chicken. Basado en el número de empresas estadounidenses que operan actualmente en Ecuador, es obvio que el país sudamericano está prácticamente americanizado.

La tasa de inflación estaba disparándose sin control como parte de un movimiento político estratégico repentino para el control de la economía ecuatoriana, y la última solución fue reemplazar el sucre por el dólar estadounidense como moneda nacional a principios del año 2000. Esa época turbulenta ya se ha ido y la implementación del dólar como unidad monetaria nacional energizo la economía a la larga.

El principal problema con el tratado de libre comercio es su estancamiento. Sin embargo, el presidente Correa ha decidido renegociar la deuda del país debido a un terremoto en la región amazónica, que destruyó los principales oleoductos, antagoniza un acuerdo de libre comercio con los EE.UU., y sin embargo, decidió frenar las concesiones para las empresas mineras.

Parte de la campaña presidencial de Correa fue de rescindir el contrato con la única base militar de Estados Unidos en América del Sur. Lo cual se convirtió en realidad cuando el contrato de la base militar estadounidense en Manta no fue renovado debido a un presunto escándalo de espionaje que implican a varios funcionarios de alto rango.

Hay diferentes argumentos que se tratan en la mesa de negociaciones para un acuerdo comercial entre Ecuador y Estados Unidos, que beneficiaría a la economía ecuatoriana, o que pondría la población ecuatoriana en una posición de desventaja económica.

Según Silkee en el 2008, las relaciones diplomáticas de Ecuador con los EE.UU. han estado en buenos términos. Sin embargo, la agitación política ha puesto en peligro las relaciones bilaterales. Aunque los EE.UU. ratifico acuerdos de libre comercio con los países vecinos de Ecuador al norte y al sur. Los esfuerzos para alcanzar un acuerdo de libre comercio con Ecuador se han postergado por una disputa con una empresa petrolera.

Desde un punto de vista político y económico de los expertos en la materia han aconsejado al gobierno de Estados Unidos no contradecir Correa, sino utilizar tácticas diplomáticas para obligarlo a mantener un mercado abierto y las políticas democráticas, tales como el mantenimiento de las preferencias comerciales de Estados Unidos para el Ecuador.

Durante el gobierno del general Guillermo Rodríguez Lara (1972-1976), la industria petrolera fue nacionalizada bajo CEPE (Corporación Estatal Petrolera Ecuatoriana). Se descubrió más petróleo en la cuenca del Amazonas, esto a su vez revolucionó la economía ecuatoriana, y dio lugar a una transición de la agricultura a la producción petrolera.

Goodman, González, 2008, afirmó que más del 50% de la economía del país sudamericano se basa principalmente en la industria del petróleo y la fabricación está reservada para el mercado local. Las empresas locales y los productos agrícolas se utilizan para el consumo interno y la exportación.

Los principales productos que representan las exportaciones generan un alto ingreso de capitales son el petróleo, flores, banano, camarones, y otros productos. En el año 2010, los productos de petróleo crudo y refinado generaron el 56% de los ingresos totales de exportación.

Ecuador es el primer exportador de banano del mundo, como se ve en los mercados de Estados Unidos y Europa, generando cerca de $2.1 billones de dólares y un importante exportador de camarón que trae $830 millones y el cacao con una producción de $402 millones.

Las exportaciones de otros productos o mejor aún nuevos en el mercado, como las flores representan $600 millones, conservas de pescado $600 millones, y la industria del automóvil contribuye con $375 millones.

El Tratado de Libre Comercio entre Ecuador y los EE.UU. no beneficiaría en última instancia a la economía ecuatoriana a corto plazo debido a que el mercado industrial de allí aún no está muy bien desarrollado.

En otras palabras, el Ecuador no tiene economías de escala, y si por ejemplo Levis entra en ese mercado, beneficiaría al consumidor, pero el productor local estaría en

desventaja. El consumidor tendrá la libertad de elección y obtener un mejor producto a un precio más alto, sin embargo, el productor estaría obligado a declararse en quiebra porque no podría producir en grandes cantidades, ni pagar a sus empleados un sueldo justo, o apoyar las operaciones de la empresa.

Lo mismo ocurriría en el mercado agrícola, digamos que el gobierno ecuatoriano permite a la industria láctea entrar en el mercado local a $1.00 por galón, el productor local está vendiendo el mismo producto por $1.50. El consumidor, por supuesto, tendría que elegir el producto más barato obligando al productor local a bajar el precio. Sin embargo, este proveedor local no sería capaz de producir en escala, y en última instancia, se vería obligado a salir en quiebra.

Como se mencionó antes, la producción ecuatoriana aún no es una economía de escala, y tomaría años para llegar a ese nivel. Una vez que se haya hecho realidad el mercado local estaría dispuesto a involucrarse en un Tratado de Libre Comercio con los Estados Unidos. Ese es el motivo principal por el cual los diferentes gobiernos no han llegado a ningún acuerdo todavía con el fin de proteger el mercado local y principalmente al productor local.

Otros sectores de la economía ecuatoriana se beneficiarían con un TLC, por ejemplo, el sector bancario se vería obligado a competir, digamos en el supuesto caso de Bank of America. ¿Por qué? Simplemente porque Bank of America ofrece un mejor servicio al cliente que obliga a los banqueros locales a competir ya sea recapacitando y rediseñando su estrategia, así como también ofreciendo al cliente mejores ofertas de préstamos a mejores tasas de interés.

El objetivo principal es profundizarse en los principales hechos y obtener la mejor respuesta para la economía ecuatoriana y las personas que están contribuyendo a su progreso. Además, para estudiar y llegar a las raíces de los problemas principales por las que este proceso ha llegado a estancarse. Se habla mucho acerca de este tema desde el punto de vista político, así como del ciudadano común, las personas

que están a favor y otros que sólo se oponen debido a la dominación yanqui.

Otro de los objetivos es llevar a cabo el análisis correcto que afecta al Tratado de Libre Comercio en lo que respecta a la producción, las importaciones, las exportaciones, el consumismo y la entrada de capitales de diferentes agentes económicos de la economía ecuatoriana. Este estudio analiza los resultados de diferentes escenarios, con y sin TLC para Ecuador utilizando un modelo aplicado de equilibrio general, que incluye datos de socios a nivel económico relacionados con el TLC de América del Sur y la Unión Europea.

Además, este estudio tiene en cuenta las interdependencias entre estos socios comerciales, así como los efectos directos e indirectos que un acuerdo con grandes y profundas repercusiones como el TLC traería a la economía ecuatoriana.

Diferentes escenarios se establecieron con el fin de analizar el impacto del TLC en la economía ecuatoriana. En los primeros escenarios, Ecuador firma el TLC, y en los otros escenarios Ecuador no firma el TLC.

El escenario en el cual Ecuador firma el TLC incluye a Colombia y Perú, lo que permite a los EE.UU. libre acceso a sus mercados. A cambio, los países andinos consolidan las preferencias arancelarias reales otorgadas por los EE.UU., además consiguen varias preferencias bajo la Ley de Preferencias Arancelarias Andinas y Erradicación de la Droga mejor conocida como ATPDEA (siglas en inglés para *Andean Trade Promotion and Drug Eradication Act*), y los EE.UU. elimina los subsidios a la producción de los bienes agrícolas.

Los escenarios en los que Ecuador no firma el TLC, pero Colombia y Perú acuerdan hacerlo tendrían estas consecuencias. Colombia y Perú permiten a los EE.UU. libre acceso a sus mercados; y a cambio estos países reciben preferencias del ATPDEA.

Ecuador pierde trato preferencial arancelario del ATPDEA y se le imponen las mismas tarifas que a Venezuela de parte de los EE.UU. En otras palabras, el Ecuador obtiene tarifas más caras que las impuestas por los EE.UU., al mismo

precio de las tarifas impuestas a Venezuela.

En la otra situación, Colombia y Perú permiten a los EE.UU. libre acceso a sus mercados. A cambio estos dos países andinos obtienen mejores preferencias del ATPDEA, y los EE.UU. eliminan los subsidios en la producción del mercado agrícola. Ecuador abandona la tarifa de trato preferencial de la ATPDEA y el país está obligado a pagar los aranceles iguales a los impuestos a Venezuela por los EE.UU.

Los resultados preliminares serán si Ecuador firma el TLC como sus socios económicos andinos Colombia y Perú. Si Ecuador no obtiene más libertad de acción de los EE.UU. que la consolidación de la ATPDEA, y suponiendo que no habría ningún tipo de fricción externa o interna. Esto implicaría un resultado inmediato de un déficit comercial, la reducción en los ingresos de los factores de producción y el colapso del bienestar social de la gente en Ecuador.

Si Ecuador firma el TLC y los países andinos obtienen mejores tarifas preferenciales y una mayor cuota de azúcar. Eso quiere decir una reducción del 50% de los aranceles equivalentes *ad valorem* impuestos por los EE.UU.

Ecuador podría incluso experimentar un efecto inmediato del déficit económico, la reducción de los ingresos de los bienes utilizados en la producción. Eso llevaría al fracaso de la asistencia social del Ecuador.

De nuevo, si Ecuador firma el TLC y permitiendo las ya mencionadas ventajas arancelarias con los otros países andinos. Como resultado, los EE.UU. reduce sus subsidios a la producción agrícola. Ecuador podría incluso terminar con un impacto negativo en la balanza comercial, una caída menor en la producción y la demanda, en los ingresos de los factores de producción, y una reducción del bienestar social de la población.

Incluso sin el TLC, Ecuador podría experimentar los mismos efectos que con un supuesto TLC, pero en menor proporción. El valor agregado de la mayoría de los sectores reflejaría una caída, el ingreso de los factores refleja la reducción en el valor agregado, la caída de la demanda agregada, así como el bienestar social. En general, la economía

Batalla de Yorktown

Fuente: Recuerdo de Virginia, 2014

no muestra un mejor rendimiento sin un TLC, debido a la caída del producto interno bruto.

Sin embargo, si se analiza lo que sucedería a corto y largo plazo, se espera que haya acontecimientos positivos con un TLC así como los mecanismos necesarios para atraer la inversión. Además, se espera que el TLC abra el camino a mejores tecnologías e innovación.

Colombia y Perú firmaron oficialmente el acuerdo de libre comercio con los EE.UU. Se suponía que sería difícil proteger algunos sectores económicos amenazados por productos de competencia importados de los Estados Unidos. En particular, el arroz, los cereales y otros productos, debido a que estos entrarían al Ecuador desde los países fronterizos que firmaron el TLC con los EE.UU.

En un análisis de los diferentes sectores, es obvio que los sectores que se beneficiarían con el TLC, a diferencia de una reducción sin el TLC, incluye los sectores que se benefician con las tarifas establecidas por el ATPDEA, como estos mercados particulares: frutas, hortalizas, flores, azúcar y la pesca.

Según el Informe del Servicio de Investigación Legislativa del 2008 (CRS *Congressional Research Service*) para el Congreso, las negociaciones diplomáticas para un acuerdo de libre comercio con Ecuador han estado paralizadas hasta nuevo aviso sobre la disputa con la compañía petrolera estadounidense. En este caso, Estados Unidos tiene que ceder y llegar a un acuerdo con el gobierno de Correa o simplemente esperar con paciencia hasta que su mandato termine y el pueblo ecuatoriano elija al próximo mandatario.

El comercio exterior ecuatoriano depende en gran medida de los EE.UU., por ejemplo, las importaciones de Ecuador desde los EE.UU. representan el 19%, y exporta el 48% a los Estados Unidos. Muchas inversiones vienen y están financiadas por las empresas estadounidenses establecidas en el Ecuador.

El impacto de un Tratado de Libre Comercio de acuerdo a Goodwin (2006), los argumentos relacionados con el "¿Por qué las Naciones se resisten frecuentemente al libre comercio?" se centran principalmente en su eficiencia, el aumento de la producción y el consumo, y por qué no en las oportunidades y los beneficios de una economía mundial bien integrada.

Una vez más, este es el caso del Gobierno de Correa en Ecuador. El presidente sostiene que un acuerdo TLC no es beneficioso para el país por muchas razones.

A esta altura, es necesario recalcar una entrevista con el presidente Rafael Correa sobre por qué "Ecuador ya no está a la venta" realizada por el periodista Greg Palast en el 2008. Correa sostuvo que el país ya no está a la venta debido a su soberanía y gobierno nacionalista.

Esto en esencia es la verdad; algunas idiosincrasias son utilizadas para fomentar el aislamiento del resto del mundo. Sin embargo, cualquier economía necesita inversión extranjera para estimular los mercados locales con el fin de sobrevivir y mantenerse competentes.

Esta es una poderosa entrevista que conduce a las raíces fundamentales de la resistencia del gobierno ecuatoriano para firmar el TLC con los EE.UU. La principal se refiere al daño

ambiental y social en la región amazónica, así como la oposición de la presencia militar estadounidense en territorio ecuatoriano.

Es esencial que se hable acerca de los principales hechos históricos de Estados Unidos para entender mejor a la primera potencia económica del mundo hasta octubre del 2014

Los españoles, franceses e ingleses colonizaron América del Norte durante los siglos XVI y XVII. Ponce de León exploró Florida en 1513, Giovanni da Verrazzano al servicio de Francia exploró el Atlántico Norte en 1524, Hernando de Soto descubrió el río Mississippi en 1542, Sir Walter Raleigh fundo la colonia de Roanoke en 1587.

El Congreso Continental de las trece colonias americanas proclamó la Declaración de Independencia del antiguo imperio británico el 4 de julio de 1776.

Las fuerzas combinadas del ejército continental al mando del general George Washington y el ejército francés dirigido por el conde de Rochambeau dieron el golpe final a las fuerzas británicas y alemanas comandadas por Lord Cornwallis, quien se rindió en la batalla de Yorktown, el 19 de octubre de 1781. El Tratado de París (3 de septiembre, 1783) selló oficialmente la guerra revolucionaria americana.

En la actualidad la economía de Ecuador se encuentra en el octavo puesto en América Latina después de Brasil, México, Argentina, Colombia, Perú, Venezuela y Chile.

En abril del 2007 se creó la Unión de Naciones Sudamericanas (UNASUR), integrada por los doce países independientes de América del Sur, con sede en Quito, y fue nombrado primer Secretario General el ex Presidente ecuatoriano Rodrigo Borja Cevallos (10 ago. 1988-10 ago.1992), pero no ejerció el cargo porque renunció por discrepancias con este organismo

La economía ecuatoriana ha logrado alcanzar una bonanza por el petróleo y gasto público, pero ha tenido en 2013 y 2014 déficit fiscales superiores al 5%, dando lugar al gasto de más de 5,000 millones de dólares por encima de los ingresos. Hubo muchos recursos invertidos en obra pública, para el beneficio de los sectores populares en educación y

salud.

El peso ecuatoriano fue renombrado a sucre en 1884, fue definido en 22.5 gramos de plata equivalente a 5 francos. En 1898 Ecuador cambia al patrón oro, con el sucre definido como 732.224 miligramos de oro fino equivalente a 2 chelines de libra esterlina.

En 1932 se suspende el oro y la tasa oficial fue fijada en 5.95 compra por dólar estadounidense. Después el precio de plata subió más del valor nominal de la mayoría de monedas de plata en la década de 1930. Esto dio lugar a varios ajustes al sistema de cambio de divisas pese a lo que el sucre siguió depreciándose.

Los controles de divisas, se levantaron en 1937 y el tipo de cambio oficial se fijó en los 13.5 sucres por dólar, se devaluó a 14.77 sucres por dólar en 1940 y se volvieron a imponer controles de cambio. El tipo de cambio oficial se convirtió en 14 en 1942 y 13.5 en 1944, 15 sucres por dólar en 1950, 18 en 1961, y 25 en 1970.

El sucre mantuvo un tipo de cambio relativamente estable contra el dólar hasta 1983, se devaluó a 42 por dólar. La depreciación toma impulso y el mercado libre lo hizo llegar a más de 800 por dólar en 1990 y casi 3,000 en 1995. Perdió 67% de su valor de cambio durante 1999 terminando en 25,000 sucres por dólar.

En enero del 2000, el Presidente Jamil Mahuad Witt (10 ago. 1998-21 ene. 2000) anunció que el dólar de los Estados Unidos sería adoptado como moneda oficial de Ecuador.

Palacio de Carondelet, Quito - Ecuador

Fuente: Wikipedia Turismo, 2015

CAPÍTULO 2

TRATADO DE LIBRE COMERCIO EN LAS AMÉRICAS

Ecuador no sólo se extiende a ambos lados de la línea equinoccial, es prácticamente único entre los países latinoamericanos y centroamericanos en general. Continua en el dialogo sobre todo en la línea de sus relaciones con los EE.UU. para llegar a un acuerdo de libre comercio común como parte de un Acuerdo de Libre Comercio más importante que integraría a todo el continente. De hecho, a pesar de las negociaciones en curso, Ecuador se destaca de sus vecinos, ya que aún no tiene una agenda común de libre comercio establecida con los Estados Unidos.

Durante las últimas dos décadas, los acuerdos comerciales con diferentes regiones del continente se han aplicado con resultados a favor y en contra, pero el país andino de Ecuador continua siendo excluido de estos acuerdos y las negociaciones hasta el día de hoy. Este capítulo examina los antecedentes de estos acuerdos comerciales y su situación actual con respecto a Ecuador para identificar posibles obstáculos para el progreso y la probabilidad del éxito.

La experiencia y visión sobre el tema económico están mejor representados por el dicho de que "la marea alta pone a

flote a todos los barcos". Se cree que tiene que ver especialmente con los acuerdos de libre comercio multilaterales que incrementan el comercio entre los países involucrados de manera que sean de beneficio mutuo. Pese a los intentos de forjar un pacto económico con Ecuador, los dos países no han logrado llegar a una solución para sobrepasar el resto de obstáculos a la inclusión del acuerdo de libre comercio con los EE.UU.

Aunque, los Estados Unidos aún representa para Ecuador el socio comercial más importante en el campo de exportación e importación. Además, algunos observadores sugieren que la aceptación de Ecuador a estos acuerdos de libre comercio son superficiales y la mayoría de las mercancías que son objeto de intercambio comercial entre los dos socios económicos ya están libres de impuestos.

Otros analistas, sin embargo, sugieren que al mantener a Ecuador fuera de estos acuerdos comerciales multilaterales, al país se le está negando de una participación equitativa con los países centroamericanos vecinos al nivel de las condiciones económicas en juego.

Inclusive Ecuador cuenta con otros recursos naturales como el oro, la madera y el cobre, así como de un sector agrícola importante. La economía del país continua dependiendo en gran escala de sus recursos petrolíferos conocidos, que representan más del 50 por ciento de sus exportaciones y aproximadamente del 40 por ciento de sus ganancias del sector público en las últimas temporadas (*USDA at work in Ecuador, 2011*).

Durante los primeros años de la década de 1990, Ecuador también experimentó una crisis económica significativa causada por la crisis bancaria, provocando una contracción del PIB de un 5.3%. Como resultado, el gobierno ecuatoriano no cumplió con el pago de la deuda externa y la pobreza en el país se incrementó sustancialmente. (Economía del Ecuador, 2012)

En respuesta a la situación, el gobierno de Estados Unidos autorizó una serie de reformas estructurales para el Ecuador para ayudar a recuperarse de su crisis económica,

incluyendo la circulación del dólar estadounidense como moneda oficial del Ecuador. Así lo describen algunos economistas a la situación durante este período.

Según Matthews, Maniam y Balasundram (2006), Ecuador pasó por una revisión completa de su sistema monetario, a través de los golpes macroeconómicos del mercado mundial, lo que finalmente obligó al país a renunciar a su moneda nacional, el sucre, y adoptar el dólar estadounidense.

Tras una serie de reformas estructurales, la economía de Ecuador mejoró y, debido en gran parte a las remesas de los expatriados, los precios altos del petróleo y las nuevas exportaciones adicionales (Economía del Ecuador, 2012). De hecho, durante el período del 2002 al 2006, la economía de Ecuador gozó de un saludable 5.2 por ciento al incremento anual promedio, lo que representa el crecimiento más agresivo en un cuarto de siglo.

Tras un crecimiento económico modesto hasta el 2007, la economía ecuatoriana se propuso alcanzar una tasa de crecimiento del 7.2% en el 2008, atribuible al aumento de los precios del petróleo a nivel mundial, así como una mayor inversión en el sector público de Ecuador.

A pesar de este prolongado crecimiento de la economía, la situación cambio cuando el presidente Rafael Correa asumió el cargo presidencial en enero del 2007. Después de la toma de mando, el nuevo presidente dejo de pagar la deuda soberana de Ecuador en diciembre del 2008 por un monto aproximado de $3.2 billones, monto que representó alrededor del 30 % de la deuda externa pública del país.

A pesar de este revés económico, en mayo del 2009, Ecuador fue lo suficientemente solvente para redimir a casi todo (91%) de sus deudas en mora (Economía del Ecuador, 2012). Las acciones posteriores de la administración del nuevo presidente han creado algunos motivos de preocupación para el futuro.

Las secuelas persistentes de la Gran Recesión del 2008, sin embargo, junto con la disminución de las remesas de los expatriados y fuertes reducciones en los precios mundiales del

petróleo han afectado adversamente la economía ecuatoriana y el crecimiento fue de apenas un 0.4% en el 2009. Para el 2010, sin embargo, el crecimiento económico se había reanudado a nivel satisfactorio y logrado un incremento del 3.6% en el 2010, así como un aumento del 6.5% en el 2011.

Últimamente, otros indicios sugieren que la economía ecuatoriana está dotada para sustentar crecimiento en el futuro. Por ejemplo, solamente en el 2011, el gobierno ecuatoriano:

1. Concreto un acuerdo de un préstamo de $2 billones de dólares con el Banco de Desarrollo de China.
2. Recibió $1 billón de dólares en virtud de un acuerdo de venta durante 2 años por intermedio de un convenio petrolífero.
3. Aseguro $571 millones de dólares de financiamiento con el Banco Exim de China para un nuevo proyecto hidroeléctrico.
4. Proclamo planes para asegurar más créditos chinos en el 2012 (Economía del Ecuador, 2012).

Aunque Estados Unidos sigue siendo el principal socio comercial de Ecuador en el campo de importación y exportación, pero desde que el gobierno ecuatoriano dejo de hacer el pago de la deuda externa en el 2008, China se ha convertido desde aquel entonces en el mayor prestamista a nivel internacional de Ecuador, lo cual da más libertad del gasto social por parte del gobierno ecuatoriano.

Por supuesto, esta libertad de gasto social, también puede conducir a gastos motivados políticamente de manera que se pretende promover oportunidades inmediatas para la reelección en lugar de la necesidad de un desarrollo económico nacional a largo plazo. Por otra parte, es difícil o incluso imposible poner un precio en el liderazgo nacional y la capacidad de imponer aranceles como más convenga en respuesta a cambios en el mercado de forma que contribuyan a su popularidad.

En conjunto, esta poderosa combinación de flexibilidad

política en el evento de una situación social dinámica quiere decir que es importante "seguir al dinero" con el fin de determinar quién se beneficia del comercio internacional y estos temas se discuten más adelante.

El estado actual de los acuerdos de libre comercio en las Américas a nivel general con las coaliciones regionales de comercio y pactos comerciales de mutuo interés han asumido un papel cada vez importante en el hemisferio occidental en los últimos tiempos.

Los EE.UU. continúan discutiendo sobre el tema de un acuerdo comercial abierto con el Gobierno ecuatoriano. En el momento de redactar el mismo, los participantes reportan el consumo, y que Ecuador también fue parte de las negociaciones del TLC Andino pero Perú logró llegar a un acuerdo mutuo unilateral con la gran potencia del norte, mientras que la posición ecuatoriana se mantuvo en proceso de negociación.

Hubo algunas claras indicaciones de un convenio *quid pro quo* en el acuerdo unilateral forjado entre los EE.UU. y Colombia. Tal es el caso de a pesar que los gobiernos de Ecuador y Perú también tuvieron un impuesto de admisión sin restricciones al mercado de Estados Unidos para una canalización de sus bienes en virtud del ALCA, el acuerdo con Colombia y Perú, asegura y aumenta la admisión libre de impuestos. A cambio, se redujo inmediatamente los aranceles colombianos en casi el 80% de los productos manufacturados estadounidenses con un acuerdo para eliminar el 18% restante que se llevara a cabo durante la próxima década.

Por otra parte, en el 2003, el Poder Ejecutivo de los Estados Unidos indicó que no habría negociaciones posteriores a tratarse con respecto a un acuerdo comercial diferente con la República de Panamá. El gobierno de Estados Unidos ratificó un consenso de asistencia y el convenio para un Acuerdo Común Estados Unidos-Uruguay a finales del 2006 (Faber, 2007).

Es evidente, sin embargo, que hay algunos "obstáculos" involucrados en estos "bloques de construcción". El Área de Libre Comercio de las Américas, con Ecuador destacándose a

pesar del doloroso proceso, pero a la larga saludable, en virtud de la ausencia completa de aranceles y pagos obligatorios que nivelan el papel que se desempeña económicamente sobre los miembros del acuerdo de libre comercio.

En realidad, los aranceles se pueden utilizar de muchas formas diferentes, dependiendo de los resultados esperados y la situación de los comerciantes involucrados. En algunos casos, las tarifas sólo están diseñadas para nivelar el campo de juego, mientras que en otros, con la intención específica de generar ganancias en las compras, y en otros casos han de considerarse como una sanción.

Por ejemplo, los aranceles son impuestos que se aplican a los productos importados del extranjero. El comercio internacional se ve afectado por los aranceles y las restricciones comerciales relacionadas; los aranceles se pueden clasificar como ingresos o como protección y un pago elevado de tarifa obligaría a que los productos nacionales sean competitivos. También es conocido como impuesto. Además, se imponen tarifas a las ganancias para recaudar fondos para el gobierno de turno, y los aranceles proteccionistas, que son más altos que los aranceles por las ganancias generadas, están diseñados para aumentar la venta de las mercancías importadas.

Como el único país andino que se mantiene sin un tratado de comercio abierto formalizado con los Estados Unidos, se trata de cuestiones esenciales ya que se refieren a la capacidad de tales acuerdos para ayudar a mejorar el nivel de vida de todos los ecuatorianos en lugar una selecta élite opulenta.

Con relación a este tema, Venkataramany y Sivakumar (2008) reportan "Los países que son actualmente parte de la planificada Área de Libre Comercio de las Américas (ALCA) tienen requisitos exclusivos y preocupaciones, pero casi todos los participantes están enfocados a reducir las remuneraciones de comercio sin restricciones a nivel de cada socio y su comunidad".

Por lo tanto, el estado actual del ALCA no puede ser en absoluto sorprendente dados los intereses en conflicto y las

cuestiones políticas dinámicas que están involucradas. La cuestión es el tiempo que los negociadores han estado trabajando para resolver sus diferencias.

De hecho, el trabajo inicial con Ecuador y su situación con respecto al ALCA a opacado directamente la idea del Tratado de Libre Comercio de América del Norte a mediados de la década de 1990 y la mayoría de los observadores en aquel tiempo y en los años siguientes han anticipado la formalización de este acuerdo para la fecha de fin de plazo establecida del 1 de enero del 2005.

Este fue el tema central de una serie de protestas que se produjeron en Ecuador en respuesta a los intentos del gobierno para forjar un acuerdo del ALCA con Estados Unidos en el 2006.

La protesta social que surgió en respuesta a las determinaciones de Ecuador para forjar un tratado de comercio libre con los Estados Unidos. De hecho, poco popular entre la mayoría de los pueblos indígenas, que creían en el esfuerzo fue un intento de extender la hegemonía estadounidense en su país aún más.

Estos son temas preocupantes en una supuestamente América Latina post-colonial, pero está claro que en el siglo 21 donde los ingresos anuales de muchas corporaciones multinacionales superan el presupuesto de todos los países centroamericanos en conjunto.

La manera en que el gobierno ecuatoriano responde a estas cuestiones probablemente determinará qué acción adecuada será requerida para aplastar la disidencia social y en qué medida será requerida la fuerza militar para lograr ese objetivo. Hasta la fecha, los informes de Ecuador suenan más como algo de Oriente Medio que en el propio terreno de los Estados Unidos.

La voluntad del gobierno ecuatoriano para emplear la fuerza militar también se puso de manifiesto durante estas manifestaciones. No sólo el gobierno respondió con la fuerza militar, suspendió los derechos constitucionales e implementó la ley marcial virtual con amplios poderes de arresto.

De acuerdo con un informe de NotiSur el cual señala que

se movilizaron las fuerzas policiales y el ejército para limpiar las carreteras bloqueadas durante el mes de marzo del 2006. Luego, la administración declaró el estado de emergencia en diferentes partes del país.

Los ecuatorianos reaccionaron a la iniciativa del ALCA con una participación joven de fuerza. Aproximadamente mil quinientos nativos y estudiantes marcharon pacíficamente por la capital, Quito, el 22 de marzo, en protesta contra un posible acuerdo con los Estados Unidos.

En un discurso al Consejo de la Organización de los Estados Americanos (OEA) y el Consejo de Asuntos Mundiales en Washington, DC el 16 de enero del 2002, el presidente George Bush hizo hincapié en el compromiso de Estados Unidos para el establecimiento de un Área de Libre Comercio de las Américas para el año 2005.

Esta no fue la primera vez que el presidente Bush se había comprometido con la iniciativa del Área de Libre Comercio para las Américas.

De hecho, en un discurso ante la OEA, en el 2001, el presidente reafirmó el compromiso de Estados Unidos al promocionar el Área de Libre Comercio para las Américas la que convertiría al continente Americano en la zona de libre comercio más grande del mundo, que abarca 34 países y 800 millones de personas. (*Bush pushes Free Trade Area, 2002*).

Sin embargo, la fecha prevista, ha llegado y pasado, y nada sustancial ha cambiado en lo provisional a excepción del creciente escepticismo que se puede alcanzar cualquier resultado positivo. El estado actual del ALCA se ha mantenido prácticamente sin cambios durante los últimos 5 años: "las conversaciones del Área de Libre Comercio de las Américas llegaron al punto de paralizarse en el 2007 y todavía están pendientes" (Venkataramany y Sivakumar, 2008).

De hecho, incluso la página de Internet del Representante Comercial de EEUU (*USTR U.S. Trade Representative* por sus siglas en inglés) se ha cerrado sin indicación de su situación futura y todo lo referente a Ecuador se ha deshabilitado aunque todavía están en la lista.

En realidad, a primera vista el concepto de "libre

comercio" suena acogedor para todo el mundo. Después de todo, más comercio equivale al aumento del desarrollo y quién puede discutir del mismo si se lo está haciendo "gratuitamente" Al respecto, Fandl (2007) informa que hay una serie de resultados satisfactorios relacionados con los acuerdos de libre comercio. Por definición, el libre comercio está destinado a aumentar las importaciones y exportaciones entre países de manera que sean mutuamente favorables.

Cuando el término "libre comercio" es utilizado para describir los acuerdos que no son necesariamente justos, sin embargo, las cosas se vuelven más perplejas. Los resultados económicos de los pactos de libre comercio en los países andinos que buscan capital en el comercio internacional son complicados debido a sus procesos multifacéticos.

Con respecto al mismo, algunos observadores sugieren que este es el caso con el ALCA, que tiene previsto fomentar la prosperidad económica de los países participantes en varios niveles, el resto del mundo lo visualiza como un esquema de distracción comercial. En un mercado cada vez más globalizado, el concepto del Acuerdo Comercial Abierto de las Américas puede haberse convertido en obsoleto antes de tener una oportunidad.

Por otra parte, algunos analistas sostienen que el Tratado para el Área de Libre Comercio de las Américas es superfluo, ya que la gran mayoría de todos los bienes ya son exentos de los impuestos de aduana establecidos por el gobierno. Por ejemplo, Kornis (2004) enfatiza que, mientras los Estados Unidos inician las conversaciones con los homólogos de América del Sur para un acuerdo de libre comercio entre Estados Unidos y los países andinos, el 86 por ciento de las importaciones combinadas estadounidenses con estos países ya están libres de impuestos aduaneros.

En el análisis "seguimiento de capital", tiene que haber algún tipo de motivación de parte de Ecuador y los demás países andinos para negociar las diferencias restantes que puedan tener con los Estados Unidos en general, y el Ministerio de Agricultura, en particular en lo que respecta al libre comercio.

Resulta que gran parte de la motivación para la formalización de estos acuerdos de libre comercio tienen que ver con la necesidad de establecerse como "primer movimiento" en el comercio internacional con los Estados Unidos.

Se han hecho esfuerzos posteriores para la negociación del potencial tratado de libre comercio con Ecuador. Los países de América del Sur de la región Andina al momento gozan de la libertad de impuestos aduaneros para la mayoría de sus exportaciones a los EE.UU. en virtud a la extensión de la Ley de Preferencias Arancelarias Andinas (*ATPA Ondean Trade Preference Act*) formalizado a principios del milenio (Kornis, 2004).

Antes de su vencimiento, sin embargo, Colombia y Estados Unidos formalizaron un tratado comercial mutuo en febrero del 2006 después de 2 años de intensas negociaciones.

El impulso para el acuerdo bilateral con Colombia por la exclusión de los países andinos se debe en parte a la incapacidad de las partes involucradas para resolver el resto de sus diferendos y no está claro si serán capaces de hacerlo en el futuro previsible.

La naturaleza política de estas negociaciones supera claramente las ventajas económicas de su entorno que podrían lograrse mediante la expansión del libre comercio. Fandl (2007), llega a la conclusión, de que los nuevos acontecimientos políticos en Bolivia y Ecuador apuntan a al dudoso logro de concretar acuerdos con aquellos países.

Mientras que Colombia logró concretar un acuerdo de libre comercio con los Estados Unidos, los obstáculos restantes para avanzar con sus homólogos andinos en general y en lo que respecta a Ecuador, en particular, tiene que ver con temas agrícolas.

Chauvin (2004) informa al respecto: "Los países de la Ley de Preferencias Arancelarias Andinas (ATPA *Andean Trade Preference Act*), que son mercados importantes para los cereales y algodón estadounidense, y al mismo tiempo con una delicadeza acerca de la entrada de sus mercados a los cultivos agrarios por intermedio de los Estados Unidos. Cristiano

Espinosa, principal negociador de Ecuador, de acuerdo a los informes, había dicho: "Los países andinos no cederán terreno en el frente agrario. Protegeremos nuestros mecanismos para neutralizar los efectos de los subsidios estadounidenses".

En la actualidad, de acuerdo con el método de la Comunidad Andina conocida como "banda de precios", los aranceles de una cantidad variable de hasta el 100 por ciento están autorizados para la aplicación a diversos productos agrícolas que se importan a los países andinos con el fin de estabilizar los precios en el caso de que caigan por debajo de los niveles predeterminados (Chauvin, 2004).

En algunos casos, la Ley de Preferencias Arancelarias Andinas se ha utilizado para el avance de otros objetivos estadounidenses para reducir el tráfico ilícito de drogas entre los países andinos y los Estados Unidos. Según Baker (2000), el *ATPA* promueve el desarrollo económico en Bolivia, Colombia, Ecuador y Perú. Fue establecido bajo la administración de George H. W. Bush de 1989 a 1993.

Su principal objetivo era proporcionar a estos países con una alternativa a la producción de coca, dándoles un mayor acceso al mercado estadounidense.

Estados Unidos ha utilizado enfoques similares en Colombia, por ejemplo, en su esfuerzo para proveer a los agricultores colombianos de una alternativa viable a los cultivos de coca. Con este fin, Burns informa, "Desde el comienzo del 2000, Estados Unidos se ha manifestado con varias mejoras de apoyo a más de 81 mil pequeñas empresas agrícolas, dando como resultado la producción de más de 100 mil hectáreas de nuevos cultivos y más de mil obras de infraestructura a menor escala, como las escuelas, caminos y puentes que se han completado en las comunidades que participan en el programa".

Para los Estados Unidos, estas cuestiones son más importantes en el ámbito político que en el económico. Por ejemplo, el comercio de Estados Unidos con las naciones del *ATPA* continúa de manera modesta, e incluso combinada.

En el 2003, todos los países del *ATPA* contabilizaron únicamente el 1 por ciento de las transferencias totales de

Estados Unidos y mucho menos del 1 por ciento del total de las importaciones estadounidenses.

A pesar de los porcentajes relativamente modestos, sin embargo, los países andinos del *ATPA* fueron responsables de proporcionar una serie de productos estratégicamente valiosos, incluyendo el cobre, así como artículos de altas ganancias como los espárragos y las flores (Kornis, 2004).

Por otra parte, los Estados Unidos han experimentado una escasez colectiva en su comercio con los países del *ATPA* desde 1991.

Por el contrario, los niveles de exportación de los Estados Unidos a los países del *ATPA* no han aumentado sustancialmente desde 1999, momento en el que se habían reducido totalmente al 28 por ciento en comparación a su valor en 1998.

Por otra parte, una verdadera "tormenta perfecta" de las condiciones económicas adversas ha obstaculizado la capacidad y el deseo de los países del *ATPA* a importar productos de los Estados Unidos.

Gran parte del comercio de exportación de *ATPA* con los Estados Unidos está vinculado al petróleo y el gas natural (33 por ciento); además, la mayor parte de la maquinaria y equipo que importan las naciones de *ATPA* de los Estados Unidos se utilizan en la industria del petróleo y el gas natural. En este tipo de industria, está claro que los EE.UU. gozan de una ventaja competitiva sobre los países exportadores de petróleo, ya que carecen de las instalaciones de refinería necesarias para incrementar su valor. Por ejemplo, Kornis informa, "Los principales productos de exportación de Estados Unidos a la región son petróleo refinado, aceites combustibles y aceites lubricantes".

En este entorno, es fácil ver cómo se pueden pasar por alto los intereses del agricultor ecuatoriano a nivel promedio. En un mundo que demanda cada vez más energía, incluso las flores más espectaculares y sabrosos espárragos, dos importantes exportaciones ecuatorianas, las cuales no tienen punto de comparación con el petróleo y el gas que Ecuador tiene para ofrecer.

Las tendencias actuales indican que estos patrones continuarán en el futuro previsible, ya que la búsqueda de combustibles alternativos viables continúa. A menos y hasta que se logre dicho objetivo, el comercio de Estados Unidos con las naciones de *ATPA* probablemente continuará caracterizándose por el uso estratégico de una política efectiva de negociaciones y los acuerdos unilaterales con los países andinos dependiendo del desarrollo de la situación política en dichos países.

Aunque no sería conveniente sugerir que el Departamento de Estado de los Estados Unidos estaría conspirando con el Ministerio de Agricultura de Ecuador para evitar el desarrollo de su homologo del sur, es razonable sugerir que los Estados Unidos sigue utilizando el análisis de costo-beneficial para determinar la mejor alternativa para proceder con la resistencia económica de Ecuador.

En algunos casos, esto puede ser interpretado como las tarifas precedentes de generación de ingresos las cuales podrían reducir la disponibilidad y afectar negativamente a los estándares de precios internacionales, temas que claramente beneficiaron al Ecuador.

Por otra parte, Ecuador dispone de ventajas significativas en su relación existente con los EE.UU. a medida que tal vez no quiere estar atado a un marco más rígido que impedía las ventajas unilaterales en la negociación con los Estados Unidos.

De hecho, Kornis añade: "El nuevo estatus libre de impuestos de petróleo y ramificaciones de las naciones de *ATPA* bajo *ATPDEA* también fue responsable de la mayor parte de la reducción de la tasa impuesta a las importaciones de Estados Unidos y los ingresos de impuestos de Estados Unidos de las naciones de *ATPA*".

Estos resultados no son del todo sorprendentes, sin embargo, ya que el *ATPA* no necesariamente está diseñado para ser un mecanismo de generación de ingresos para los Estados Unidos, sino más bien un marco en el que los objetivos políticos se pueden lograr a través de diversos programas de incentivos. El *ATPA* también se diferencia de

otros acuerdos comerciales de Estados Unidos debido a que uno de sus principales objetivos es reducir el comercio ilícito al menos del tráfico de drogas, entre las naciones de la *ATPA* y los Estados Unidos.

La ubicuidad de las drogas ilícitas en estos países representa una amenaza continua que afecta a la seguridad general de la gran nación del norte, Sheppard mantiene y sugiere que esta sea una actividad de valor agregado que Estados Unidos podría prescindir.

Como ya se ha señalado, desde 1991, la mayor parte de las mercancías exportadas desde Ecuador entran a los Estados Unidos libres de derechos de aduana en virtud de la Ley de Preferencias Arancelarias Andinas. Como también se señaló, el *ATPA* ha sido utilizado por los Estados Unidos en su guerra global contra el terrorismo y las drogas de diversas maneras con resultados mixtos en los países participantes.

Al mismo tiempo, los Estados Unidos no ha recibido ningún tratamiento preferencial para sus mercancías importadas a las naciones de la *ATPA*; Estados Unidos hace recibir otra recompensa en términos de estrategia internacional y la necesidad de establecer un acuerdo hegemónico con estos países.

El valor estratégico de los productos derivados del petróleo y gas natural que van hacia los Estados Unidos y las relaciones bilaterales con estos países son, por lo tanto, probablemente mucho más importantes que su impacto económico, pero estos niveles van en aumento. Algún indicio de estos resultados se puede discernir por la actividad comercial reciente.

Actualmente, las exportaciones de Ecuador totalizaron $22.29 billones en el 2011, un aumento de
$17.49 billones en el 2010. Las principales exportaciones del país son el petróleo, el banano, flores cortadas, camarones, cacao, café, madera y pescado (Economía del Ecuador, 2012), con sus principales socios de productos de exportación comercial como los Estados Unidos, Panamá, Perú, Venezuela, Chile y Rusia, como se muestra en la Tabla 2.1.

Tabla 2.1: Principales Socios de Exportación del Ecuador

País	**2009 % de Exportaciones**
Estados Unidos	**38.3%**
Panamá	**10.1%**
Perú	**6.3%**
Venezuela	**5.3%**
Chile	**5%**
Rusia	**4.6%**

Fuente: Datos tabulares, Economía del Ecuador, 2012

En la actualidad, las importaciones de Ecuador incluyen materiales industriales, combustibles y lubricantes, bienes de consumo no durables y totalizaron $23.58 billones en el 2011, un aumento considerable con respecto a sus niveles del 2010 de $19.3 billones. Los principales socios de importación de Ecuador son Estados Unidos, China, Colombia, Panamá, Perú, Brasil y Corea del Sur como se establece en la Tabla 2.2.

Tabla 2.2: Principales Socios de Importación del Ecuador

Country	**2009 % de Importaciones**
Estados Unidos	**27.9%**
China	**10.3%**
Colombia	**9.2%**
Panamá	**4.6%**
Perú	**4.6%**
Brasil	**4.3%**
Corea del Sur	**4.1%**

Fuente: Datos tabulares, Economía del Ecuador, 2012

Como se ha señalado en el capítulo de introducción, sin embargo, el aumento del comercio con los EE.UU. no se ha traducido necesariamente en una amplia aceptación de los acuerdos de libre comercio por el pueblo ecuatoriano. De

hecho, este tema fue el centro de una serie de protestas que se produjeron en Ecuador, en respuesta a los intentos del gobierno para forjar un acuerdo ALCA con Estados Unidos en el 2006.

De hecho, la protesta social que surgió en respuesta a la lucha del Ecuador para forjar un acuerdo comercial sin restricciones con los Estados Unidos dio a conocer que muchos indígenas creen que el esfuerzo fue un intento de extender la hegemonía estadounidense en su país aún más.

Parecería que las preocupaciones del gobierno de Estados Unidos relacionadas con las drogas no logran tomar en cuenta estos problemas locales y regionales con una visión etnocéntrica. Está claro que el tema general que proviene de la literatura relevante es una preocupación de alto nivel para las amenazas a la seguridad nacional de conformidad a la Ley de Seguridad Nacional que prevaleció sobre todo lo demás.

Los países andinos representan una situación única; sin embargo, todavía hay una orientación general hacia la codificación del Tratado de Libre Comercio de las Américas y este objetivo probablemente continuará siendo el centro principal de las relaciones y convenios Estados Unidos-América Latina en el futuro (Gillgannon, 2004).

Sin embargo, hay algunos indicios de que hay mucho más involucrado en estas negociaciones en curso que muchos estadounidenses se dan cuenta, o les importa, pero que todavía tienen un efecto potencial del comercio entre los EE.UU. y Ecuador basado en cuestiones primordiales que involucran la seguridad nacional.

Tal vez no es tanto una cuestión de la opinión pública estadounidense al no preocuparse por estas cuestiones. Puede que más bien sea un indicio de cuán potencialmente cuestiones de seguridad nacional de Estados Unidos son actualmente disruptivas y muy probablemente en el futuro con respecto al comercio con los países de América del Sur de la región andina en general y Ecuador en particular.

Por un lado, Gillgannon sugiere que el público estadounidense no puede estar en conocimiento debido a la falta de información sobre estos temas: "Como de costumbre,

la prensa escrita y la televisión estadounidense se han desempeñado con una pobre campaña informativa de estas negociaciones comerciales y el público estadounidense permanece desinformado y apático".

Aunque, el postula una justificación potencial en apoyo de esta escasez de cobertura oportuna acerca de las negociaciones comerciales entre Estados Unidos y Ecuador. Estas razones anteriores, sin embargo, no toman en cuenta los beneficios de la seguridad nacional que están establecidos de conformidad con la Ley de Seguridad Nacional.

En realidad, los intereses de seguridad nacional de Estados Unidos que están involucrados en el Ecuador son ampliamente percibidos como los esfuerzos para extender la hegemonía estadounidense en la región centroamericana aún más y la imagen del estadounidense se está generalizando una vez más. En este sentido, Gillgannon añade, "los sudamericanos temen a los convenios comerciales con Asia y la Unión Europea para contrarrestar las presiones en favor de la supremacía del comercio y el control de su autonomía nacional por parte de los Estados Unidos".

Aunque estas negociaciones han sido el punto central y el tema a las disposiciones de la Ley de Seguridad Nacional. Algún indicio de cómo se han percibido estos esfuerzos en Ecuador se puede destacar de la observación de Gillgannon, "el Ministro de Defensa Donald Rumsfeld se reunió con sus homólogos del Hemisferio Occidental en Quito, Ecuador, en busca de la solidaridad hemisférica para la guerra de Estados Unidos contra el terrorismo".

Por otra parte, hay algunos indicios de que los EE.UU. no sólo están dispuestos a maniobrar militarmente si es necesario, está francamente ansioso por hacerlo, en un esfuerzo para encausar su guerra global contra el terrorismo en su propio hemisferio. Estas recomendaciones no parecen ser tan benignas, sin embargo, cuando se toma en cuenta al secretario de Defensa, los otros comentarios son importantes.

Estas consideraciones son importantes al evaluar el futuro de las relaciones comerciales entre Estados Unidos y Ecuador. Es razonable sugerir que si las cosas se perciben

como haberse pasado de la mano en cualquier región del Hemisferio Occidental, EE.UU. esta listo para intervenir amparado en la Doctrina Monroe de larga trayectoria y de las disposiciones más recientes de la Ley de Seguridad Nacional.

Para muchos estadounidenses, estas temáticas son simplemente desconocidas debido a la escasez de los informes de los medios de comunicación de Estados Unidos, y es razonable sugerir que algunos estadounidenses simplemente no saben ni les importa lo suficiente acerca de estas cuestiones que tienen un impacto en las futuras negociaciones entre estos dos países.

Como Gillgannon señala, "No hemos descifrado las señales de los tiempos. Alabamos a las abstracciones de la justicia y la paz, pero no hacemos el trabajo sucio en los complicados detalles de los nuevos temas mundiales de intercambio y comercio".

Claramente, estos son temas preocupantes en un supuesta América Latina post-colonial, pero está claro que en el siglo 21 cuando los ingresos anuales de muchas corporaciones multinacionales, las cuales superan el presupuesto de todos los países centroamericanos, la manera en que el gobierno ecuatoriano responde a estas cuestiones probablemente determinará qué curso de acción será necesario para reprimir la disidencia social y hasta qué medida la fuerza militar será necesaria para lograr ese objetivo.

La reciente condena a Siria por el nuevo presidente de Egipto deja en claro que las cosas están cambiando en el Medio Oriente, y una reacción árabe similar a un efecto domino podría ocurrir en la esfera de influencia de Estados Unidos. Hasta la fecha, los informes de Ecuador suenan más como algo de Oriente Medio que la esfera de influencia de los Estados Unidos.

La voluntad del Gobierno ecuatoriano para emplear la fuerza militar también se puso de manifiesto durante estas manifestaciones. No sólo el gobierno respondió con la fuerza militar, suspendió los derechos constitucionales del país e implementó la ley marcial virtual con amplios poderes de arresto, una medida muy inti-popular.

En este medio ambiente, el comercio entre los EE.UU. y Ecuador puede pasar a segundo plano para tratar de resistir la influencia estadounidense en la región mientras se busca al mismo tiempo la mejora de las futuras relaciones comerciales.

El hecho de que los ecuatorianos estén saliendo a las calles tomando gran riesgo personal y protestar contra el llamado "libre comercio" que significaría más ganancias sugiere que nada es realmente "gratis" cuando se trata de lidiar con los Estados Unidos y la Ley de Seguridad Nacional, de acuerdo a Clark (2012).

Cuando los miembros de la comunidad ecuatoriana tienen encuentros con el sistema legal, deben lidiar no sólo con las oficinas judiciales y los procedimientos legales, pero con los contradictorios valores indígenas.

También, cuando bloquean carreteras como parte de un levantamiento nacional contra un acuerdo de libre comercio, deben negociar internamente, como miembros de la comunidad también pueden sufrir de un paro cívico, así como con los actores regionales y nacionales, tanto dentro y fuera del estado. Estos acontecimientos sugieren que las negociaciones entre los EE.UU. y Ecuador pueden requerir de más tiempo e involucrar más compensaciones que muchos observadores pudieran creer.

Washington, DC., Estados Unidos

Fuente: Wikipedia Turismo, 2015

CAPÍTULO 3

ANÁLISIS DE POSIBLES ACUERDOS COMERCIALES

En este estudio se utilizó un esquema combinado de revisión que emplea elementos cualitativos y cuantitativos. El componente cualitativo consistió en una evaluación de la literatura con jurado y nivel académico significativo en relación al Acuerdo de Libre Comercio de las Américas y sus implicaciones para el Ecuador, sus vecinos de América Central y los Estados Unidos, concentrándose en los recursos disponibles más recientes.

Este elemento cualitativo es coherente con la orientación proporcionada por un gran número de investigadores sociales que hacen hincapié en la necesidad de revisar la literatura existente para determinar lo que se conoce acerca de un tema, así como las brechas que existen (Silverman, 2005).

El uso de ambos componentes cualitativos y cuantitativos es coherente con la orientación proporcionada por Neuman (2003). Él aconseja: "Tanto la investigación cualitativa como cuantitativa utilizan varias técnicas específicas de investigación (por ejemplo, encuesta, entrevista y análisis histórico), sin embargo, hay mucha coincidencia entre el tipo de datos y el estilo de la investigación. La mayoría de los investigadores del estilo cuantitativo examinan los datos

cualitativos y viceversa".

Similitudes-Diferencias, Análisis Cuantitativo-Cualitativo de Ecuador

Similitudes

El análisis de estos comprende la inferencia – quiere decir que ambos lleguen a una decisión basada en la prueba.
Ambos implican un proceso sistemático.
Ambos implican la comparación, ya sea internamente o con la evidencia relacionada de otros lugares.
Ambos tratan de evitar errores, falsas conclusiones e inferencias engañosas y buscar una descripción y explicaciones válidas.

Diferencias

El análisis cuantitativo es altamente estandarizado con poca diferencia entre los proyectos; sin embargo, el análisis cualitativo tiene muchas maneras de abordar los temas posibles.
El análisis cuantitativo se lleva a cabo al final de los datos; el análisis cualitativo se lleva a cabo durante la recolección de datos.
El análisis cuantitativo tiende a poner a prueba las hipótesis a través de la manipulación de números que representan hechos.
El análisis cualitativo es menos abstracto y no asume que la vida real se puede medir numéricamente.

Prominencia

Desde la perspectiva cualitativa, el hecho de que la Oficina del Representante Comercial de Estados Unidos haya deshabilitado los enlaces de los documentos pertinentes a las negociaciones estadounidenses y ecuatorianas es digno de mención, ya que refleja cuestiones de seguridad internacional

las cuales están involucradas (por ejemplo, la Ley de Seguridad Nacional).

Los niveles históricos de los aranceles como evidencia son palpables, sin embargo, estos temas pueden ser sintetizados bajo la teoría del equilibrio general.

Se procederá con la investigación en varios niveles.
Por motivo de falta de documentos del Gobierno de Estados Unidos relacionadas con las negociaciones en curso con los funcionarios del comercio ecuatoriano, es razonable sugerir que la situación continúa dinámica y sujeta a anular y clasificar lo concerniente a la seguridad.

El concepto de equilibrio define un sistema completo en el que todas las piezas se comportan de manera predecible. Por lo tanto, la teoría del equilibrio general es un sistema con muchas de las características de los sistemas físicos y, por lo tanto, pueden ser examinados con muchas de las mismas herramientas de investigación utilizadas por intermedio de sistemas físicos. Además, ofrece el poder de predicción, el objetivo final de la investigación científica (Kirchhoff, 1999).

Hay varias similitudes muy claras entre los métodos cualitativos y cuantitativos que pueden contribuir de manera sustantiva a los procesos de investigación, si se tienen en cuenta sus diferencias.

De hecho, de acuerdo a Jenson (2004), "Es a nivel de metodología, que se define como un plan teóricamente informado de acción en relación al campo empírico, que la distinción entre la investigación cualitativa y cuantitativa es más evidente".

Un estudio tanto cuantitativo como cualitativo puede emplear los mismos métodos, y ambos comparten el objetivo común de desarrollar un mapa preciso de los conceptos fundamentales a los fenómenos empíricos. En otras palabras, "Es la unión específica de una selección empírica necesaria del 'microcosmos' con una teória del 'macrocosmos' que distingue sobre todo a la investigación cuantitativa y cualitativa como tradiciones y paradigmas" (Jenson, 2004).

El componente cuantitativo consistió en un análisis de los posibles resultados de diferentes oportunidades, con el

acuerdo de libre comercio y sin acuerdo de libre comercio con un modelo aplicado de equilibrio general, que incluye datos de los agentes económicos relacionados con el TLC en América Central y del Sur, así como de Estados Unidos y la Unión Europea.

Además, este análisis tiene en cuenta las interdependencias entre estos socios comerciales, así como los efectos directos e indirectos que un acuerdo con grandes y profundas repercusiones, como el TLC, traería sobre la economía ecuatoriana en términos del efecto del aumento del comercio en la economía nacional de hoy, así como en el futuro a medida que se crean nuevos puestos de trabajo y se establecen proyectos empresariales sostenibles.

En un mundo perfecto, el aumento del comercio entre Estados Unidos y Ecuador tendría un resultado positivo, pero las exigencias del tráfico de drogas que fluye a los EE.UU. de los países andinos implica que los EE.UU. están dispuestos a asumir una pérdida económica cada vez al favorecer el proceso de la guerra global contra el terrorismo, de la cual el tráfico de drogas representa un componente importante.

En consecuencia, el Gobierno de Ecuador puede aceptar las ofertas de Estados Unidos, sean las que sean en la actualidad, o no comprometerse a las mismas, pero los EE.UU. probablemente continuarán ejerciendo una gran influencia en el área de manera que no son cuantificables.

La CIA (*Central Intelligence Agency*) es la Agencia Central de Inteligencia estadounidense con un largo historial de operaciones clandestinas en los países andinos, y es razonable sugerir que permanecen vigentes en la actualidad.

El pueblo ecuatoriano, sin embargo, no está sentado sin hacer nada mientras estas presiones están ejecutándose en su contra, y nuevamente, el hecho de que están dispuestos a correr el riesgo de ser arrestados, heridos e incluso la muerte para protestar estas acciones se basa en el reconocimiento generalizado de que el aumento del comercio, de hecho, será beneficioso para su país.

Por ejemplo, aunque el comercio entre los Estados Unidos y los países de América Latina disminuyó tras el final

de la Segunda Guerra Mundial durante la década de 1980, mientras que por la década de 1990 se experimentó un crecimiento impresionante en el comercio internacional en esta región.

Según Hirst (2004), la tasa de comercio de Estados Unidos con los países de América Latina aumentó significativamente más rápido que el crecimiento concomitante de los Estados Unidos en el comercio mundial: "América Latina se había convertido en el mercado de mayor crecimiento para las exportaciones estadounidenses. Los Estados Unidos, por fin, se dieron cuenta de Latinoamérica. El comercio entre algunos países de América Latina entraron en auge, sobre todo dentro de las subregiones como el Cono Sur de Sudamérica, Venezuela y Colombia, los países de América Central".

El impacto de este aumento del comercio con los EE.UU. se manifiesta de varias maneras, incluyendo una serie de acuerdos de libre comercio de la región que siguen vigentes hoy en día. Por ejemplo, Hirst informa, "El establecimiento de zonas libres de comercio formales facilitó el crecimiento del comercio y otras relaciones económicas".

Estos acuerdos de libre comercio fueron los siguientes:

1. Tratado de Libre Comercio de Norteamérica (TLCAN): este acuerdo incluye México, Estados Unidos y Canadá.
2. El Mercado Común del Cono Sur (más bien y comúnmente conocido como MERCOSUR): este acuerdo es entre Argentina, Brasil, Paraguay y Uruguay.
3. La Comunidad Andina, sus miembros son Bolivia, Colombia, Ecuador, Perú y Venezuela.
4. El Mercado Común Centroamericano (MCCA).
5. La Comunidad del Caribe (CARICOM).

El efecto colectivo de estos acuerdos de libre comercio se ha pronunciado y todos estos países, con la única excepción

de Ecuador, han finalizado sus negociaciones con los Estados Unidos con respecto a los tratados de libre comercio vigentes.

El estancamiento que existe hoy en día debe ser visualizado con estos acontecimientos en mente, ya que tienen el poder de compensar cualquier ganancia potencial que puede ser ejecutado por cualquier socio comercial en el futuro.

A pesar de que Estados Unidos sigue siendo el mayor socio de importación y exportación de Ecuador, el estado sudamericano tiene mucho más en juego económicamente.

En este sentido, Hirst enfatiza, "La inversión extranjera directa y de cartera de EE.UU. en grandes cantidades fluyó hacia América Latina y el Caribe, financiando la expansión de las actividades económicas. La velocidad de las transacciones de inversión de cartera, sin embargo, también expuso a estos y otros países a la volatilidad financiera marcada y pánicos financieros recurrentes".

De particular interés es la observación de Hirst, que independientemente del resultado de estas eventualidades para Ecuador, el escenario se había fijado para lo que vendría después del 11 de septiembre del 2001.

Según Hirst, "La transformación en las relaciones económicas internacionales del hemisferio y específicamente en las relaciones económicas de los Estados Unidos con el resto del hemisferio ya era más difícil a principios del siglo XXI".

El acuerdo de la Ley de Preferencias Arancelarias Andinas, ATPA (*Andean Trade Preference Act*), se inició en diciembre de 1991 entre los Estados Unidos y cuatro países andinos, Bolivia, Colombia, Ecuador y Perú con el objetivo de apoyar a estos países andinos en su lucha contra la producción y tráfico de drogas mediante la ampliación de sus alternativas económicas (Relyea, 2011).

La Ley de Preferencias Arancelarias Andinas y Erradicación de la Droga, ATPDEA (*Andean Trade Promotion and Drug Eradication Act*), ofrece garantías libres de aranceles para ciertos productos excluidos anteriormente del ATPA. Los posibles escenarios en los que Ecuador no firma el TLC, pero Colombia y Perú acuerdan hacerlo y, alternativamente,

Colombia y Perú permiten a los EE.UU. libre acceso a sus mercados.

Las organizaciones que entran en el mercado global tienen varias estrategias a su disposición: la exportación, licencias, empresas conjuntas, propiedad directa, importadores, y compensación. Como los sistemas de logística, los cuales mejoraron el consumo y la producción, lo que dio lugar a la separación geográfica.

Escenarios Posibles del Tratado de Libre Comercio Ecuatoriano

Primer Escenario

Colombia y Perú permiten a los EE.UU. libre acceso a sus mercados.

Implicaciones Potenciales

A cambio de este acceso gratuito, dichos países reciben preferencias del ATPDEA. Ecuador pierde tratamiento preferencial arancelario del ATPDEA y se le imponen las mismas tarifas que a Venezuela de parte de los EE.UU. En otras palabras, al Ecuador se le impone tarifas más elevadas por los EE.UU., al mismo precio de las tarifas de Venezuela por intermedio de los EE.UU.

Medidas a Examinarse

1. Proyección de los niveles de exportación de Estados Unidos a Colombia, Perú y Ecuador bajo este escenario.
2. El nivel de los nuevos aranceles para Ecuador en línea con los de Venezuela.
3. El impacto de tasas arancelarias más elevadas para Ecuador.

Segundo Escenario

Colombia y Perú permiten a los EE.UU. libre acceso a sus

mercados.

Implicaciones Potenciales

A cambio de este acceso gratuito, estos dos países andinos aseguran mejores preferencias del ATPDEA, y los EE.UU. eliminan los subsidios en la producción de bienes agrícolas. Ecuador renuncia a la tarifa de trato preferencial de la ATPDEA y el país está obligado a pagar aranceles iguales a los impuestos a Venezuela por los EE. UU.

Medidas a Examinarse

1. Proyección de los niveles de exportación de Estados Unidos a Colombia, Perú y Ecuador bajo este escenario.
2. El nivel de nuevos niveles arancelarios para Ecuador en línea con la de Venezuela.
3. El impacto de trato preferencial a Colombia y Perú bajo el ATPDEA.

Lógica de Apoyo al Esquema de Investigación

Según Ackerman, Nadal, Benniti, Gallagher y Salas (2004), "El nivel de sofisticación metódica que está implicada en la predicción de los efectos económicos potenciales de un acuerdo comercial depende de la escala de actividad económica entre los socios comerciales propuestos".

La teoría del equilibrio general ofrece un marco útil para desarrollar lo anterior de manera diferente y que pasaría de acuerdo a escenarios basados en la tendencia de los sistemas económicos que buscan el equilibrio.

Por ejemplo, según Kirchhoff (1999), "Fundamentalmente, la economía neoclásica propone que una economía de Mercado que toma la forma del equilibrio entre la oferta y la demanda. El mecanismo equilibrador es el precio. Aunque el mercado no siempre puede estar en equilibrio, la tendencia del sistema económico es moverse hacia el equilibrio".

Morrison (1999) advierte que, "En la teoría neoclásica del equilibrio general, se supone que un subastador ajusta los precios a fin de lograr la igualdad entre la oferta y la demanda de forma simultánea en todos los mercados".

La teoría del equilibrio general del comercio internacional, obviamente, se refiere a la interacción de varios mercados; en consecuencia, la extensión de la teoría del equilibrio general de la economía nacional se puede diferenciar entre las teorías del comercio nacional e internacional (Marsh, 1991).

Por lo tanto, la teoría del equilibrio general ofrece un marco útil que permite evaluar los elementos cuantificables en cada uno de los escenarios descritos anteriormente, utilizando las técnicas de recolección de datos y la base de datos del estudio que se describe a continuación.

Técnicas de Recolección de Datos y Base de Datos del Estudio

Basado en el marco de lo establecido, las medidas cuantitativas correspondientes a los dos escenarios diferentes fueron examinadas utilizando los archivos de los datos de rendimiento económico publicados por los gobiernos de los Estados Unidos y Ecuador, así como por analistas comerciales a nivel internacional y las pertinentes organizaciones no gubernamentales.

Los datos cualitativos consisten de recursos revisados por expertos y académicos publicados en los últimos 10 años que se obtuvieron de las bibliotecas públicas y universitarias, así como recursos confiables de investigación en el Internet.

Además, también se consultó a los recursos de organizacionales gubernamentales y no gubernamentales para desarrollar los diferentes escenarios utilizando los datos comerciales como punto de referencia para los países de interés.

Los datos cualitativos y cuantitativos se sintetizaron utilizando el "proceso de traducción recíproca" en el que cada

estudio y con la ayuda de recursos fomentaron al análisis posterior.

Limitaciones del Esquema de Investigación

A pesar de su capacidad para modelar los mercados económicos, la teoría del equilibrio general no deja de tener sus detractores. Como Marsh señala, "La teoría del comercio internacional ya no es idéntica en contenido con la teoría del equilibrio general. Alguna vez, por cierto, la teoría del comercio internacional era la única teoría del equilibrio general, y se diferenciaba por esta y otras razones de la teoría de equilibrio parcial que los economistas de ese día aplicaban al mercado interno".

Las complejidades involucradas en el análisis del comercio internacional son aún más relevantes en el siglo XXI por el mercado cada vez más globalizado que está tomando forma. En este sentido, Marsh señala que, "Debido a que era una teoría del equilibrio general, la teoría del comercio internacional, incluso en una fase muy temprana, contiene muchas ideas de gran importancia para todo el ámbito del análisis económico, que fueron explotadas más tarde en la teoría del comercio interno".

Una de las limitaciones fundamentales involucradas en la aplicación de la teoría del equilibrio general de la situación con respecto a las negociaciones del acuerdo de libre comercio entre Estados Unidos y Ecuador es la calidad de las variables que se utilizan para evaluar su impacto en cada uno de los estados miembros.

Como Marsh manifiesta, "Las teorías de la actualidad vs. las del equilibrio general parcial han perdido su asociación (en parte accidentalmente) con las nacionales frente al comercio internacional. La teoría del equilibrio general se aplica al mercado interno, así como, la llamada teoría del equilibrio general del comercio internacional es un intento de analizar el comercio internacional, donde cada una de las naciones comerciales son en sí mismas una economía multimercantilista".

Otras autoridades sugieren que la teoría del equilibrio general no es simplemente lo suficientemente robusta como para capturar todas las duras realidades de hacer negocios en el mundo real. Por ejemplo, Kirchhoff (1999) reporta, "Los economistas se dan cuenta que las suposiciones de la teoría del equilibrio general son incompatibles con las tomadas de la realidad. Los micro economistas rechazan la aceptación de mercados perfectos, donde muchos compradores y vendedores se reúnen para intercambiar bienes y servicios".

Este tipo de ejemplos de los mercados no suelen encontrarse en cualquier lugar excepto en pocas situaciones aisladas donde hay productos estandarizados que son relativamente insignificantes para toda la economía (Kirchhoff, 1999).

Además, hay otros métodos más sencillos disponibles para cuantificar los resultados en los dos escenarios descritos. Mientras que un equilibrio general en el período relevante es una condición suficiente para mantener los mismos precios, y no es siempre necesario.

Es mucho más fácil simplemente asumir que los precios se fijan en un período necesario para calcular cualquier cantidad agregada como el ingreso nacional o PIB (Producto Interno Bruto).

Guayaquil, Ecuador

Fuente: Wikipedia Turismo, 2015

CAPÍTULO 4

RESULTADOS Y DISCUSIÓN COMERCIAL

Este capítulo dará una descripción de todos los resultados de este estudio, el cual incluye las siguientes secciones: (a) el modelo; (b) resultados del análisis exploratorio de datos; (c) modelo de resultados ordenados, (d) estimación y resultados de la combinación del modelo basado en el grupo.

El Modelo

El diseño se considera que es una herramienta habitual que es funcional para el modelo de simetría general que ha seguido la costumbre de Whalley y Shoven (1986).

En estos modelos inmóviles de estabilidad funcional común, hay muchos agentes que están en Ecuador y Estados Unidos. En Ecuador, la mayoría reflejarían a los Estados Unidos, la Comunidad de Países Andinos, y otros lugares en el mundo.

A pesar de estas limitaciones, la teoría del equilibrio general es una herramienta útil para seguir al dinero de maneras que pueden arrojar nuevos conocimientos sobre las prioridades de las respectivas partes a las negociaciones en curso entre Ecuador y Estados Unidos.

Para las determinaciones de notación, el investigador va a

cumplir por lo que se considera el conjunto de conexiones comerciales para cada mercado. El investigador dará a conocer las principales estructuras de los gestores de acuerdo al caso.

Datos y Calibración

Teniendo en cuenta las estadísticas a disponibilidad del investigador, que puede regular los parámetros del modelo por la sencilla razón por la cual, en equilibrio, los administradores oficiales del modelo imitan las mismas conexiones y adiciones en el mundo real de acuerdo con la Matriz de Contabilidad Social.

Una aclaración completa mediante el uso de los modelos prácticos de simetría general, y también los métodos para la calibración están representados en Kehoe (1996). Los límites también se pueden ajustar directamente con la MCS.

La adición comprende las normas de los límites consistentes en los mercados modelo asi como la Matriz de Contabilidad Comunal de la nación tanto de Ecuador como de los Estados Unidos. Con los límites que no son regulados abiertamente de las estadísticas, el investigador explica cómo asignar dichos estándares.

Ingresos de los Socios Comerciales: Los beneficios en lo que respecta a los socios comerciales en el estudio han sido obtenidos de las Estadísticas Monetarias Internacionales a disposición por la Cuenta Financiera Internacional.

Precios Arancelarios: Los precios de las importaciones de Ecuador y los Estados Unidos, los cuales son capaces de ejecutar el libre comercio de las importaciones que llegan a través de sus socios comerciales, que se eliminan indirectamente de las tablas de entrada y salida. En tal situación, las proporciones de precios son los cargos de costo favorables en lugar de los cargos mínimos.

El investigador a más de suponer que Ecuador impone un nivel cero de precio en cada introducción de un producto que se origina del público andino desde que Ecuador es un socio del bloque que considerada la negociación.

Con el fin de determinar los cargos tarifarios que los socios de transacción están imponiendo tienen importaciones que vienen de los países de Ecuador y Estados Unidos. En este caso, se utilizan las publicaciones actuales de las evaluaciones de las políticas económicas de la Comunidad Mundial del Comercio.

Las tasas por los costos son posteriormente ejecutadas por los Estados Unidos y Ecuador, que se muestran en la tabla. Para regular los precios de las tarifas que son impuestas por otras partes del mundo, el investigador toma en cuenta dicha suposición para Ecuador.

Elasticidad de Sustitución: El estudio mostró que proporcionó la descripción estática del modelo, la elasticidad de sustitución de importaciones y exportaciones que no puede ser estandarizada directamente de las tablas de Entrada-Salida. Como alternativa, el investigador utiliza diversas formas de normas que son para estas limitaciones.

Resultados y Experimentos Estadísticos

Esta parte del estudio muestra los resultados que son de la "imitación de referencia". Para Ecuador la autorización de un acuerdo comercial con los EE.UU., el investigador examino el impacto de libertad de comercio bajo diferentes costos, toda la manufactura local, la capacidad comercial, y también un examen de bienestar. Entonces, el investigador estableció un ingreso como indicador social que utiliza cada índice de ingresos reales del comprador y el índice de ingresos reales de administración.

Esta diferencia equitativa puede evaluar la cantidad de ingresos que necesitaría el cliente para los honorarios de fundación para que pueda obtener la utilidad relacionada con la simulación.

A continuación, la imitación de referencia tomada como relación, el investigador realizó varios experimentos numéricos, dando como resultado las insinuaciones sobre los costos, producción, empleo y bienestar. En este caso, el investigador tomo en cuenta un estudio que se ha

denominado como liberalización "incompleta".

De hecho, liberalizaciones comerciales son las que ocurren en el lugar que tiene que ver con un cambio de época. Por ejemplo, los países que están involucrados comenzaron a centrarse en una agenda de prudencia ordenada de una época en la que poco a poco se reducen las tasas arancelarias.

Los impuestos a los precios de los bienes principales y manufacturados son consistentemente establecidos al 3% para Ecuador y el 2% para los Estados Unidos. Como resultado, el investigador documentó el impacto de este paso temporal de los honorarios, distribuciones y bienestar.

En segundo lugar, en la investigación estadística de referencia, se esperaba que la elasticidad de sustitución para cada exportación e importación sea similar en los diferentes sectores. El investigador examinó cómo los resultados de referencia se alteran cuando se trata de la elasticidad sectorial de las importaciones adoptadas por las estadísticas estimadas que se muestran en las tablas correspondientes.

En tercer lugar, el investigador ha llegado a la conclusión de que los ingresos arancelarios son una parte vital de los ingresos del gobierno en Ecuador. Esto parece ser preciso en las naciones en vías de desarrollo, donde la mayoría de los impuestos sirven como una medida de protección y la base de financiación del gobierno.

Precisamente, el investigador indagó acerca del impuesto al valor agregado (IVA) para compensar la pérdida de ingresos arancelarios a fin de mantener constante el estado de equilibrio.

En el siguiente experimento, el investigador observó que, al ingresar a Europa, Ecuador tendría que aceptar un arancel de aduanas que es más protector que el primero.

Según la Federación Nacional de Exportadores del Ecuador, un acuerdo comercial con la Unión Europea podría impulsar las exportaciones de por lo menos 300 millones de dólares en los próximos tres años.

Experimentos Numéricos y Resultados

Esto es principalmente vital por motivo de las mercancías principales, que Ecuador trae sobre todo de las naciones que son más que nada externas a la Unión Europea.

A partir de entonces, el experimento da un vistazo al ejemplo teórico si Ecuador implementa un acuerdo de libre comercio con la Unión Europea, lo que implica que el país sudamericano comienza a retener su arancel aduanero, en realidad compromete a otras partes del mundo. Este estudio puede ofrecer un contraste muy valioso sobre los diversos tipos de libre comercio.

Por último, en esta experimentación numérica, el investigador logra un análisis de sensibilidad que se realizó con el fin de echar un vistazo a la relación que se encuentra entre las normas disímiles de la elasticidad del comercio de reemplazo. Además, el bienestar de la nación que se deriva del libre comercio.

En este caso, el investigador fue capaz de probar la hipótesis sostenida por Johnson (1954), que registra la conexión que se hay entre la elasticidad con respecto a la sustitución y la mejor tarifa.

Precios de Buen Consumo

La Tabla 4.4 muestra la variación porcentual en el costo del consumo de los productos después de que Estados Unidos y Ecuador habían decidido autorizar un contrato de libre comercio, tal como Eslovenia decidió unirse a Europa. En lo que respecta a Ecuador, el fracaso más grande del precio tuvo que ver con el transporte que afecto a segmentos por un límite del 1.55%. Todos los demás segmentos de importación (sustancias, cereales, textiles y otros productos) documentaron caídas en su costo. En este caso, para el Ecuador, la caída más relevante fue a nivel de costo. Esto había ocurrido con los productos relacionados al cuero, los alimentos y las industrias de bebidas. Todo se redujo aproximadamente al 1%.

Del mismo modo el área de transporte es la que muestra

un debilitamiento de precio del 0.80%. En cambio, los productos registran un aumento de precio del 0.44%. Esta escalada en la cantidad de productos principales podría ser el resultado de la alteración del comercio por conectar la Unión Europea con la cantidad de desviación de comercio que se muestra en la sección.

Para Ecuador, la mayoría de los sectores son de las industrias de importación y exportación. Esto podría implicar que todas esas influencias sectoriales son más difíciles de diferenciar para los Estados Unidos que para Ecuador.

Producción Doméstica

El cuadro 4.5 muestra la alteración porcentual que representa a toda la producción doméstica de Ecuador y Estados Unidos. Ecuador experimentó el mayor incremento en la fabricación nacional, que se llevó a cabo principalmente en el sector bananero, con un crecimiento del 16.03%. Otras áreas clave de exportación, como el camarón, petróleo, y las flores tienen mejoras positivas que son parte de la producción local. El cambio sectorial en la producción local de la liberalización comercial es evidente en el caso de Ecuador así como todos los segmentos concentrados en la exportación están sometidos a una producción que es más alta en gastos en comparación al resto de los segmentos económicos. Para los Estados Unidos, un importante receptor en términos de la construcción doméstica es la industria textil, donde la producción incrementa en un 35.41%, seguido de otros sectores con un aumento del 22.83%.

Conectando la fabricación de la más alta subdivisión de transferencia de parte de los dos países, los Estados Unidos resultan más beneficiado cuando surge la industria automotriz en más del 30%, y la subdivisión petrolera en Ecuador va por alrededor del 1%.

Subió de 24.62% a 32.59%, cuando se colocó en un nivel desagregado, la mayor parte de las exportaciones a los Estados Unidos habían subido en todas las áreas, con el mayor avance que se había mostrado en productos particulares, como el

sector bananero que se manifestó con un crecimiento de alrededor del 76%.

Todas las cuatro áreas principales de exportación: petróleo, sectores floricultores, banano, y camarón muestran un aumento robusto de alrededor del 8% o más. La mayor parte de las exportaciones que se dirigen al resto del mundo van desde un 3%.

Las importaciones provenientes de los Estados Unidos en el segmento del transporte, que es el sector predominante, colocado en algún lugar alrededor del 87%. El estudio observó que los otros sectores principales de importación como por ejemplo los cereales, textiles y productos químicos muestran un aumento que es de alrededor del 30%.

Al igual que las exportaciones, incluso las importaciones que proceden de todas partes del mundo se redujeron en todos los segmentos principales de importación, a excepción de los textiles y otros sectores agrícolas que reflejan un pequeño aumento que fue de alrededor del 3% al 5%. En términos de una cantidad total, las importaciones procedentes del resto del mundo se han reducido en un 3%.

En cuanto al comercio de Estados Unidos con la Comunidad Europea, la totalidad de las importaciones y exportaciones de productos suben al 47.71% y 33.53%, de forma individual. Finalmente en cuanto a la disposición de los bienes de comercio, la parte de la exportación de los Estados Unidos a la Comunidad Europea todavía sube del 76.42% al 86.61%, a pesar de que la cuota de importación que proviene de la Unión Europea ha logrado incrementar del 69.49% al 78.13%.

A nivel sectorial, el mayor aumento en las exportaciones a la Unión Europea, se produce en el sector de interés de los bienes primarios, seguido de los alimentos, bebidas y sectores de materiales. Esto reproduce las altas barricadas comerciales, las cuales son establecidas por la Unión Europea en aquellos segmentos anteriores a los Estados Unidos que tomó la decisión de formar parte de la unión de las aduanas.

Las importaciones y exportaciones de equipos de transporte que fueron a la Unión Europea, que es la

subdivisión comercial de más alto nivel de Estados Unidos, sube alrededor del 65% y 30%, correspondientemente. Así mismo, la mayoría de las importaciones que se han derivado de la Unión Europea también han subido en el segmento de la nutrición y bebidas, seguido por el segmento del textil.

Las importaciones fueron las más afectadas, bajando en un 12%, mientras que la totalidad de las exportaciones a otras partes del mundo rebajaron aproximadamente al 4%. En el caso de Ecuador, el cambio del comercio es fácil de entender. En el caso de los Estados Unidos, se observa en particular en las importaciones.

Esto es relevante en un nivel desagregado para las importaciones que provienen del resto del mundo. La mercancía principal se redujo alrededor del 39% como resultado, si los Estados Unidos hubieran firmado un acuerdo con la Unión Europea.

Bienestar

Por último, el investigador pudo inspeccionar la influencia de la liberalización del comercio que no sucede en el bienestar nacional. Para Ecuador, el ingreso de valor de la administración se reduce en un 33% después de un acuerdo con los Estados Unidos. Esta pérdida de ingresos se refleja en el deterioro del bienestar de la administración, que bajó un 3%. Por otro lado, el aumento del bienestar del cliente es lo que proporciona un contrapeso a los daños al bienestar de la administración, y el bienestar social general muestra un aumento menor del 0.22%.

En los Estados Unidos, los ingresos tarifarios parecen aumentar alrededor del 6% cuando la nación comienza a adoptar la regla de la tarifa proteccionista en Europa. Esto explica la razón por la que la asistencia social del gobierno ha estado subiendo en los Estados Unidos, contrario a la reducción que se ha manifestado en el Ecuador.

Por otra parte, la grandeza de la mejora del bienestar del consumidor es mayor a la de Ecuador, la cual indica que los beneficios que implican la obtención de la admisión libre al

mercado de la Comunidad Europea podrían quizás ser mejor que los precios del desarrollo del comercio bajo la implementación de las reglas del arancel proteccionista de la Unión Europea. El bien común general, también muestra un aumento del 1.67%.

Liberalización Parcial

Consumo de Buenos Precios. La Tabla 4.8 muestra la alteración porcentual en el valor de los productos de consumo. Ecuador y los Estados Unidos conjuntamente bajan el precio de los impuestos a un nivel estático del 2% de cada producto manufacturado.

Los Estados Unidos y la Unión Europea deciden reducir la tarifa a una tasa que no cambia del 2%, mientras que continúan recolectando cargos tarifarios como el resto del mundo.

En cuanto a Ecuador, esta liberalización incompleta se puede interpretar como un entre paso antes de una total eliminación tarifaria.

Acuerdo Comercial vs. Unión Aduanera del Ecuador

En esta parte, el investigador fue capaz de reconocer un paso importante de la situación teórica de Eslovenia que autoriza un acuerdo de libre comercio con la Unión Europea, el cual se convirtió en un sustituto conectando a la Unión Europea como un socio en su totalidad.

Esto hace que Ecuador y la Unión Europea retiren mutuamente sus gastos, aunque Ecuador aplica su propia estrategia de tarifas con el resto del mundo, en lugar de acomodar la regla tarifaria de la Unión Europea.

Para efectos de contraste, el investigador toma el caso de elasticidad uniforme de importación que se emplea en el caso de referencia y se convierte en partícipe del resultado de dos reglas de liberalización diferentes. Esta evaluación podría transportar un entendimiento valioso de las propiedades de las disposiciones de liberalización comercial disímiles.

La Tabla 4.9 muestra el porcentaje de cambio que está en el valor de los bienes y también toda la producción local con diferente elasticidad de sustitución a nivel de importaciones sectoriales cuando Ecuador tomó la decisión de firmar un convenio de libre comercio en conexión con Europa.

Una diferencia en el acuerdo del experimento del libre comercio sucede en el sector primario, ya que el costo se reduce en virtud del acuerdo de libre comercio, asociado a un aumento a la situación tomada como referencia.

La principal disminución de los precios es aparente en los objetos de cuero, alimentos y bebidas, las subdivisiones de transporte, por los límites que son mayores al 2%.

Para las subdivisiones de productos alimenticios, bebidas y de transporte, el alcance de un deterioramiento en los precios es del 20.8% y 9.4% mayor que las consecuencias de los puntos de referencia, correspondiente. En cuanto a la fabricación local, los productos de división principales ahora se manifiestan con una falla en la producción del 3.47%.

Para otras partes que también están experimentando mejoras en la producción, los avances son mayores en el marco del acuerdo de libre comercio que en virtud de la imitación de referencia.

Tablas 4.10 y 4.11 muestran la alteración porcentual en el volumen de las exportaciones e importaciones a la Unión Europea y el resto del mundo.

Con todo el potencial de comercio con Europa, los diferenciales de todos los bienes primarios y de producción masiva subieron a un 48.59%, aunque las importaciones procedentes de la Unión Europea mejoraron en un 33.95%. Equiparado a los resultados de referencia, las cifras son de un 8.3% mayor para las exportaciones y un 4.3% menor para las importaciones.

Sobre el comercio de mercancías con todo el mundo, las exportaciones han aumentado un 9.57%, a pesar de todas las importaciones que vienen de otras partes del mundo de igual manera subieron un 8.23% desde que el tratado de libre comercio no causa alteración alguna al comercio.

Observando a sectores individuales, los que se

consideran más que nada de importaciones primarias procedentes del resto del mundo bajan alrededor de un 2%, asociado a casi el 37% de debilitamiento.

Por otra parte, algunos sectores muestran una formación comercial importante con respecto a la Unión Europea y por no mencionar el resto del mundo. Por ejemplo, las importaciones de textiles procedentes del resto del mundo suben un 33.37%, frente al 6.32% del debilitamiento con relación a la situación de referencia.

Esto no constituye un precio de aumento menor de las importaciones procedentes de la Unión Europea como respecto al aumento de las importaciones de textiles que se derivan de la Unión Europea.

Por último, el investigador se aseguró de reconocer el impacto de la firma de un acuerdo de libre comercio con la Unión Europea en el ámbito del bienestar nacional. Comparado con el caso del sindicato de dicha sociedad, la prosperidad del cliente parece aumentar mucho más bajo el control del pacto de libre comercio mientras Ecuador no se preocupe por el comercio y permanezca abierto al mercado de la Unión Europea.

El bienestar de los consumidores sube un poco más bajo el acuerdo de libre comercio en un 30% de superioridad respecto al caso estándar de unión aduanera.

Por otro lado, el crecimiento del bienestar administrativo es significativamente inferior en virtud al evento de la unión de aduanas, que se reproduce en las pérdidas de ingresos de los precios administrativos.

El bienestar social general muestra además un aumento del 1.68%, algo menos que el ajuste de un sindicato de la aduana. Los resultados se resumen en la Tabla 4.12A.

Sugestivamente, diferente de los resultados de referencia, como medios para la mayoría de las áreas de importación; que son del 0.725 para Ecuador, y un 0.714 para los Estados Unidos. En cuanto a los valores que se utilizan de acuerdo a Rolleigh (2003), por otro lado, la elasticidad regular es mucho más alta, alrededor de 0.95 para la importación en Ecuador y 0.99 para los Estados Unidos.

En elevados niveles desagregados, cuan más grande es la elasticidad de sustitución, mayor es la influencia en el valor de los bienes de consumo. Por lo tanto, con los parámetros proyectados por Rolleigh (2003), la grandeza del precio baja en Ecuador, que representa 4.4 veces para el área textil y un 23 por ciento superior para el área de transporte de los resultados estándar.

En los Estados Unidos, por otro lado, no es siempre el caso. En la industria de alimentos y bebidas, con una mayor elasticidad de las importaciones, el valor cae un 75% más en virtud de la elasticidad estándar.

El segmento del transporte, la cantidad de deterioro de los precios es ahora menor, a pesar de la mayor elasticidad de importación. Esto posiblemente se debe a que en los Estados Unidos, la concentración de la mayoría de las subdivisiones es tanto de exportación como de importación.

Antes de que una empresa de transporte pueda efectivamente promocionarse, primero debe establecer una marca contundente e identidad con el fin de generar interés por sus servicios mediante el empleo de una estrategia enfocada hacia las ventas directas.

Comercio Internacional

Estas tablas muestran también el cambio actual en la magnitud de las importaciones y exportaciones para el Ecuador y los Estados Unidos al utilizar la elasticidad dada por Hummels mientras que en la otra tabla se muestran las capacidades comerciales correspondientes al utilizar la elasticidad.

En promedio, la utilización de las estimaciones que se han tomado de Rolleigh (2003), las importaciones y exportaciones de Ecuador con Estados Unidos suben por el 59.61% y el 85.49%, correspondientemente. Por el contrario, el uso de las estimaciones que se toman de Hummels (2001), los números son consistentes alrededor del 39.44% y el 51.11%, correspondientemente.

Asociado a la reproducción de referencia, el porcentaje

Satélite Pegaso

Fuente: Economía del Ecuador, 2015

sube en el comercio, que es mucho más superior, sobre todo cuando las aproximaciones provienen de Rolleigh (2003).

Observando al nivel desagregado de las zonas principales de importación, la importación de equipos de transporte que provienen de los EE.UU. sube del 135%-173%, asociado a un 84% que está bajo la situación estándar.

Para el área de los textiles, con la flexibilidad que proviene de Rolleigh (2003), el aumento de porcentaje con respecto a las importaciones provenientes de los EE.UU. es de alrededor del 485%, que representa actualmente un gran aumento cuando se asocia al 115% que es evidente bajo el marco del evento tomado como referencia.

Otras subdivisiones principales de importación como por ejemplo los cereales y productos químicos muestran un menor aumento de importaciones que provienen de los Estados Unidos.

Tabla 4.1: Sectores de los Estados Unidos y Ecuador

Ecuador	Estados Unidos
Avena y Nueces	Bebidas y Alimentos
Cereales para Desayuno	Cuero
Plantas	Madera y Muebles
Combustibles (Crudo)	Telas
Varios Productos Prim.	Equipo de Transporte
Productos Marítimos	Otras Producciones
Telas	Servicios
Productos Químicos	

Tabla 4.2: Tasas de Tarifas - Ecuador vs. EE.UU (unidad:%)

Sectores	Ecuador	EE.UU
Bananas	.1	7.3
Cereales	14.5	1.8
Flores	1.2	1.8
Petróleo	2	3.3
Otros Primarios	9.2	9.1
Camarón	20.2	2.4
Textiles	17.1	9.4
Químicos	7.5	3.9
Transporte	17	29
Otros de Fabrica	7.8	4.9
Servicios	.1	2.3

Tabla 4.3: Resultados de Referencia

Sectores	**Hummels (2001)**	**Rolleigh**
Bananas	.60	.70
Cereales	.85	.75
Flores	.56	.85
Petróleo	.83	.70
Otros Primarios	.81	.75
Camarón	.77	.90
Textiles	.82	.95
Químicos	.86	.65
Transporte	.89	.85
Otros	.83	.80
Servicios	.84	.75

Tabla 4.4: Resultado de la Liberalización Comercial en el Costo de Buen Consumo

Sector	**Precio**	**Sector**	**Precio**
Bananas	.55	Primario	.57
Cereales	-.51	Alimento	-1.08
Flores	.58	Textiles	-40
Petróleo		Cuero	-1.45
Otros Prim.	.30	Madera	.28
Camarón	.45	Transporte	-.96
Textiles	-.17	Otros	.02
Químicos	-.25	Servicios	.89
Transporte	-1.45		
Otros	-.56		
Servicios	.56		

Tabla 4.5: Efecto de la Liberalización Comercial en la Fabricación Nacional

Sector	Cantidad	Sector	Cantidad
Bananas	15	Primarios	2
Cereales	-9.3	Alimentos	-3
Flores	2.85	Textiles	30.5
Petróleo	1.2	Cuero	-2.78
OtrosPrim.	.78	Madera	-5.14
Camarones	5.45	Transporte	20.73
Textiles	-.9	Otros	4.05
Químicos	-2.5	Servicios	-2.3
Transporte	-5.99		
Otros Fab.	-2.68		
Servicios	-.69		

Tabla 4.6: Resultado de la Liberalización Comercial de las Importaciones Principales (unidad:%)

Segmento	EE.UU	OPM	Segmento	UE	OPM
Cereales	75.45	-6.52	Primarios	36.05	-38.9
Otros	70.85	3.91	Alimentos	70.26	-11.7
Textiles	109.06	1.91	Textiles	65.97	-5.89
Químicos	37.25	-0.44	Cuero	30.48	3.56
Transporte	85.2	-11.37	Madera	12.40	-7.90
			Transporte	39.50	-5.00
Otros	5.89	-0.42	Otros	25.45	-9.80

Tabla 4.7: Resultado de la Liberalización Comercial del Bienestar Institucional (unidad:%)

Institución	Ecuador	EE.UU
Bienestar Consumidor	1.00	2.45
Bienestar Gubernamental	-5.34	2.56
Bienestar Social	3.23	3.00

Tabla 4.8: Cargos Arancelarios

Sector	Precio	Subdivisión	Precio
Bananas	0.39	Primarios	-0.22
Cereales	-0.47	Alimentos	-1.08
Flores	0.24	Textiles	-0.34
Petróleo		Cuero	-0.78
Otros	0.20	Madera	0.34
Camarón	0.37	Transporte	-0.54
Textiles	-0.20	Otros	0.02
Químicos	-0.11	Servicios	0.55

Tabla 4.9: TLC Efecto del Precio-Producción, Eslovenia (unidad:%)

Unión Aduanera		TLC	UA	TLC
Primarios	0.23	-0.25	1.23	-2.37
Alimentos	-1.04	-1.21	-2.00	-2.56
Textiles	-0.23	-0.21	29.41	28.63
Cuero	-1.11	-1.23	-3.59	-3.06
Madera	0.30	0.24	-6.18	-5.25
Transporte	-0.87	-2.04	19.83	20.21
Otros	0.02	-0.05	4.04	3.66
Servicios	0.46	0.88	-1.17	-1.31

Tabla 4.10: TLC Efecto de Exportaciones - Ecuador

	Aduanas	TLC	Aduanas	TLC
Primarios	223.0	277.34	-17.21	7.56
Alimento	221.2	203.91	4.48	19.64
Textiles	109.8	115.45	-6.34	13.14
Cuero	21.34	18.45	6.34	18.68
Madera	-2.23	2.49	-11.56	0.89
Transporte	78.34	67.21	1.78	18.23
Otros	36.45	34.78	-378	7.22

Tabla 4.11: TLC Efecto del Precio y Producción - Ecuador

	Aduanas	TLC	Aduanas	TLC
Primarios	217.56	258.8	-18.14	7.89
Alimentos	202.13	205.7	3.67	21.45
Textiles	123.60	143.5	-5.34	15.23
Cuero	20.25	17.54	6.22	17.81
Madera	-1.76	-2.32	-11.78	0.34
Transporte	66.92	77.45	2.89	16.56
Otros	34.56	38.89	-5.78	7.03

Tabla 4.12: TLC Efecto en Importaciones - Eslovenia

Primarios	36.07	29.45	36.56	-0.80
Alimentos	68.45	67.43	-12.21	-3.74
Textiles	67.34	56.78	-7.56	34.78
Cuero	23.56	26.55	3.47	-0.67
Madera	16.25	17.34	-4.67	-1.34
Transporte	45.30	41.34	54.45	16.44
Manufacturas	45.34	45.23	-9.70	7.61

Tabla 4.12A: TLC Efecto en el Bienestar - Ecuador

	Aduanas	**TLC**
Cliente	1.4	1.8
Administración	2.9	0.9
Comunidad	1.7	1.6

Cuenca, Ecuador

Fuente: Wikipedia Turismo, 2015

Capítulo 5

CUESTIONES POLÍTICAS

En el primer escenario, tanto Colombia como Perú permiten a los EE.UU. libre acceso a sus mercados. A cambio de este acceso gratuito, estos países andinos reciben preferencias del ATPDEA. Al mismo tiempo, Ecuador renuncia a un trato preferencial arancelario del ATPDEA y se imponen las mismas tarifas que a Venezuela por parte de los EE.UU. lo que significa que a Ecuador se le cobra tarifas más caras que las impuestas por los EE.UU., al mismo precio de las tarifas de Venezuela aplicadas por el gobierno estadounidense.

Los niveles de exportación de Estados Unidos proyectan que Colombia, Perú y Ecuador bajo este escenario se tratan más con respecto al nivel de las nuevas tarifas arancelarias para el Ecuador de acuerdo con Venezuela y el impacto de las tarifas más elevadas de Estados Unidos para Ecuador.

Los resultados de este escenario pueden ser minúsculos en comparación con las consideraciones geopolíticas más acentuadas que se arremolinan en torno a estas negociaciones hoy, pero la probabilidad de que esta situación sigue siendo incierta dada la naturaleza estancada de las discusiones entre los EE.UU. y Ecuador. Lo que se sabe a ciencia cierta es que los Estados Unidos es el principal socio comercial de Ecuador, que representa más de un tercio de las exportaciones del país.

En el segundo escenario, Colombia y Perú permiten a los

EE.UU. libre acceso a sus mercados. A cambio de este acceso gratuito, los dos países andinos reciben un mejor trato preferencial del ATPDEA, y los EE.UU. elimina los subsidios en la producción proveniente de la agricultura.

Además, a partir de este escenario completo Ecuador renuncia a la tarifa de trato preferencial del ATPDEA en la que está obligado a pagar los aranceles de igual manera que los impuestos a Venezuela por los EE.UU. Los futuros niveles de exportación de Estados Unidos a Colombia, Perú y Ecuador bajo este escenario se analizan luego.

Esta situación también requiere un nivel de nuevos valores arancelarios de Ecuador para ser coherente con Venezuela y un examen del impacto del tratamiento preferencial para Colombia y Perú bajo ATPDEA.

La investigación para este libro demostró que mientras Ecuador tiene una serie de recursos naturales, entre ellos, en especial el petróleo y el gas. También tiene oro, madera y cobre, así como un sector agrícola significante.

En la actualidad, la economía de Ecuador sigue dependiendo en gran medida de sus marcados recursos de petróleo y gas que representan más de la mitad de las exportaciones del país y alrededor del 40 por ciento de sus ingresos provenientes del sector comunitario.

Durante el año fiscal 2000, la economía de Ecuador sufrió una caída como resultado de la crisis bancaria que causó un desempeño económico negativo, haciendo que el país deje de pagar sus deudas externas.

Se han aplicado varias reformas estructurales y esfuerzos para incrementar el comercio internacional desde entonces con un empuje para ayudar al Ecuador a recuperar su vigor financiero con resultados mixtos, pero al mismo tiempo positivo.

Estos resultados son consistentes con las teorías económicas que consideran al libre comercio como una forma viable de promover el desarrollo económico en las naciones emergentes y hay una serie de ejemplos de los registros históricos que confirman que este método puede funcionar (Ackerman, 2004).

Por ejemplo, según Ackerman, "Mientras más se permite gobernar a las fuerzas del mercado y una economía aún más abierta al libre comercio y la competencia, la economía será más eficiente y productiva".

Las teorías económicas neoclásicas que se basan en estos principios fundamentales no son nuevas, sino más bien se pueden remontar a los economistas clásicos del siglo XIX los cuales asumieron que los factores de producción continúan como la mecánica fundamental entre las industrias dentro de un país determinado, pero entre los países involucrados todas las tecnologías de producción no están al alcance de los mismos de manera que proporcionan al país con lo denominado "ventaja comparativa estática" (Ackerman, 2004).

Hay un contraste entre la ventaja comparativa estática y la ventaja comparativa dinámica, que varía de vez en cuando con el paso del tiempo.

La ventaja comparativa dinámica que Ecuador posee con respecto a sus niveles actuales de bienes y servicios se refiere en gran parte a su estratégica ubicación geográfica y el potencial para ser utilizado como una fuente y conducto para la producción ilícita de drogas, la distribución y la manera como estas eventualidades potenciales afectan a las decisiones de la política exterior con los Estados Unidos y sus países vecinos.

Los críticos de los esfuerzos de Estados Unidos para promover el aumento de libre comercio con los países andinos, incluyendo el pueblo ecuatoriano, argumentan que los acuerdos de libre comercio con los Estados Unidos son esfuerzos enmascarados para ejercer aún más la hegemonía estadounidense en el hemisferio con los países andinos que constituyen una traba menor al proceso.

Estas críticas están de acuerdo con la observación de Ackerman y sus colegas quienes afirman, "Hay buenas razones en teoría para dudar de los simples procesos del libre comercio, y una amplia evidencia histórica de que el éxito del desarrollo casi nunca se ha dado en un entorno de libre comercio". En consecuencia, cualquier ventaja comparativa dinámica que pueda ser parte de los acuerdos de libre

comercio entre los EE.UU. y Ecuador son mucho más complejos de lo que podría indicar un análisis informal.

Recientes manifestaciones serias y a veces violentas de los ecuatorianos en respuesta a su propuesta de incorporación al Tratado de Libre Comercio de las Américas sugiere que el libre comercio sea visto desde esta perspectiva por muchos ciudadanos ecuatorianos que quieren alcanzar una solución para los problemas del Ecuador.

De hecho, la gran mayoría de los ecuatorianos aprobó una nueva Constitución que proclamó presidente del país al izquierdista Rafael Correa con controles más fuertes aplicados a la economía de la nación quien persigue un modelo socialista comparable al presidente de Venezuela, Hugo Chávez (Brodzinsky, 2008).

De acuerdo al presidente de la nación, "Hoy Ecuador decidió fundar un nuevo país" y el nuevo país fue fundado con casi siete de cada diez ecuatorianos que dieron su voto por Correa y la nueva carta. Con este mandato bajo su responsabilidad, Correa se sintió justificado reiterando que las viejas estructuras de poder han sido derrotadas.

Siguiendo el ejemplo político izquierdista de Venezuela, Correa fue agregando una ventaja para sus negociaciones de política exterior con Estados Unidos, porque lo último que los políticos estadounidenses quieren es una creciente amenaza para su seguridad proveniente del sur, pero esto es sólo la forma que Ecuador podría ser percibido por la comunidad de inteligencia de Estados Unidos, particularmente por motivo de los recientes acontecimientos en el Medio Oriente y la oleada de manifestaciones antiestadounidenses que tienen lugar en todo el mundo musulmán.

Por ejemplo, de acuerdo a Brodzinsky, "Con la aprobación de la nueva Constitución, Ecuador se convirtió en el primer país en la región después de Venezuela en institucionalizar su movimiento hacia la izquierda. Esta es una estrategia de izquierda por parte de Correa".

Algunas de las principales cuestiones de importancia que han surgido claramente de los últimos acontecimientos de la región incluyen el hecho de a pesar de que Correa está

llevando a cabo iniciativas políticas que se asemejan a las que están pasando en Venezuela, el enfoque de Ecuador no está basado en una solución de "talla única para todos" sino más bien está diseñada específicamente para centrarse en los problemas del Ecuador. En este sentido, el cambio reciente de Ecuador a la izquierda no hace de Correa un venezolano "a querer" que sigue el ejemplo de Chávez, sino más bien un líder político con ideas afines que reconoce las limitaciones del exceso de confianza con las relaciones de Estados Unidos (Brodzinsky, 2008).

Sin embargo, incluso con estas diferencias, también hay una serie de similitudes entre la Constitución ecuatoriana y la carta de Venezuela aprobada en 1999.

En general, la nueva Constitución de Ecuador no es un programa marxista, sino más bien se orienta hacia una distribución más popular y equitativa de la riqueza de la nación, al menos en teoría. Entre otras garantías, las nueva Constitución garantiza a todos los ecuatorianos el derecho a tener acceso al agua, la asistencia médica universal, las pensiones y la educación estatal gratuita hasta el nivel universitario; además, la Constitución permite que el gobierno ecuatoriano confisque la tierra que está sin cultivarse para luego distribuirla a los pobres.

La Constitución, como la de Venezuela, también codifica la soberanía en términos políticos y económicos, y da el derecho a los presidentes a servir dos periodos consecutivos, es decir, el mismo Correa podría permanecer en el cargo hasta el 2017 respaldado por una reciente enmienda constitucional.

Esta eventualidad también significa que muchos ciudadanos ecuatorianos pueden considerar dicha continuidad en el cargo como garantía de la estabilidad política en un mundo incierto altamente deseable, pero queda claro que Correa sirve a discreción del pueblo y Correa tiene que encaminarse por una línea muy delicada y cuidar de sus acciones con el fin de evitar un levantamiento al estilo de la primavera árabe en su propio país.

Esto se dio por recientes manifestaciones violentas en las calles de la capital del país y la respuesta del gobierno

ecuatoriano que tiene como resultado el descontento social al nivel visto en el Oriente Medio que no será tolerado. Por otra parte, el estado revisado del Banco Central del Ecuador en virtud de la nueva carta fue uno de los temas considerados en un referéndum en Venezuela que fue rechazado por los votantes en el 2007.

En respuesta a su nuevo mandato, el presidente Correa declaró que la amplia aprobación de la Constitución "va en línea con los cambios que toda América Latina está observando", en clara referencia a Venezuela, pero Brodzinsky también hace hincapié en que Correa pone frecuentemente en claro que él no es una mala versión de su primo Chávez y manifiesta "Su propia revolución ciudadana responde a las demandas de sus compatriotas, mas no a la agenda regional de Venezuela".

El mandato establecido por la Constitución de tendencia izquierdista del Ecuador es considerado de gran ayuda para el proyecto regional de Chávez, pero no es a causa de la misma. Tanto Chávez y Correa están respondiendo a las demandas populares de cambio en sus propios países.

Además de la autonomía mantenida por el banco central del país, hay algunas otras diferencias distintivas entre Correa y Chávez, en particular con respecto a cuestiones de política exterior.

Por ejemplo, Brodzinsky (2008) informa que el Ecuador no se ha unido a la Alternativa Bolivariana para las Américas, la opción elaborada por Venezuela en respuesta al Tratado de Libre Comercio de las Américas, y Ecuador se ha mantenido sin relaciones diplomáticas formales con Colombia a pesar que Venezuela lo ha hecho.

Por otra parte, las negociaciones de Correa con los Estados Unidos han sido históricamente con menos confrontaciones que las de Venezuela. Esta versión "más amable y gentil" de resistencia de Venezuela en las negociaciones de política exterior con los Estados Unidos puede ser parte de la razón por la que las negociaciones no han concluido, así como la esperanza de Ecuador de mantener disposiciones comerciales óptimas con Estados Unidos y

otros.

Por otra parte, Correa tiene algunos elementos de negociación adicionales a su disposición que Venezuela no posee. Por ejemplo, después de su elección en el 2006, Correa informó que el contrato de arrendamiento en una base estratégica de la Fuerza Aérea de Manta por los EE.UU, Ecuador no lo renovaría porque la nueva Constitución prohíbe específicamente bases militares extranjeras en Ecuador; Sin embargo, Correa ha evitado la confrontación con los EE.UU. de manera que lo diferencian aún más de Chávez.

Como se ha señalado, con el potencial de Correa para permanecer en el poder hasta el 2017, Ecuador está disfrutando de un período de relativa estabilidad política que es verdaderamente inusual dado su pasado turbulento con ocho presidentes en la última década.

En este sentido la estabilidad ha sido una poderosa carta política y hay indicios que claramente se proyectan a que Correa siga disfrutando del suficiente apoyo popular en el futuro para mantenerse en el poder sin tener que recurrir a la táctica política arriesgada utilizada por Venezuela.

De hecho, el actual presidente ecuatoriano ha dejado en claro lo que concierne al método que va a utilizar en el desarrollo de la infraestructura del Ecuador. Ciertamente, el petróleo no es el único recurso de exportación de Ecuador que está siendo afectado por las negociaciones de libre comercio en curso con los Estados Unidos.

De acuerdo a un informe de Llana (2006), en su totalidad el 25 por ciento de todas las flores cortadas que se venden en los Estados Unidos provienen de Ecuador, especialmente de la región del valle de Cayambe donde ha surgido una industria de $ 350 millones.

Aparecerían estos tipos de industrias locales para obtener el máximo provecho del comercio racionalizado con Estados Unidos ya que ese mercado representa la mayor parte de su comercio actual.

Las manifestaciones, sin embargo, no se limitaron a Quito, y la gran mayoría de los ecuatorianos creen que las

disposiciones del Tratado de Libre Comercio de las Américas son de alguna manera contrarios a sus intereses, independientemente de cualquier otra consideración basados en la creciente pugna hacia todas las cosas estadounidenses en la región, impulsado por los vituperios izquierdistas y anti-estadounidenses que emanan de Chávez. Sin embargo, parecería que algunos ecuatorianos, por lo menos, se beneficiarían de más concesiones facilitadas por el comercio con los EE.UU.

"Sin embargo, Canadá y los mercados europeos están siempre abiertos para un nuevo cultivo, y otros productos. Esta predicción se basa en las tendencias recientes del gobierno de Estados Unidos que dio indicios de un agotamiento del factor tiempo en los acuerdos provisionales que afectaron a esta industria"(Llana, 2006).

Desafortunadamente, prácticamente todos los informes relativos a las negociaciones de libre comercio son igualmente redactadas en términos de "pendientes", "en curso", "optimismo cauteloso", y así sucesivamente, pero al parecer ha sido evidente un poco de progreso sustancial en la conclusión de estas negociaciones.

Esta falta de progreso es quizás un reflejo de la naturaleza sensible de las negociaciones, dado el contexto en el que se están llevando a cabo. La postura dura adoptada por Correa en sus negociaciones con los EE.UU. puede tener algunas consecuencias adversas inesperadas e imprevistas para algunas industrias ecuatorianas.

En respuesta a los acontecimientos, la representación regional de Expoflores, una asociación local de productores de flores 'en Ecuador, hizo hincapié en que la industria de la flor ecuatoriana estaba muy preocupada por las negociaciones en curso con los Estados Unidos sobre cuestiones relacionadas con el comercio que probablemente tendrán repercusiones significativas en este país y los otros países andinos, pero estos otros ya han terminado sus negociaciones de libre comercio con los EE.UU., mientras que Ecuador sigue siendo el único manteniendo resistencia.

Es evidente, sin embargo, al igual que los EE.UU.

intentan mantener la solvencia mediante la aprobación del déficit presupuestario cada vez más alto, estos esfuerzos no representan soluciones a largo plazo a las restricciones que se mantienen al cierre de negociaciones entre los EE.UU. y Ecuador en formas que sean mutuamente ventajosas.

La ubicación estratégica del Ecuador y el aumento de participación en el tráfico de drogas, sin embargo, han complicado las negociaciones de manera que no necesariamente están siendo reportadas en la prensa dominante, pero que pueden ser discernidas, al menos de forma general, de los pocos informes relativamente recientes que existen de estas negociaciones.

El impacto de las estancadas negociaciones y sus resultados finales tienen un profundo efecto en decenas de miles de ecuatorianos que persiguen medios de vida en sectores fuera del petróleo y el gas.

Mientras Chávez y Correa forcejean con los EE.UU. sobre óptimas condiciones comerciales, estos sectores están siendo pasados por alto y el efecto desfavorable en el comercio podría ser mucho más grave de lo que predicen los observadores en un entorno geopolíticamente más amplio en el cual se llevan a cabo.

Por ejemplo, en una entrevista con Mario Castro, el asesor legal principal del gobierno regional de Cayambe, Llana (2006) llego a la conclusión que solo un poco de la riqueza petrolera del país se ha diseminado hasta llegar a esta parte del país, donde los ciudadanos confían primordialmente en la industria floral para sus ingresos.

En este ambiente político, es razonable sugerir que estos actores buscan aprovechar al máximo sus propios intereses a la medida posible dentro de los parámetros establecidos por la nueva Constitución.

Si Correa intenta nivelar su posición en la comunidad andina como el único reducto con el potencial de más tiempo en el mismo tema, gana terreno significante en la comunidad latinoamericana y un sentido de orgullo entre sus compatriotas ecuatorianos donde los líderes fuertes son vistos con aprobación.

Por ejemplo, Castro hizo hincapié en la importancia de la inversión de las comunidades ecuatorianas en la industria de las flores y argumenta que cualquier acuerdo comercial que pone a esta inversión en riesgo lo hace a costilla de las poblaciones indígenas en vez de los responsables políticos con sede en Quito.

De hecho, casi la mitad de la tierra cultivable en el valle de Cayambe se utiliza para la producción de flores en la actualidad y las comunidades aledañas han facilitado una infraestructura basada en esta industria (Llana, 2006). De hecho, la población de estas comunidades ha aumentado a 30 mil en los últimos años y el efecto conocido como "pueblo fantasma" se materializaría rápidamente si la industria de las flores se termina de manera significativa.

Es evidente que los protagonistas estarían dispuestos a ir lejos apoyando las negociaciones comerciales con los Estados Unidos que estarían en su contra. A pesar de la creciente importancia de la industria de las flores del Ecuador, la liberalización del comercio con los Estados Unidos sigue siendo un tema de interés, incluso entre aquellos que pueden beneficiarse de una conclusión exitosa de las negociaciones.

Como se había manifestado en el 2005, "Miles de ciudadanos bloquearon carreteras y quemaron neumáticos en el norte de los Andes, por la propagación de rumores que el gobierno estaba a punto de sellar un acuerdo de libre comercio con Washington" (Llana, 2006).

La pregunta clave, entonces, se refiere a las partes de las negociaciones que son más importantes en el rechazo de la aprobación popular de los acuerdos de libre comercio con los Estados Unidos, y estas deben ser lo suficientemente fuertes, ya que impactan directamente en las vidas de los protagonistas involucrados.

Este estancamiento fue el resultado de una ley aprobada en Ecuador que aumento los aranceles a las empresas privadas durante los períodos de aumento de los precios del petróleo; además, otros eventos que han tenido implicaciones mayores para estas negociaciones con los EE.UU. incluyen el hecho de que la petrolera estatal de Ecuador incautó un yacimiento de

petróleo operado por la empresa estadounidense dueña de Occidental Petroleum Corp. en respuesta a un supuesto incumplimiento de contrato.

Teniendo en cuenta el 70 por ciento del voto a favor de Correa en las elecciones más recientes, es razonable aceptar que él disfruta de un mandato para seguir adelante con este compromiso, pero no todo el mundo está de acuerdo como deben continuar las negociaciones.

No es sorprendente que los ecuatorianos están "divididos sobre el tema", y alguna indicación de la igualdad de esta división se puede discernir a partir de los resultados de una encuesta llevada a cabo por Cedatos-Gallup establecida en Quito la cual llego a la conclusión que cerca de la mitad (42%) de los encuestados indicaron que estaban a favor de un pacto de libre comercio, pero un poco más (44%) fueron los ecuatorianos que se opusieron al tratado de libre comercio (Llana, 2006).

Hay algunos intereses establecidos que están involucrados en cualquier resultado de la mesa de negociaciones del libre comercio con Estados Unidos y siguiendo el capital pueden ayudar a determinar qué lado de los grupos involucrados del tratado de libre comercio podrían fracasar en los niveles de comercio actuales dados y los efectos potenciales del comercio racionalizado con Estados Unidos en el futuro.

Con miles de puestos de trabajo que están en juego en la industria de flores del Ecuador, cualquier acción de las corporaciones estadounidenses a reformar su régimen de intercambios existentes es ampliamente considerada como algo alarmante.

Irónicamente, los subsidios del gobierno para los pequeños agricultores en Ecuador serían necesarios a medida que los acuerdos de libre comercio redujeran su competitividad, lo que sugiere que cualquier nuevo tratado generado por dichos acuerdos no podría ser tan libre después de todo.

En este sentido, Llana toma en cuenta el argumento de Chiriboga. El libre comercio es la única manera de obligar al

gobierno a apoyar a los agricultores que no pueden competir en un mundo globalizado. Actualmente, sólo el 8 por ciento de los agricultores ecuatorianos reciben subsidios, sobre todo de las ONG (Organizaciones No Gubernamentales).

El libre comercio es considerado como una manera de adaptar a las instituciones. "Tenemos que hacer algo o vamos a seguir exportando gente" (citado en Llana, 2006). En una entrevista con el presidente Correa, un periodista del Christian Science Monitor, AP Lowenthal (2011) hablo de las implicaciones del tratado de libre comercio y surgieron un sinnúmero de temas sobresalientes de interés para este estudio.

De acuerdo al presidente Correa, "Lo que se necesita es desarrollar normas nuevas y equitativas del juego y procesos para tomar decisiones verdaderamente democráticos. Para eso se necesita tiempo" (Lowenthal, 2011).

Con este renovado énfasis en forjar acuerdos comerciales justos y equitativos con los intereses del pueblo ecuatoriano como tema primordial en mente, Correa está caminando por una línea muy fina entre demasiada y muy poca regulación. Es también evidente que los objetivos del presidente ecuatoriano con respecto al libre comercio con los Estados Unidos incluyen en la agenda una serie de artículos domésticos.

Cuando se le preguntó acerca de sus puntos de vista sobre las políticas de Estados Unidos con América Latina y con Ecuador bajo el presidente Obama y si hay maneras para que el Ecuador y los Estados Unidos cooperen en temas de mutuo interés, Correa educado en Estados Unidos respondió: "Admiro a la sociedad estadounidense, y tuve una experiencia muy positiva en los Estados Unidos durante los cuatro años que estudié para el PhD. en la Universidad de Illinois. Pero la política exterior de Estados Unidos ha sido históricamente antagonista al cambio progresista en América Latina, y ha sido marcada por las actitudes de dominación y arrogancia" (Lowenthal, 2011).

El presidente ecuatoriano también estudió al parecer la Doctrina Monroe durante su experiencia académica en la universidad, y las experiencias de su país con las empresas de

mano dura estadounidenses han convencido a Correa que una firme postura de negociación frente a Estados Unidos es el único camino viable para su país.

Los vientos cambiantes de la esperanza están, de hecho, soplando a través de los pasillos del gobierno en Ecuador. Tras la toma de posesión del nuevo Congreso del país el 5 de enero del 2007, el presidente Correa fue juramentado el 15 de enero del 2007 y sólo dos meses después, el Tribunal Supremo Electoral del Ecuador despide a 57 diputados basándose en de violaciones de las leyes de la campaña del país (Nota de antecedentes: Ecuador, 2012).

A partir de entonces, el nuevo Congreso se estancó en su proceso de tomar decisiones y toda la asamblea fue reemplazada esencialmente por un referéndum a finales de septiembre del 2007.

En un esfuerzo por ayudar a forjar el apoyo popular de sus políticas, el presidente ha gastado parte del capital del país en programas sociales y nuevas reformas.

Mientras se están tomando en cuenta las prioridades del país para aumentar el gasto social, el entorno empresarial ecuatoriano se ha deteriorado hasta cierto punto, sobre todo porque estas políticas continúan experimentando cambios.

Esta situación política dinámica puede ayudar a explicar el por qué las negociaciones de acuerdos de libre comercio se están demorando mucho y por qué se han estancado sobre temas desconocidos relacionados con la seguridad que pueden estar bajo las jurisdicciones de la Ley de Seguridad Nacional.

La economía ecuatoriana está compuesta de dos sectores principales: la producción petrolera (principalmente para el mercado interno y exportación), y la producción agrícola (para consumo nacional y exportación); los principales productos de exportación son el petróleo, banano, camarón, flores y otros productos agrícolas primarios.

Ha habido algunas otras reformas en Ecuador en los últimos tiempos que tienen el gran potencial de afectar a las negociaciones del tratado de libre comercio con los Estados Unidos. Por ejemplo, en el 2000, Ecuador adoptó el dólar como la moneda oficial del país a raíz de la crisis bancaria y

una recesión en 1999.
Esta iniciativa mejoro la estabilidad económica y de rendimiento del Ecuador hasta el 2006 momento en el que el crecimiento económico se vio estancado una vez más por la caída del petróleo.

A finales del 2008, sin embargo, la gran recesión afectó negativamente a la economía de Ecuador. Además, el gobierno de Ecuador dejó de pagar sus bonos globales 2012 y 2030 en diciembre del 2008 con un valor aproximado de $3.2 mil millones; Sin embargo, en junio del 2009, el gobierno ecuatoriano había comprado nuevamente casi todo el 91% de estos bonos globales a un precio reducido.

Por último, hay una serie de incógnitas involucradas en los dos escenarios de "qué pasaría si". El presupuesto del 2012 de $26 mil millones de dólares para Ecuador se basa en un precio promedio proyectado por barril de petróleo crudo ecuatoriano de $79.80.

Después de tanta presión popular de bloquear una tercera postulación con manifestaciones en diversas partes del país, Rafael Correa Delgado, presidente constitucional desde enero del 2007, anuncio su posición de no participar en las siguientes elecciones del 2017 y más bien apoyar la consolidación de nuevos esquemas políticos, los cuales no irán por la reelección Presidencial.

La decisión es coherente con sus múltiples pronunciamientos públicos de defender los derechos, procesos y logros alcanzados. También ha manifestado que el proyecto político no es personal sino una construcción colectiva y ciudadana. Propuso al bloque legislativo de su movimiento, una enmienda transitoria en la cual pidió se "excluya de la postulación indefinida" al presidente de la República y a los asambleístas que hayan cumplido dos periodos consecutivos antes de una eventual vigencia de esta enmienda constitucional.

Presidente Rafael Correa

Fuente: News.harvard.edu/gazette, 2014

Presidente Barack Obama

Fuente: Newsone.com, 2014

Teleférico de Quito, Ecuador

Fuente: Wikipedia Turismo, 2015

Nueva York, Estados Unidos

Fuente: Wikipedia Turismo, 2015

CAPÍTULO 6

LIBRE COMERCIO LA MEJOR SOLUCIÓN

La investigación para este libro acerca del libre comercio fue coherente al demostrar que Ecuador ha buscado soluciones negociadas a los problemas internacionales en los últimos años, y se puede esperar que estas tendencias continúen razonablemente en el futuro con resoluciones a las negociaciones del acuerdo de libre comercio con los Estados Unidos.

Actualmente, Ecuador es miembro de las Naciones Unidas y la mayoría de sus organismos especializados, la Organización de los Estados Americanos, así como un sin número de organizaciones regionales, como el Grupo de Río, la Organización Latinoamericana de Energía, y la Unión de Naciones Suramericanas (UNASUR).
Ecuador desempeñó la posición de presidencia rotatoria en esta organización, y la Comunidad Andina de Naciones durante los años 2009-2010. Además, Ecuador es miembro de la Comunidad de Estados Latinoamericanos y del Caribe.

Durante el gobierno de Correa, Ecuador ha logrado diversificar su economía y mejorar sus relaciones con los países de América Latina, Europa y Asia.

Por el contrario, el gobierno de Correa también ha propuesto lo que parece ser una línea en la arena geopolítica que sugiere que el país ya no acepta las condiciones

comerciales que considere insatisfactorias para los intereses a largo plazo del país, pero la volatilidad de la industria petrolera y el naciente estado de la diversificación de la economía ecuatoriana, así como la probabilidad de que cualquiera de los escenarios que desempeñan un papel en el mundo real continua siendo incierto.

Lo que se puede concluir razonablemente es que Ecuador no puede ser otra Venezuela, pero se perfila para ser un competidor regional que pueda ejercer una influencia muy por encima de su peso económico debido a su estratégica ubicación geográfica y el potencial para perfilarse como plataforma de las organizaciones terroristas en el mismo territorio Americano.

En el análisis final, un Tratado de Libre Comercio de las Américas, que incluye a todos los países de la región, entre ellos Ecuador, no podría llevarse a cabo de ninguna manera mientras el presidente Correa este en el poder, y continuara en la presidencia por otros 5 años.

Esto quiere decir que a pesar de la estabilidad política, la prevista permanencia en el poder ofrece particularmente en vista del turbulento pasado político de Ecuador, el primer escenario y dos resultados que afectarán positivamente a unos interesados más que otros.

Con los ciudadanos del país igualmente divididos sobre el deseo de las disposiciones del Tratado de Libre Comercio de las Américas, es razonable llegar a la conclusión de que no importa cual escenario se siga.

Correa complacerá a algunos constituyentes mientras que defraudara a otros. La pregunta clave entonces será cuál de estos grupos de interés tiene más en juego y la mejor manera de balancear sus necesidades con los otros sectores de la economía ecuatoriana.

Históricamente hablando esta sobreentendido que las grandes corporaciones petroleras multinacionales prevalecen, pero la línea en la arena geopolítica establecida por Correa puede ayudar a excluir esta eventualidad y promover prácticas comerciales que beneficien a los sectores históricamente marginados.

Las elecciones en Estados Unidos fijadas para noviembre del 2012, reeligieron a Obama y Biden con el 51% de los votos. El pueblo ecuatoriano decidió el resultado político del país en las urnas el 17 de febrero del 2013. Como resultado, Correa fue reelegido debido a que el país ha experimentado un gran progreso bajo su liderazgo. A pesar de que se opone a tal acuerdo con Estados Unidos, la mayoría de la población apoya su agenda política.

El libro se centró en la perspectiva agroeconómica para determinar la implicación de un acuerdo económico de tal magnitud, las ventajas, desventajas, y la mejor decisión para la comunidad ecuatoriana y de los EE.UU.

El Tratado de Libre Comercio entre Ecuador y los EE.UU., no beneficiará en última instancia a la economía ecuatoriana a corto plazo debido a que la industria no está muy bien desarrollada todavía. Eventualmente tomara por lo menos 20 años para conseguir la infraestructura adecuada para proyectar a Ecuador en esa dirección.

Los elementos de la mezcla del mercado juegan un papel importante en el desarrollo de las conversaciones del Tratado de Libre Comercio para un acuerdo entre los principales involucrados a nivel económico.

El precio es el principal determinante de si un país importa o exporta un producto. Precio mundial es el valor que prevalece en los mercados mundiales y es el valor acordado por medio del cual podemos vender o comprar productos.

Precio domestico es establecido a nivel local sin comercio. Haciendo caso omiso de los costos de transporte, si el precio mundial es mayor que el precio domestico antes que el comercio, por lo tanto, el producto está listo para la exportación. Si el precio mundial es menor que el precio domestico antes que el comercio, entonces importaríamos dicho producto.

El comercio permite a cualquier persona comprar productos a un precio más barato que en los mercados internacionales y vender aquellos a un precio más alto que si estuviéramos restringidos al precio doméstico. Los precios internos caen por los bienes que importamos y los

consumidores se benefician mientras que los productores se ven perjudicados como se demuestra por una reducción en las compras.

Los aranceles, por definición, son impuestos a las empresas que tienen bienes importados. Los aranceles tradicionalmente aumentan el precio domestico por encima del precio mundial por el valor de la tarifa. El aumento en el precio domestico dará lugar a una disminución de la cantidad nacional en demanda, y un aumento en la cantidad doméstica proveída.

Las cuotas son restricciones de la máxima cantidad de bienes que se pueden importar, y tienen un efecto similar al de los aranceles. Estas cuotas restringen la cantidad disponible a los consumidores domésticos y tienden a aumentar el precio dando como resultado un efecto negativo similar al de un arancel.

La diferencia principal es la distribución de los excedentes. Un arancel aumenta los ingresos del gobierno, mientras que la cuota de importación crea excedente.

El Tratado de Libre Comercio de Norteamérica ha incrementado el comercio y la inversión entre Canadá, México y Estados Unidos por la eliminación gradual de los aranceles, cuotas de importación y las barreras a la propiedad extranjera.

La economía de Canadá depende mucho más del comercio internacional de los Estados Unidos. Los canadienses pagan más impuestos y son menos remunerados que sus vecinos estadounidenses. Corporaciones canadienses han invertido fuertemente en el extranjero, especialmente en Estados Unidos. Canadian Pacific cuenta con casi un tercio de sus vías de ferrocarril en los Estados Unidos y es uno de los mayores propietarios de hoteles de lujo al sur de la frontera.

Antes de NAFTA ya había el sector de las maquiladoras. Establecidas por el gobierno, las maquiladoras son fábricas extranjeras las cuales pueden importar partes y materiales a México libres de impuestos, siempre y cuando se utilicen para fabricar productos para la exportación.

México es el segundo socio comercial de Estados Unidos, como resultado del comercio bilateral entre los países.

Se pronosticaba que México sería la décima economía más grande en el 2015.

Los Estados Unidos continúan siendo la mayor economía del mundo, con un PIB de $15.7 trillones de dólares y el más grande para la gestión de la inversión extranjera. Los Estados Unidos también es líder en el campo del conocimiento, servicios financieros, y tecnología informática, estas posiciones serán importantes en los próximos años.

Hoy en día, las multinacionales estadounidenses operan en todo el mundo, muchas disfrutando de posiciones dominantes en sus respectivas industrias, seguidas por China, Alemania, Corea del Sur y Japón, con un PIB de más de 1 trillón de dólares correspondiente.

Las diferentes regiones se especializan en aquellos productos que se pudieran producir de manera más eficiente, el exceso de producción podría ser enviado económicamente a otras áreas que producen y consumen, y otros bienes que no se produjeron localmente fueron importados.

Este concepto de comercio, cuando se aplica a los mercados mundiales, ayuda a explicar el alto nivel de comercio internacional que se ejecuta en la actualidad. Sistemas logísticos eficientes permiten a las empresas mundiales aprovechar el hecho de que las tierras, y las personas que viven allí no son igualmente productivas, y la logística es la esencia misma del comercio. Esto contribuye a un nivel económico de vida más alto.

Cuando una organización logística opera a nivel mundial, la mejor empresa es generalmente una en la que las funciones de planificación y control están centralizadas y las funciones de la operación están descentralizadas.

Los diferentes sistemas tienen que estar en su lugar para la introducción rápida del producto, las necesidades del mercado concentradas, entrega inmediata, diversos servicios, canales innovadores, capital, inventario, crédito, inversión de trabajo en la construcción y equipo.

Los ecuatorianos son gente amable, quienes acogen a los extranjeros de todo el mundo, porque a ellos les encanta aprender su lengua y cultura, y aún mejor si hay inversión de

capital para darle vida a la economía local.

Hay aproximadamente 10 mil expatriados estadounidenses que viven en Ecuador y alrededor de 700 mil ecuatorianos que residen en los Estados Unidos. Por lo tanto, un acuerdo de libre comercio beneficiaría a ambos países.

Los EE.UU. es una potencia tecnológica que exporta automóviles, maquinaria y otros productos manufacturados a Ecuador. Sin embargo, China es la nueva fábrica a nivel mundial, y la India es la agencia de servicio al cliente para el mundo.

Los ecuatorianos tienen un cariño especial por el producto "Hecho en EE.UU", y están dispuestos a pagar cualquier cosa para conseguir dichos artículos debido a la buena calidad.

Por otro lado, Ecuador se beneficia de sus exportaciones libres de impuestos, tales como, el petróleo crudo, camarón, productos agrícolas y materias primas que van a los EE.UU., y da la bienvenida a la inversión estadounidense liderada por los sectores manufactureros y de venta al por mayor. El Tratado de Libre Comercio entre Ecuador y Estados Unidos le daría al ciudadano ecuatoriano común una mejor oportunidad para comprar, digamos, un vehículo que se vende en Detroit por $20,000 en lugar de $40,000 en el mercado ecuatoriano. Si el precio de la entrada de automóviles fuera lo mismo que en Detroit, los ecuatorianos aumentarían la demanda de más vehículos, que a la vez beneficiarían ambas economías.

En Ecuador se paga el Impuesto a Consumos Especiales (ICE) que varía de acuerdo al costo del vehículo y puede llegar hasta el 35%. Tambien, hay que pagar gastos de nacionalización, el 12% del Impuesto al Valor Agregado (IVA) y el 40% del *ad valórem.*

Los denunciantes, sobre la intromisión del gobierno de Estados Unidos en los asuntos privados de sus ciudadanos, casos Assange (fundador australiano de WikiLeaks) y Snowden (analista de la Agencia de Seguridad Nacional de los Estados Unidos), ambos en busca de asilo político con el gobierno ecuatoriano y ruso, son acontecimientos recientes que han demorado y comprometido el acuerdo comercial con

Estados Unidos.

Una vuelta rápida de la historia que espero no se repita durante este proceso de negociaciones. Aunque el movimiento de independencia en América del Sur logró su objetivo primordial, ninguno de los líderes disfrutó de los beneficios de la victoria.

La invasión francesa de España dio a las colonias americanas la oportunidad de rebelarse. El golpe final para España tuvo lugar en la batalla de Ayacucho, cuando Antonio José de Sucre lideraba el ejército patriota que derrotó al ejército realista.

Cuando los Libertadores, Bolívar y San Martín, se reunieron en Guayaquil, el histórico encuentro no fue un éxito, y posteriormente, San Martín se retiró de todos sus cargos. Bolívar, luego regresó a mantener unida a su creación de la Gran Colombia, pero resultó ser una tarea imposible cuando su salud comenzó a fallar.

La Gran Colombia se vino abajo con la secesión de Ecuador y Colombia debido a las diferencias políticas. Tanto Bolívar como San Martín murieron en el exilio. Es triste darse cuenta que los padres de la libertad sudamericana terminaron en la desolación después de sacrificar todo en el nombre de una nación poderosa.

El sueño de Bolívar fue de unir a toda la América española bajo una sola bandera para estar al mismo nivel de las potencias extranjeras. Sin embargo, su sueño de panamericanismo se desvaneció con él. En 1830, en su lecho de muerte, Bolívar escribió: "América es ingobernable. Los que han servido a la revolución han arado en el mar".

El ex presidente de Venezuela, Hugo Chávez falleció en marzo del 2013, después de 14 años en el poder. Nicolás Maduro, asumió el cargo presidencial, y el país se encuentra actualmente en el caos político y económico, así como, una revuelta feroz por la oposición.

El presidente Rafael Correa visitó los Estados Unidos para dar discursos en Harvard, la Universidad de Yale y el Instituto de Tecnología de Massachusetts durante la segunda semana de abril del 2014. Correa habló de su agenda política y

Cóndor Fabricado en Ecuador

Fuente: Economía del Ecuador, 2012

sus logros durante su mandato como presidente. También aseguró el programa de intercambio de estudiantes.

Entre 2007 y 2013, manifestó Correa, 1.13 millones de ecuatorianos salieron de la pobreza, y la incidencia de la pobreza extrema se redujo del 16.9 por ciento al 8.6 por ciento, mientras que el crecimiento promedio de un 4.2 por ciento al año. El país también ostenta la tasa de desempleo más baja en la región, un 4.1 por ciento, y de acuerdo al informe de desarrollo de la ONU es uno de los tres países con mayor movilidad ascendente (Harvard Gazette, 2014).

Un terremoto de magnitud 7.8 dejó más de 600 muertos en la zona norte de la costa ecuatoriana el 16 de abril de 2016. Las próximas eleciones están programadas para el 19 de febrero de 2017.

Ahora bien, es sólo cuestión de ambos gobiernos para discutir los términos del Tratado de Libre Comercio que ha estado en la mesa de negociación durante tanto tiempo. Las negociaciones están actualmente suspendidas debido a las diferencias políticas diseñadas para beneficiar a los pocos que controlan la economía nacional.

MAPAS

Mapa 1.1: República del Ecuador

Fuente: Economía del Ecuador, 2012

Mapa 1.2: Estados Unidos de América

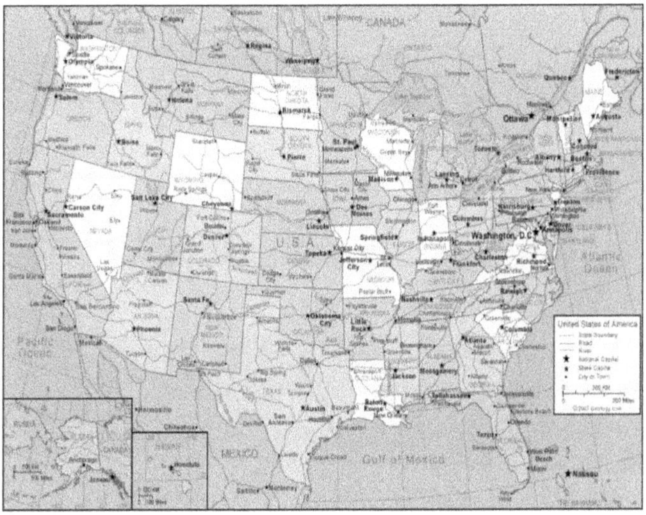

Fuente: Geology.com, 2014

BIBLIOGRAFÍA SELECTA

Ackerman, I., Nadal, F., Benetti, A., Gallagher, C., Salas, C. (2004). *The flawed foundations of general equilibrium theory: Critical essays on economic theory*. New York: Routledge.

Baker, J. C. (2003). *Financing international trade.* Westport, CT: Praeger.

Bush pushes Free Trade Area. (2002). *The New American*, 18(3), 1.

Chauvin, L. O. (2004). Farm tariffs, intellectual property rights seen as main obstacles to U.S.-Andean FTA. *International Trade Daily, Bureau of National Affairs.*

Ecuador Economy. (2012). *U.S. government: CIA world factbook.* https://www.cia.gov/library/publications.

Ecuador: Militant opposition to Andean Free Trade Agreement blockades several regions, moves to capital. (2006). *NotiSur.*

Faber, E. M. (2007). Pregnancy discrimination in Latin America: The exclusion of "employment discrimination" from the definition of "Labor Laws" in the Central American Free Trade Agreement. *Columbia Journal of Gender and Law*, 16(1), 37-44

Fandl, K. J. (2007). Bilateral agreements and fair trade practices: A policy analysis of the Colombia-U.S. Free Trade Agreement (2006). *Yale Human Rights and Development Law Journal*, 10, 111-115.

Gillgannon, M. (2004). Globalization proceeds without practical ethical guides. *National Catholic Reporter*, 41(8), 19-21.

Goodman, A., Gonzalez, J. (2008). *Democracy Now. Ecuadorian President Rafael Correa on the Lawsuit Against Chevron, Eradicating Foreign Debt and Why He Says "Ecuador is No Longer for Sale."* ww.democracy.org.

Goodwin, N. (2006). *Global Development and Environmental Institutions.* http://www.eoearth.org/article/Global_economy.

Hirst, M. (2004). *The United States and Brazil: A long road of unmet expectations.* New York: Routledge.

Hummels, D. (2001). "Toward a Geography of Trade Costs", *GTAP Working Papers* 1162, Center for Global Trade Analysis, Purdue University.

Jensen, K. B. (2004). *A handbook of media and communication research: Qualitative and quantitative methodologies.* London: Routledge

Kirchhoff, B. A. (1999). *Entrepreneurship and dynamic capitalism: The economics of business firm formation and growth.* Westport, CT: Praeger Publishers.

Kornis, M. (2004). The expanded Andean Trade Preferences Act and a U.S. free trade agreement. *International Economic Review*, 19-22.

Maniam., B. (2005). The Global Perspective of Free Trade Area of the Americas. *The Business Review, Cambridge, 21*(16), 12-23.

Matthews, D. A. & Maniam, B. (2006). Reassessing the case of Ecuador's Dollarization. *Journal of Economics and Economic*

Research, 7(3), 22-26.

Neuman, W. L. (2003). *Social research methods: Qualitative and quantitative approaches, 5th ed.* New York: Allyn & Bacon.

O'Neill, R. (2014). *HARVARD Gazette.* http://news.harvard.edu/gazette.

Palast, G. (2008). *Democracy Now.* www.democracy.org.

Relyea, C. W., Liliana, S., Fish, K. E. & Shankar, G. S. (2011). Laissez faire and international trade: A critique of the proposed United States - Colombia Free Trade Agreement. *Journal of International Business Research*, 10(1), 37-39.

Rolleigh, M. (2003), "Plant Heterogeneity and Applied General Equilibrium Models of Trade: Lessons from the Canada-US Free Trade Agreement", Mimeo, University of Minnesota.

Schott, J. (2004). *Assessing US FTA Policy. Institute for International Economics.* Retrieved from: www.iie.com.

Seelke, C.R. (2008). *CRS Report for Congress. Ecuador: Political and Economic Situation and U.S. Relations.* Retrieved from: www.fas.org/sgp/crs/RS21687.

USDA at work in Ecuador. (2011). *U.S. Department of Agriculture.* www.fas.usda.gov/icd/tim/FINAL.

Wikipedia Turismo. (2015). www.wikipedia

ACERCA DEL AUTOR

"SOLDADO BALANCEA CARRERA, EDUCACIÓN PARA TRIUNFAR"

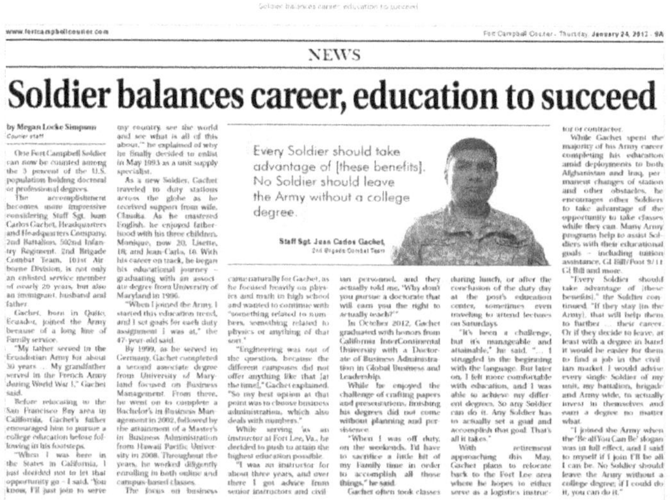

Soldier balances career, education to succeed

www.fortcampbellcourier.com — Fort Campbell Courier · Thursday, January 24, [illegible] · 9A

NEWS

Soldier balances career, education to succeed

by Megan Locke Simpson
Courier staff

One Fort Campbell Soldier can now be counted among the 3 percent of the U.S. population holding doctoral or professional degrees.

The accomplishment becomes more impressive considering Staff Sgt. Juan Carlos Gachet, Headquarters and Headquarters Company, 2nd Battalion, 502nd Infantry Regiment, 2nd Brigade Combat Team, 101st Airborne Division, is not only an enlisted service member of nearly 20 years, but also an immigrant, husband and father.

Gachet, born in Quito, Ecuador, joined the Army because of a long line of Family service.

"My father served in the Ecuadorian Army for about 30 years … My grandfather served in the French Army during World War I," Gachet said.

Before relocating to the San Francisco Bay area in California, Gachet's father encouraged him to pursue a college education before following in his footsteps.

"When I was here in the States in California, I just decided not to let that opportunity go – I said, 'You know, I'll just join to serve my country, see the world and see what is all of this about,'" he explained of why he finally decided to enlist in May 1993 as a unit supply specialist.

As a new Soldier, Gachet traveled to duty stations across the globe as he received support from wife, Claudia. As he mastered English, he enjoyed fatherhood with his three children, Monique, now 20, Lisette, 18, and Juan-Carlos, 10. With his career on track, he began his educational journey – graduating with an associate degree from University of Maryland in 1996.

"When I joined the Army, I started this education trend, and I set goals for each duty assignment I was at," the 47-year-old said.

By 1999, as he served in Germany, Gachet completed a second associate degree from University of Maryland focused on Business Management. From there, he went on to complete a Bachelor's in Business Management in 2002, followed by the attainment of a Master's in Business Administration from Hawaii Pacific University in 2008. Throughout the years, he worked diligently enrolling in both online and campus-based classes.

The focus on business came naturally for Gachet, as he focused heavily on physics and math in high school and wanted to continue with "something related to numbers, something related to physics or anything of that sort."

"Engineering was out of the question, because the different campuses did not offer anything like that [at the time]," Gachet explained. "So my best option at that point was to choose business administration, which also deals with numbers."

While serving as an instructor at Fort Lee, Va., he decided to push to attain the highest education possible.

"I was an instructor for about three years, and over there I got advice from senior instructors and civilian personnel, and they actually told me, 'Why don't you pursue a doctorate that will earn you the right to actually teach?'"

In October 2012, Gachet graduated with honors from California InterContinental University with a Doctorate of Business Administration in Global Business and Leadership.

While he enjoyed the challenge of crafting papers and presentations, finishing his degrees did not come without planning and persistence.

"When I was off duty, on the weekends, I'd have to sacrifice a little bit of my Family time in order to accomplish all those things," he said.

Gachet often took classes during lunch, or after the conclusion of the duty day at the post's education center, sometimes even traveling to attend lectures on Saturdays.

"It's been a challenge, but it's manageable and attainable," he said. "… I struggled in the beginning with the language. But later on, I felt more comfortable with education, and I was able to achieve my different degrees. So any Soldier can do it. Any Soldier has to actually set a goal and accomplish that goal. That's all it takes."

With retirement approaching this May, Gachet plans to relocate back to the Fort Lee area where he hopes to either serve as a logistics instructor or contractor.

While Gachet spent the majority of his Army career completing his education amid deployments to both Afghanistan and Iraq, permanent changes of station and other obstacles, he encourages other Soldiers to take advantage of the opportunity to take classes while they can. Many Army programs help to assist Soldiers with their educational goals – including tuition assistance, GI Bill/Post 9/11 GI Bill and more.

"Every Soldier should take advantage of [these benefits]," the Soldier continued. "If they stay [in the Army], that will help them to further … their career. Or if they decide to leave, at least with a degree in hand it would be easier for them to find a job in the civilian market. I would advise every single Soldier of my unit, my battalion, brigade and Army wide, to actually invest in themselves and earn a degree no matter what.

"I joined the Army when the 'Be all You Can Be' slogan was in full effect, and I said to myself if I join I'll be all I can be. No Soldier should leave the Army without a college degree; if I could do it, you can do it."

Every Soldier should take advantage of [these benefits]. No Soldier should leave the Army without a college degree.

Staff Sgt. Juan Carlos Gachet,
2nd Brigade Combat Team

Fuente: Fort Campbell Courier, 2013

Uno de los soldados de Fort Campbell ya cuenta como parte del 3 porciento de la población estadounidense que posee doctorados o títulos profesionales.

Este logro es aún más impresionante considerando que el Sgto. Juan Carlos Gachet, de la Compañía del Cuartel General, 2do Batallón, 502vo Regimiento de Infantería, 2da Brigada de Combate, 101va División Aérea, no es solamente

un miembro enlistado del servicio militar
de casi 20 años, pero también un inmigrante, esposo y padre.

Gachet, oriundo de Quito-Ecuador, se enlistó en las filas del Ejército por tradición militar de su familia.
"Mi padre sirvió en el Ejército Ecuatoriano por casi 30 años...Mi abuelo sirvió en el Ejército Francés durante la Primera Guerra Mundial". Dijo Gachet.

Antes de establecerse en el área de la bahía de San Francisco en California, el padre de Gachet le motivó a que continúe con su educación universitaria antes de seguir sus pasos.

"Cuando estaba ya en Los Estados en California, decidí no desperdiciar esa oportunidad, me dije, me enlistaré para servir mi país, ver el mundo y saber de qué se trata", explico la razón por la cual finalmente decidió enlistarse en mayo de 1993 como especialista de logística.

Como nuevo integrante, Gachet viajó a bases militares en diversas partes del mundo con el apoyo de su esposa Claudia. Aprendiendo y dominando el idioma inglés, disfrutó como padre con sus tres hijos, Monique de 20; Lisette de 18; y Jean-Carlo de 16 en la actualidad. Con su carrera establecida, empezó a recorrer el sendero educativo graduándose con un título asociado de la Universidad de Maryland en 1996.
"Cuando me enlisté en las filas del Ejército, empecé con este afán educativo y establecí objetivos para cada puesto que pase", dijo el hombre de 47 años de edad.

Por el año de 1999, durante su paso por Alemania, Gachet completó su segundo título asociado de la Universidad de Maryland enfocado en Administración de Empresas. Luego, adquirió la Licenciatura en Administración de Empresas en el 2002, seguido por el Magíster en Administración de Empresas con la Universidad de Hawaii Pacific en el 2008. A través de los años, trabajó diligentemente matriculándose en clases virtuales y tradicionales.

Administración de Empresas fue algo natural para Gachet, ya que su concentración principal era en física y matemáticas durante la vida colegial y decidió continuar con "algo relacionado con los números, física o algo al respecto".

"La ingeniería quedó descartada, ya que los diferentes centros educacionales no ofrecían algo similar en aquella época", lo manifestó Gachet. "Por lo tanto mi mejor opción fue escoger administración de empresas, la cual también tiene que ver con números".

Durante su trabajo como instructor en Fort Lee, Va., tuvo que ver con el empuje de obtener lo máximo posible en educación.

"Fui instructor por casi tres años, y allí otros instructores más experimentados y personal civil me habían manifestado 'por qué no continuas con el doctorado que te daría pleno derecho a enseñar'".

En Octubre del 2012, Gachet se graduó con honores en la Universidad California InterContinental con un Doctorado en Administración de Empresas en Comercio Internacional y Liderazgo.

Al disfrutar el desafío terminando tareas y preparando presentaciones, la culminación de sus títulos no llegó sin planificación y persistencia.

"Cuando estaba libre durante los fines de semana, tenía que sacrificar un poco de tiempo con la familia para lograr todas estas cosas", manifestó.

Gachet a menudo tomó clases durante la hora de almuerzo, o después de la conclusión del día de trabajo en el centro educativo de la base, a veces inclusive viajando a otros lugares para asistir a lecturas los sábados.

"A sido un desafío, pero posible de obtenerlo", lo manifestó. Al principio tuve estragos con el idioma, pero luego, me sentí más tranquilo con la educación, y pude lograr mis diferentes títulos. Por lo tanto cualquier soldado lo puede lograr. El soldado tiene que en realidad ponerse un objetivo y lograr dicho objetivo. Eso es todo.

Cuando Gachet paso la mayor parte de su carrera militar completando su educación en medio de movilizaciones tanto a Irak como Afganistán, cambios permanentes de bases y otros obstáculos, el motiva a otros soldados a que aprovechen de las oportunidades de tomar clases cuando puedan. Muchos programas del Ejército ayudan a los soldados con sus

objetivos educativos - incluyendo ayuda financiera por los diferentes cursos, GI Bill, Post 9/11 GI Bill y más.

"Todo soldado tiene que aprovechar de estos beneficios," continuo el militar". Si ellos permanecen en el Ejército, eso les ayudara para avanzar...en su carrera. O si deciden salir, por lo menos con un título en mano será más fácil encontrar un empleo en el mercado laboral civil. Aconsejaría a cada soldado de mi unidad, de mi batallón, brigada y todo el Ejército, a invertir en ellos y obtener un título sin importar lo que sea".

"Me enliste en las filas del Ejército cuando el eslogan de 'Se Todo lo que Puedes Ser' estaba en pleno auge, y me dije si me enlisto seré todo lo que pueda ser. Ningún soldado tiene que salir del Ejército sin un título universitario; si yo lo pude, tú lo puedes" (Megan Locke Simpson, 2013).

Tradición Militar

Juan Carlos Gachet Castro, oriundo de Quito, Ecuador, se enlistó en el Ejército debido a una gran tradición familiar de servicio y aventura militar. Su padre, Juan Edmundo Gachet Valencia, sirvió en el Ejército Ecuatoriano por casi 30 años.

Durante la Primera Guerra Mundial, Francia enlistó en sus filas a todos los primogénitos de los ciudadanos franceses y por lo tanto Augusto Gachet viajó a Europa y fue parte de esta conflagración mundial. En aquel entonces Philippe Pétain estaba al mando de su unidad militar.

Andrés Augusto Gachet Baca

Veterano de la Primera Guerra Mundial. 1914-1919.
"Distinguido ciudadano nacido en Ecuador, registrado en la embajada francesa, cumplió su servicio militar en Francia entregando su contingente para la libertad del mundo, casado felizmente con Laura M. Valencia, distinguida mujer quiteña,

tuvo 5 distinguidos hijos y 5 bellas, hermosas y espirituales damitas, muchos nietos y bisnietos y miles de recuerdos nobles e imborrables. Tristeza y nostalgia por no tenerle conmigo, su nombre: Andrés Augusto Gachet Baca" (Gabriel Gachet).

Juan Edmundo Gachet Valencia

Servicio Activo: 1956-1981.
Padre querido y ciudadano ilustre nacido en Quito - Ecuador, sirvió en la Caballería, Infantería Mecanizada como tanquista, en el Batallón de Logística y Cuerpo de Ingenieros en Transportes como conductor. Felizmente casado con Bertha Castro de Gachet, tienen tres hijos: Juan Carlos, Augusto, Roberto, y siete nietos. Su cuota de la libertad del país es muy apreciada.

Juan Carlos Gachet Castro

Servicio Activo: 1993-2013.
Originario de Quito, Ecuador. Completó el entrenamiento básico en Fort Jackson, Carolina del Sur, y el entrenamiento individual avanzado en Fort Lee, Virginia. Después de su graduación, se le asignó la Especialidad Ocupacional Militar 92Y (Logística). Ejerció sus funciones en Italia, Alemania, Texas, Hawái, Virginia y Kentucky. Participó en dos frentes de combate, una vez en Irak y Afganistán en dos ocasiones. Continuó su servicio civil en el Ministerio de Veteranos de Guerra en la sección Fiscal y actualmente es contratista en la base militar Lee.

Residente de Chester Sirve en Afganistán

KABUL, Afganistán (21 de febrero, 2011) -. Un residente de Chester, Virginia está sirviendo actualmente como el sargento de abastecimientos para el Elemento Nacional de Apoyo a los Estados Unidos localizado en el aeropuerto internacional de Kabul.

El Sargento. Juan Gachet del Ejército de EE.UU. es responsable de más de $500,000 dólares en equipo y materiales, con apoyo directo a los miembros del servicio que entran en Afganistán a través de Kabul. Además de sus tareas habituales, también ayuda en el campo administrativo de las tropas que entran y salen, y brinda asesoramiento a los soldados que son desplegados por primera vez a la región.

"Es mi deber servir a mi país", dijo Gachet. "Estoy aquí para apoyar a las tropas estadounidenses estacionadas en KAIA, que salen del perímetro de la base a diario para brindar la seguridad a la población civil".

Cuando se le preguntó acerca de por qué se unió al Ejército de Estados Unidos, dijo que era "tradición familiar. "Corre por mis venas", agregó Gachet. Sólo quería mantener la tradición familiar, la cual empezó con mi abuelo durante la Primera Guerra Mundial".

Gachet cree que necesita ayudar a los afganos en la ardua terea de la reconstrucción de su país. "Mi motivación para proporcionar un ambiente seguro para Afganistán proviene de mi familia", explicó Gachet. "Es difícil estar lejos pero con su apoyo y saber qué puedo hacer esa diferencia, soy capaz de seguir y comprometerme con la misión al 100 por ciento" (Relato de Peter Lee, Marina de EE.UU., Afganistán).

Kabul, Afganistán – Sgto. del Ejército estadounidense, Juan Gachet (derecha) participa en una media maratón en el Aeropuerto Internacional de Kabul durante un evento de recreación moral.

Kabul, Afganistán – Con más de 50 libras en equipo militar, el Sgto. Juan Gachet (derecha) marcha 22 km".

Instituto Nacional Mejía

La trayectoria por el Instituto Nacional Mejía de Quito, Ecuador. Su lema "*per aspera ad astra*" latín que significa "por el camino áspero a las estrellas" establece un hito importante y la inspiración por mis esfuerzos académicos que me llevaron al mundo de la educación postsecundaria.

Apuesto a que todo el mundo extraña la vida colegial hasta cierto punto. Seis años, los mejores de mi vida, los pasé en el Mejía, donde la transición de niñez a juventud tuvo lugar. Tres años dedicados a la secundaria, al aprendizaje de las materias tradicionales, y los tres últimos aprendiendo Matemáticas y Física como materias principales.

Mejía es el centro de enseñanza a nivel secundario más grande. Varios líderes políticos y celebridades ecuatorianas experimentaron y compartieron sus actividades de aprendizaje.

Instituto Nacional Mejía. 21 de Agosto, 2015

Juan Carlos Gachet

Hace 32 años nos graduamos, no hubo incorporación ya que solo tres salimos en las listas oficiales, pero nos graduamos aunque sea un par de meses después. Fuimos partícipes de las protestas en contra de la injusticia social de aquel tiempo por lo tanto el gobierno suspendió las clases indefinidamente y este acontecimiento fue uno de los ingredientes para que no se realice la ceremonia oficial de incorporación, pero el último día de clases de 1983 hicimos el juramento al pie del monumento a José Mejía Lequerica, nuestro patrono, de reunirnos en el futuro después de unas décadas.

Aquel juramento se hizo realidad en diciembre del 2014, 31 años después al encontrarnos en las gradas del Patrón Mejía y de allí nace el sentimiento del acto simbólico de incorporación que llegó a concretarse el 21 de agosto del 2015.

California Intercontinental University, Clase del 2012

Sólo el tres por ciento de los estadounidenses han logrado obtener títulos de doctorado o profesionales, lo que convierte a un soldado de Fort Campbell, el sargento. Juan Carlos Gachet en élite estadounidense. Gachet, un soldado de 20 años, es también un esposo, padre e inmigrante. Los logros del hombre de 47 años de edad no son sólo honorables; son prueba de que la educación está al alcance de todo el personal militar.

Gachet comenzó con su educación superior en la década de 1990, y se graduó de la Universidad de Maryland con un grado asociado. Estas metas tuvieron un efecto dominó rápido, y así Gachet completó un segundo grado asociado en 1999, esta vez en Administración de Empresas, y continuó su educación en dicho campo. Recibió su Licenciatura en Administración de Empresas en el 2002 y una Maestría en Administración de Empresas en el 2008.

Él atribuye su éxito a su propia determinación y al apoyo de su familia, así como también la asistencia económica proporcionada por el Ejército. Los sentimientos de Gachet son exactamente la razón por la que CalUniversity apoya a las tropas y motiva al personal militar a utilizar sus beneficios académicos del VA y *GI Bill* en nuestra universidad (Angelo Lioudakis, 2014).

Asistentes a la Ceremonia de Graduación

Dr. Finian Tan, President
Rax Ribadia, Director Finance
Dr. Troy Roland, Chief Academic Officer
Dr. Fathiah Inserto, Dean of Academics
Randy Saban, Director of Admissions
Chris Ho, Manager – Student Affairs
Verna Zak, Staff Accountant
Andrea Clarke, Manager – Student Services
Thai Le, IT & Designs Project
Eupeterson Lewis, Military Education Liaison
Terry Ruiz, Senior Admissions Advisor
Marc Mu, Senior Admissions Advisor
Michael Winn, Admissions Advisor
Ray Soriano, Admissions Advisor
Dr. Jennie Wong, EdD Instructor
Dr. Kamalnayan Inamdar, EdD Instructor
Dr. Kenneth Phillips, EdD Instructor
Benjamin Perez, MSc Instructor

Graduados

Dr. Scott Vowels, DBA-EBM
Dr. Juan Gachet, DBA-GBL
Dr. Kim Erwin, DBA-GBL
Dr. Marshall Cowell, DBA-GBL
Dr. Augustus Ynares, DBA-HCML
Dr. Blanche Shenosky, DBA-HCML
Dr. Julie Ann Carson, DBA-HCML
Dr. Monserrate Jurgensen, DBA-HCML
Dr. Sulma Gandhi, DBA-HCML
Dr. Tracy McClelland, DBA-HCML
Dr. Andrea Clarke, DBA-HCML
Dr. Hartinder Singh Johal, DBA-ISERM

Dr. Thomson Lukose, DBA-ISERM
Dr. Wajdi Aghnatios, DBA-ISERM
Sr. Fabio Sotelo, BBA-IME

Sábado, 6 de octubre, 2012

Estimado Juan:

"Permítame felicitarlo por su excelente presentación de la mañana del viernes pasado. Estuvo muy informativa e interesante. Usted aprobó la Defensa Oral de manera ejemplar".

Muchos saludos.

Dr. Dennis Toutant
EdD Instructor

"Fue un placer conocerlo y ver a su familia en la ceremonia de investidura. Como lo había mencionado, ¡¡ me fascino su discurso!! Como lo manifestó, los padres tenemos que guiar a nuestros hijos pero nosotros debemos dar el ejemplo. Eso fue maravilloso.

Nuevamente, ¡¡felicitaciones por graduarse con nosotros en CalU, Dr. Juan Gachet!!"

Gracias.

Maritza Sandoval
Asesora de Servicios Estudiantiles

Familia

Padres, Juan E. Gachet y Bertha; hermanos, Augusto y Roberto Gachet. **E**sposa, Claudia Gachet; hijas, Monique y Lisette; hijo, Jean Carlo.

Mis padres Juan E. Gachet y Bertha Castro han dedicado con mucho cariño gran parte de su vida a sus hijos; Juan Carlos, Andrés Augusto y Luis Roberto, quienes a su vez correspondieron a dicho sacrificio en su formación académica.

Augusto es Profesor de Cultura Física y Roberto es Economista. Nuestros padres en esencia han sido una inspiración a lo largo de todos estos años.

Mi esposa, Claudia Gachet, Profesora de Educación Preescolar, ha dado todo por nuestros hijos, Monique, Lisette, y Jean Carlo.

Ellos han respondido a dicho sacrificio y Monique se

graduó de Licenciada de Ciencias en Contabilidad, Lisette logró el título de Licenciada en Ciencias de Informática en Sistemas, y Jean Carlo en mayo del 2016 alcanzó su primer título universitario en Ciencias de Administración de Empresas.

Próxima Publicación

Anécdotas del Glorioso Instituto Nacional Mejía

CAPÍTULO 1

INSTITUTO NACIONAL MEJÍA

Instituto Nacional Mejía, 1983

La trayectoria por el Instituto Nacional Mejía de Quito, Ecuador. Su lema "*per aspera ad astra*" latín que significa "por el camino áspero a las estrellas" establece un hito importante y la inspiración por mis esfuerzos académicos que me llevaron al mundo de la educación postsecundaria.

Apuesto a que todo el mundo extraña la vida colegial

hasta cierto punto. Seis años, los mejores de mi vida, los pasé en el Mejía, donde la transición de niñez a juventud tuvo lugar. Tres años dedicados a la secundaria, al aprendizaje de las materias tradicionales, y los tres últimos aprendiendo Matemáticas y Física como materias principales.

Mejía es el centro de enseñanza a nivel secundario más grande. Varios líderes políticos y celebridades ecuatorianas experimentaron y compartieron sus actividades de aprendizaje.

Juan Carlos Gachet

Hace 32 años nos graduamos, no hubo incorporación ya que solo tres salimos en las listas oficiales, pero nos graduamos aunque sea un par de meses después. Fuimos partícipes de las protestas en contra de la injusticia social de aquel tiempo por lo tanto el gobierno suspendió las clases indefinidamente y este acontecimiento fue uno de los ingredientes para que no se realice la ceremonia oficial de incorporación, pero el último día de clases de 1983 hicimos el juramento al pie del monumento a José Mejía Lequerica, nuestro patrono, de reunirnos en el futuro después de unas décadas.

Aquel juramento se hizo realidad en diciembre del 2014, 31 años después al encontrarnos en las gradas del Patrón Mejía y de allí nace el sentimiento del acto simbólico de incorporación que llegó a concretarse el 21 de agosto del 2015.

Quito, 21 de agosto, 2015.
Discurso:

Sr. Rector del Instituto Nacional Mejía, autoridades, compañeros, invitados y familias.
Ha llegado el gran día para cristalizar este evento de incorporación simbólica que tanto habíamos esperado desde diciembre del 2014.
Gracias por su participación en las diferentes reuniones.

Instituto Nacional Mejía. 21 de agosto, 2015

50 fuimos los integrantes del 6to curso 6ta sección y otros compañeros que egresaron de otras secciones de la promoción 82-83.

Hoy día somos 14, pero el resto de nuestros compañeros están presentes de corazón como es el caso de Oscar Albán, Raúl Ibarra, Freddy Arias, Muñoz, Álvaro Villavicencio y López quienes hicieron todo lo posible por venir, pero las circunstancias de la vida no lo permitieron por la obligación laboral, o por su residencia en el exterior tanto en Estados Unidos, España e Italia y pido un minuto de silencio por los compañeros fallecidos, QEPD.

Fuimos estudiantes del Glorioso Instituto Nacional Mejía, ¡el Partenón de la sabiduría! En 1983 o posteriormente egresamos del mismo como bachilleres de la República y con el paso de los años hemos cumplido metas, académicas, laborales y de familia.

Ya somos cincuentones, padres de familia, solterones empedernidos, arquitectos, ingenieros, doctores, catedráticos universitarios, corredores de bienes raíces, licenciados en restauración colonial, marinos, soldados, chefs, fiscalizadores, comerciantes. De todo un poco, pero ¡¡¡más que nada Mejías!!!

¡¡¡No son seis años es toda la vida!!!
¡¡¡ADELANTE VA EL MEJÍA!!!
¡¡¡TRES MEJÍAS POR UN PATRÓN MEJÍA!!!
¡¡¡TODA LA VIDA!!!
¡¡¡MEJÍA...MEJÍA...PATRÓN MEJÍA!!!
¡¡¡HASTA LA VICTORIA SIEMPRE!!!

Participantes del Evento del 21 de agosto, 2015

Raúl Xavier Domínguez Egas
Néstor Marcelo Esparza Cuadrado
Vicente Patricio Espinoza Boada
Juan Carlos Gachet Castro
José Antonio Lovato Villarreal
Juan Manuel Muñoz Valverde
Celso Geovanny Palacios Arturo
Jorge Rodríguez Santamaría
Orlando Marcelo Rojas Maldonado
Pablo Vinicio Sánchez Sánchez
Ricardo Iván Silva Álvarez
Hugo Patricio Tobar Buitrón
Bolermán Boanerges Tobar Duque
Byron Javier Trujillo Solórzano

Charlotte, North Carolina
21 de agosto, 2015

Enoch Ibarra

Estimados y recordados compañeros definitivamente uno siente mucho más lo que ha perdido cuando siente cuanto le hace falta, este sentimiento es el que me ha llegado estos días al ver a algunos de ustedes después de tanto tiempo, lamento más que nada no haber conservado la amistad de mis compañeros de colegio, pensaba que el tiempo nunca pasa, y ya pasaron 32 años. Salí del colegio con una meta fija, quería ingresar a la Fuerza Aérea Ecuatoriana, no lo pude hacer, me fallaron los ojos o no tuve un padrino ja...ja..., luego de esto,

cambio drásticamente mi meta y me perdí de mantener una buena amistad con muchos de ustedes.

Gracias a la iniciativa de Juan Carlos he podido volver a ver a algunos de ustedes, cuantos recuerdos inundan mi mente al regresar a esas aulas donde pasamos hermosos momentos, siempre les dijo a mis hijos (tengo 3 por supuesto, el primero se llama Enoch Raúl, el segundo José Eduardo y el último mi Benjamín, se llama Matthew Benjamín) que la época del colegio es la etapa más linda que yo he vivido, estábamos aprendiendo a ser hombres, podíamos hacer las cosas que un hombre hace, pero seguimos siendo unos jóvenes que disfrutamos mucho y que pasamos lindos momentos, entre estudios, huelgas, paseos…

Hoy cuando estén en la reunión en nuestra incorporación, estaré desde lo lejos sintiendo la misma o mayor emoción, espero que ustedes la sientan, yo sé que lo van a disfrutar, he visto como han disfrutado cuando se han reunido, he visto que seguimos manteniendo la chispa, la alegría de aquellos años, por eso el éxito de esta reunión está garantizada.

Sólo quería hacerme presente con unas pocas palabras, quisiera decir tantas cosas, pero no quiero cansarlos, habrá otra oportunidad de hacerlo, sólo quería decirles que estoy muy feliz de que esta reunión se lleve a cabo y decirles que por eso siento una profunda tristeza no poder compartir con ustedes, espero hacerlo muy pronto y lograr reunirnos otra vez, porque esto no debe terminar, más bien debe comenzar aquí y debemos reunirnos más seguido, todavía tenemos mucho tiempo.

Un fuerte abrazo a todos.

Enoch, Negro, Kalule, Ibarra

"Al oír un fuerte grito, se asustó San Pedro un día, pero al mirar hacia Quito, alegre dijo el viejito, QUE VIVA EL PATRÓN MEJÍA".

Y para los que no se acuerden, el himno del colegio, comienza

Saludamos la luz de tu nombre
Juventud de la tierra, Ecuador

Los sagrados derechos del hombre
Te iluminan con su resplandor...

Quito, 21 de agosto, 2015

Marcelo Rojas Maldonado

Compañeros egresados en el año 1983, tengo el gran gusto y satisfacción de dirigirme a ustedes para darle mis sinceros agradecimientos, puesto que con la colaboración y presencia se realizó el extraordinario magno evento de incorporación algo que se veía casi imposible por lo que quiero agradecer a mi esposa María Elena, a mi hija Dianita, a mi hijo Israel y especialmente a mi hijo David que estudia en Argentina y a la distancia me decía papi tú puedes lograr que se realice tu incorporación ya que tu cuando te lo propones lo consigues y lo sabes hacer excelentemente bien.

Este impulso anímico, emocional es lo que hizo que me proponga a colaborar y empezar a realizar un bosquejo del programa, presupuestos, fechas, etc. Y es así que se convocó a una reunión el 11 de julio del 2015 en la que Pato Espinosa estuvo muy emocionado y dijo el día que llegue a ser realidad nuestra incorporación pondré una botella de whisky, luego la otra el 25 de julio y la última el 15 de agosto para terminar y realizar nuestra incorporación el 21 de agosto del 2015.

Verdaderamente que fue extraordinaria,
¿Por qué? Porque estuvieron autoridades del colegio,
¿Por qué? Era la incorporación de estudiantes que habíamos egresado hace 32 años,
¿Por qué? No fuimos incorporados y no fuimos investidos como bachilleres de la república
¿Por qué? Somos unos verdaderos Mejías
¿Por qué? Adelante va el Mejía y adelante siempre irá...

Es decir que hay muchas razones de ser extraordinaria nuestra incorporación, además debemos pensar que quizá somos los más que nos organizamos a más largo tiempo para realizar esta ceremonia que fue organizada por nosotros con la colaboración de las autoridades de la institución.

También fue extraordinaria porque estuvieron los invitados selectos como fueron los familiares más allegados, cercanos y debo indicar que hubo más paparazzi que invitados jajaja...

En fin estas y muchas razones para agradecer a mis compañeros que hicieron posible que se lleve a cabo este programa, además es extraordinario porque el festejo y el programa social fue en la casa de Mejía Lequerica en la que tuvimos condecoración, unos presentes, recuerdos de CDs, palabras, experiencias, brindis, buffet etc., y también la botella de whisky de Patricio Espinosa quien ha tenido pablas de gallero la botella que ofreció, llevo y nos tomamos hasta la última gota que terminamos llevando hasta el parqueadero.

Gracias compañeros por todo lo que hicieron y lo más importante fue su presencia ya que debo confesar que en la incorporación llegue a quebrantarme de emoción al ver que se estaba haciendo realidad algo que quizá hace años atrás no lo pensé ni me imagine.

Deberás que el evento magno que se realizo fue extraordinariamente excelente por lo que no me canso de agradecerles y decirle que un Mejía nace, crece y muere siendo Mejía.

Un abrazo del compañero que les quiere y aprecia.

North Brunswick, New Jersey
26 de octubre, 2015

Álvaro Gabriel Villavicencio Tumipamba

Nací en un barrio muy popular de la ciudad de Quito, San Roque, La Libertad. Un barrio muy querido, donde jugábamos fútbol en la calle hasta altas horas de la noche yo lo disfrutaba porque esa era mi pasión el fútbol. Entré al Instituto Nacional Mejía por ser uno de los mejores colegios de Quito. Éramos panas con Eddy Chango, Oswaldo Jorge Raza, y otros amigos.

El fútbol y los estudios me acompañaron en ese bello colegio...ahh...y también las chicas chicas estudiantes de los colegios cercanos al nuestro que nos quedaban viendo además

por llevar puesto ese impecable uniforme con el sello distintivo de Patrón Mejía, que orgullo salir a la calle, íbamos caminando y orgullosos nos veían.
Muchos y muy buenos profesores que junto a mi Madre que está en los cielos nos enseñaron antes de lo intelectual a ser personas de bien y buenos seres humanos.

Yo terminé el colegio como Físico-Matemático, pero en el Patrón Mejía todos somos amigos... ¡LOS AMIGOS SON PARA SIEMPRE!
Me gradué, luego pude ir a la Universidad Central para seguir la carrera de Ingeniería Civil. Cuando me gradué, empecé a construir viviendas de manera particular. Luego estudié una Licenciatura en Matemáticas y continúe hasta obtener el título de Master en Educación Superior.

Tengo 2 hijos que iniciaron sus estudios en Ecuador pero ya llevan 10 años por acá y ahora lo están continuando acá en EEUU, Rutgers University. La carrera de medicina. Yo también estoy por acá desde hace 9 años. Actualmente estoy trabajando en una compañía japonesa que se llama Shiseido, donde hago el control de la calidad de los productos en un laboratorio hasta la presente fecha.

No quiero olvidarme de dar gracias a Dios por todo las cosas que él ha hecho en mi vida y con mis hijos que en un país extranjero y tan difícil hemos visto su mano ponderosa.

Como una anécdota un poco dura y a la vez graciosa fue cuando al salir del colegio estaban afuera en las manifestaciones pero yo salía normal por la puerta principal por donde es la fachada del colegio Mejía esas gradas y cuando salí me doy cuenta que todavía continuaban las manifestaciones, lo único que hice es subirme las gradas hasta la puerta principal y quedarme de curioso hasta ver qué pasaba a ver que hacía, y ver los policías cómo lanzaban las bombas y yo ahí mirando con otros compañeros desde arriba la puerta principal de la entrada en la parte donde es el rectorado, se ve hacia el rectorado hasta que hubo un momento que de pronto los policías vinieron en gran cantidad hasta entrar al colegio y subir las gradas en eso me puse nervioso y quise entrar por la puerta principal y todos mis compañeros entraron menos yo,

y la puerta se cerró ya no que la cerraron inmediatamente, yo dije ya me quedé solito ya me llevaron estos policías sufría y estaba muy apenado porque les veía subir a los policías por las gradas, yo golpeaba la puerta para que me abran y ya casi lloraba, yo no sé si estaría llorando pero de pronto alguien me abrió la puerta y creo que fue el portero que estaba a cargo de las puertas que le dio pena y sólo estiró la mano y me metió, y el policía ya no me alcanzó a llevar.

Esas son unas de las cosas que me he reído muchas veces y les he contado a mis hijos de que a veces no vale ser muy curioso hasta en nuestra propia casa.

Ibarra, 26 de octubre, 2015

Oscar Omar Albán Villena

Un saludo cordial a todos mis compañeros Mejías de la Gloriosa Sexta sección FF.MM, La vida estudiantil en las aulas, patios e instalaciones del colegio transcurrieron llenos de alegrías, tristezas, esfuerzo, estudio, quien de nosotros no se deleitó con la capacidad de ese gran maestro de anatomía Tarzán Toscano y al mismo tiempo sufrió memorizando cada parte del cuerpo humano, conocimiento sin el cual no se podía aprobar esa asignatura.

Quien no sufrió aprendiéndose de cabo a rabo el libro de filosofía del "Bombillo" Lara, que acostumbraba a evaluar de forma minuciosa al estudiante, y realizaba preguntas que solo quien había indagado entre líneas a lo largo del texto era capaz de contestar, la respuesta estaba tan escondida en el libro que generalmente releyendo una y otra vez y ni aun así se hallaba la misma.

Quien no sufrió conatos de paro cardiaco cuando nuestro matemático "El Toro" Custode escribía en el pizarrón un problema de Factoreo y miraba a su alrededor a todos nosotros buscando cuál sería su próxima víctima del llamado "Tiro al Blanco" es decir estrellar la cabeza en un blanco dibujado en el pizarrón previamente dibujado por el maestro, el famoso "guaracazo" o el golpe, como castigo por no haber

sido capaz de resolver el ejercicio; en fin, muchas cosas que todos pasamos y sentimos en carne propia.

Anécdotas propiamente dichas no recuerdo ninguna, pero si podría compartirles mi experiencia personal y fue en el último año cuando había que elaborar la Monografía, tesis o trabajo final de titulación. El suscrito era buen estudiante o al menos de los que están por arriba del promedio, competíamos en notas con los más aplicados A. Lovato, Hugo Tobar y otros más estudiosos que yo, les comento para ello , que a partir de que entre a conformar en Quinto Curso la Banda de Guerra del "Glorioso", mi rendimiento académico bajo ostensiblemente, una vez en sexto curso trate de recuperar el mismo y para demostrarlo, tome la "sabia" decisión de escoger como opción de titulación para consolidar la Nota de grado la opción de la monografía en lugar de la de Alfabetización que había implementado el Ministerio de Educación , por qué?, porque se tenía la percepción de que a Alfabetización "iban solo los vagos". En efecto, poco a poco fui dándome cuenta de lo equivocado que estaba; al ver los listados de quienes habíamos escogido hacer la monografía en lugar del programa de alfabetización, me di cuenta de que no habían ni 5 estudiantes, quienes eran, Hugo Tobar, Oscar Albán y algún otro que si hubo se me escapa de la memoria.

Como les relato nunca en mi vida estudiantil me había arrepentido de haber tomado tan desacertadamente una decisión y porque les comento esto, porque el director o tutor del trabajo de graduación de bachiller, que nos habían asignado era nada más y nada menos que el "Cuco" de Física , el Sr. Ingeniero Víctor Hugo Olalla que en esos momentos ostentaba el cargo de decano de la Facultad de Ingeniería de la Universidad Central del Ecuador, Me aprobaron un tema que nunca habíamos tratado a fondo "Péndulo Físico"; allí inicio mi calvario. Paulatinamente nos dimos cuenta de que mientras Hugo Tobar y yo investigábamos, pasábamos interminables horas en la biblioteca, avanzábamos lentamente con nuestra monografía, esperábamos por horas a que nos atienda nuestro tutor tanto en el colegio, como en su oficina en la Universidad Central.

Escuchábamos arrepentidos como el resto de compañeros entre ellos recuerdo a Marcelo Esparza, "Pugitos" Domínguez, Freddy. Arias, etc. Como contaban sus "experiencias" en el programa de alfabetización, ellos relataban entre otras cosas lo bien que pasaban en los diferentes recintos educativos donde ejercían esta noble labor de hacer conocer y capacitar a los iletrados , la misma que no solo se limitaba a eso sino también les daba la oportunidad de congeniar con las alumnas de los colegios femeninos "Las Chivas , Bomberas, Limoneras, las del Fernández Madrid, Colegio Espejo, entre otros y uno que otro colegio aniñado" que también estaban en el programa.

Muchos de ellos comentaban de los bailes que organizaban, de las fiestas y reuniones que tenían, incluso que muchos de ellos ya se habían conseguido una nueva "pelada". Al escuchar aquello una sana envidia y arrepentimiento nos embargaba el ser. Lo único que hacíamos al oír estas experiencias era mirarnos con Hugo Tobar y decirnos entre murmullos "que hemos hecho", donde nos metimos "nos tocó bailar con la más fea".

A lo hecho pecho dijimos, la única opción que teníamos era seguir con el desarrollo de nuestra monografía y esperar que la virgen nos ampare a ver si el Ingeniero nos aprobaba el trabajo final y esperar también que no nos masacrara en la defensa del mismo.

El arrepentimiento era de todos los días, nosotros viendo la claridad del día estudiando, investigando y mecanografiando, mientras los demás la pasaban de "agache" en el programa de alfabetización, todo por el orgullo de decir yo soy buen estudiante y tomo el camino más difícil, al final que sucedió, mientras la mayoría de alfabetizadores aseguraron su nota de 10/10 nosotros a duras penas sacamos una nota regular. Me quedo como experiencia que a veces más valen los resultados que los medios para lograrlo.

Han quedado muchas experiencias atrás, las mismas que nos han servido hasta nuestros días, incursione en un campo que ni por idea se me paso por la mente, en 1987 me gradué de subteniente de la Policía Militar Aduanera , el mismo

Febres Cordero fue a la Graduación, luego he incursionado en la docencia universitaria y actualmente estoy postulando para un Doctorado en Administración de Negocios, siempre agradezco al Colegio Mejía y siempre llevo sus enseñanzas en todos los actos de mi vida , constantemente estoy consciente de que por el camino áspero se llega a la cumbre.
Sexto Curso Sexta Sección
Promoción 82-83 Diurno.

Quito, 1 de noviembre, 2015

Patricio Tobar

Hola compañeros es muy placentero tener amigos como ustedes que se esfuerzan por estar en contacto.
Después que salimos fui a la Politécnica y al poco tiempo falleció mi papi que era Maestro, me dieron la oportunidad en el colegio donde él trabajaba de quedarme dando matemáticas. Desde ese momento me dediqué a la docencia.

Me cambie a la U. Central para respaldar el nombramiento y saque la Licenciatura en Matemáticas posteriormente saque otra Licenciatura en Administración y Supervisión Educativa luego una Maestría en Docencia de Matemáticas, lo que me permitió escribir 3 libros de matemáticas en EDINUN para educación secundaria y trabajar en la Universidad Central, hasta que llegó el Presidente Correa con sus leyes...Posteriormente una Maestría en Educación y Desarrollo Social, pues estábamos trabajando en el colegio Militar N°10 del Pintado y la idea era de enseñar a los líderes comunitarios en el Oriente. Actualmente estoy terminando la maestría que el estado tiene con la Universidad de Barcelona.

Una de las anécdotas que me acuerdo es que estando en sexto curso jugaba Ecuador - Argentina por las eliminatorias al mundial y al Glorioso Mejía le escogieron para integrar la barra Eveready cuyo sitio era en Tribuna del Estadio Atahualpa (no habían ya entradas para ese partido). Con el Inspector que era el Lic. Egas (Guaguazo) decidimos la forma

para entregar esas entradas y solicitamos 10 compañeros para que nos ayuden en una actividad. Nadie quería voluntariamente y tuvimos que rogarles para que nos acompañen. Al final logramos juntar esos compañeros y casi les da un patatús cuando se les dijo que van a ir al estadio al ver ese partido. Regrese a las 23h00 y tuve que amanecerme estudiando porque al otro día teníamos examen de física con el Ing. Olalla.

Así mismo en sexto curso integraba la selección de fútbol del colegio, era capitán y en el partido contra la Academia Militar Ecuador me expulsaron y no pudimos llegar a la final, entonces el entrenador Hugo Ruales no me pasaba la nota de educación física, y sí que me asusté porque ya en todo estaba pasado...

No olvides en primer curso cuando nos llevaron al atletismo al estadio y la paliza que nos metieron esos grandotes del Colegio Militar y cómo al otro día los Mejías fuimos hasta la Orellana al Colegio Militar a protestar este maltrato...

En el paseo de sexto curso a Esmeraldas hicimos las gestiones para que nos presten el bus grande un Botar que tenía el Colegio. Nos acompañó el Anibital Altamirano. Quién no recuerda cómo el coco Freire (Micra) se quedó dormido debajo del carro...Nos habíamos hospedado en Esmeraldas y fuimos a Atacames al regreso no pudimos pasar por un deslave teniendo que quedarnos en la playa haciendo una fogata a la que se sumaron muchos ex Mejías, mientras ustedes disfrutaban esos momentos junto con el Anibital nos enlodábamos pasando el derrumbe y haciendo trasbordos hasta Esmeraldas a dejar la cosas en una sola habitación para evitar el pago de un día de hospedaje. Al regresar el bus se había enterrado en la arena y tuvimos que sacarlo buscando un tractor.

Uno de los momentos más bonitos fue cuando les lleve a jugar al Politécnico a Oscar Albán y Alcides Barahona, disfrutamos esos momentos saliendo del colegio nos íbamos a entrenar, y cuando el colegio formaba líderes estudiantiles con la presencia de personajes brillantes que se habían graduado

en el Patrón.

Tengo 3 hijos, dos mujercitas y un varón que también egresó del patrón con quién compartimos desfiles de la banda, él en la de estudiantes y yo en la de egresados.

Ojalá el Bolermán nos cuente cómo siempre pasaba y era yo el que se quedaba a los supletorios.

El Geovanny también nos recuerde cómo alguna vez tocaron en el grupo folklórico que tenía.

Es fabuloso recordar esa parte de la vida que nos permitió formarnos para ser buenas personas.

Gracias amigos por hacer posible estos recuerdos. ¡Un abrazo!

Quito, 29 de noviembre, 2015

Byron Trujillo

Bueno, hasta que empecé a escribir, vengo de una familia con tradición de Mejías, fue mi abuelo, mi padre y mis hermanos alumnos del Glorioso Patrón Mejía. Yo no podía ser la excepción. Mis días dentro del colegio empezaron cuando mis hermanos mayores me llevaban a los entrenamientos de Judo, puesto que ellos fueron seleccionados no solo del Colegio si no de Pichincha y hasta del Ecuador. Al pasar de la Primaria a la Secundaria mis padres me inscribieron en el curso vacacional que el colegio ofrecía.

El primer curso lo pase en el edificio sur, donde antes había sido de los milicos, del Cuerpo de Ingenieros del Ejecito, ahora entiendo la forma de su fachada.

Aprendí a ser lo que todo "cachorro" en su momento debía hacer, primero estudiar y luego a protestar, recuerdo que alguna vez entraron los compañeros de los cursos superiores a sacarnos para una manifestación que había habido, una vez en la calle comenzamos a grita y al lanzar piedras, palos o lo que encontrábamos, luego vino la policía y nos dispersó, nos iban siguiendo hasta que encontramos una tienda cuya puerta estaba casi cerrada y no tuvimos más que ocultarnos allí hasta que todo pase, bueno toco hacer el gasto y luego de un buen rato la dueña nos mandó de su local. Recuerdo que unos

policías golpearon las puertas y nosotros desde adentro gritamos no hay quien venda. Otra anécdota que recuerdo fue que cuando salíamos a protestar por el "alto costo de la vida", los "chapas" nos seguían y nosotros corrimos hacia San Juan por las empinadas calles, los chapas nunca nos pudieron coger porque casualmente eran los más gordos los que nos seguían...ahí conocí los famosos motes de San Juan.

Dentro de las aulas la vida en clases no se pasaba con muchos sobresaltos, en primer curso recuerdo a un profesor que nos daba taller de carpintería, aprendimos a usar las herramientas me pareció bastante interesante, a un profesor que nos daba geografía cuyo cuaderno se llenó con dibujos de la tierra de la luna, de los planetas dibujos que copie de un cuaderno de dibujo que tenía mi padre desde su época de estudiante; en segundo curso el Lcdo. Almeida que nos daba dibujo técnico, recuerdo que la nota de mi examen era 19/20 y que para tener la nota de 20 el me hizo una pregunta por un punto, si contestaba bien tenía 20 y si no había un compañero que se encargaba de darnos un correazo por cada pregunta mal contestada...nunca saque 20.
Ya en tercer curso recuerdo del profesor que nos daba anatomía, del recordado Toro Custode y muchos más.

De los cursos superiores recuerdo al Dr. Flores, "El Loco Flores", que a pesar de haberse jubilado el siguió dictando clases y que su sueldo donaba para implementar el laboratorio de Química, del recuerdo la frase que nos decía en clases y cuando tomaba exámenes..."que el 10% era conocimiento y el 90% era sentido común"... Y que eso apliquemos en toda nuestra vida estudiantil y profesional...Palabras sabias.

Le recuerdo con admiración al "Indio Olalla", buen profesor de Física, me enseñó a quererle a la física, no sé por qué razón alguna vez nos contó que el al igual que el "Chinche Duque" el colegio por primera vez en la vida institucional, había otorgado una medalla de oro a cada uno por ser los mejores egresados de su promoción, que habían sido toda la vida estudiantil rivales inclusive en las aulas universitarias.

Recuerdo al entrenador de básquet el "Indio" Juan

Escalante, que me llamo para la tarde como aspirante a formar parte de la tercera categoría del básquet del colegio, a esa edad todavía tenía que pedir permiso en mi casa para poder asistir a los entrenamientos, nunca me dieron permiso, porque "tenía que estudiar", pero al siguiente año me dieron permiso. Me toco combinar el estudio con el deporte, lo cual lo hice, la tradición era que un buen estudiante era también un buen deportista, lo cual había sido cierto. Como olvidar las infaltables finales en el Coliseo Julio Cesar Hidalgo, cuando la selección entraba al coliseo todo el público se alborotaba, era algo indescriptible…En mi último año en la selección y que quedamos campeones intercolegiales por 13 años consecutivos, recuerdo la calle de honor que se formó desde el coliseo hasta el colegio, donde nos esperaban las autoridades para rendirnos un merecido homenaje…

Con mucha razón se dice que la vida de colegio es la más hermosa, que la mejor época es la de estudiante secundario, se la vive con tanta intensidad, como se añora esa época,… Pero en fin la vida continúa.

Bueno la segunda parte estará en la siguiente edición…saludos muy cordiales y un fuerte abrazo a todos mis compañeros, a los que estuvieron en la ceremonia de graduación y a los que no pudieron estar.

Sexto Curso Sexta Sección, Físico Matemático
Año Lectivo 1982-1983

Guido Fabián Aguilar Maldonado
Oscar Omar Albán Villena
Freddy Kelvin Arias Toscano
José Ramiro Ayala Carrillo
Hernán Alcides Barahona Proaño
Jaime Leonardo Brito Zaldumbide
Franklin Patricio Cárdenas Miranda
Eddy Patricio Chango López
Ernesto Efraín Chávez Ordoñez
Raúl Xavier Domínguez Egas
Walter Eduardo Egas Palma
Daniel Rómulo Enríquez Caicedo
Néstor Marcelo Esparza Cuadrado
Vicente Patricio Espinoza Boada

Jorge Humberto Freire Arellano
Juan Carlos Gachet Castro
Carlos Alberto Gómez del Salto
Jaime Oswaldo Guerrero Ben alcázar
Raúl Enoch Ibarra Rivadeneira
Washington Alberto Jiménez Duque
Víctor Eduardo Lara Chiriboga
Mario Eduardo López Guamán
José Antonio Lovato Villarreal
Jorge Fabián Mariño Erazo
Sergio Segundo Moreno Torres
Marco Antonio Muñoz Rodríguez
Juan Manuel Muñoz Valverde
Celso Geovanny Palacios Arturo
Juan Genaro Parra Galarraga
Kléver Arturo Proaño Velásquez
Jorge Roberto Pungil Carvajal
José Ignacio Quiroga Arévalo
Jorge Rodríguez Santamaría
Orlando Marcelo Rojas Maldonado
Gonzalo Vidal Román Martínez
Ramiro Javier Salvador Pérez
Pablo Vinicio Sánchez Sánchez
Guillermo Ulises Sanipatin Torres
Daniel Hermógenes Silva Almache
Ricardo Iván Silva Álvarez
Ángel Marcelo Silva Villagrán
Miguel Oswaldo Solórzano Tapia
Efrén Enrique Suarez Yépez
Luis Fernando Sulca Dueñas
Hugo Patricio Tobar Buitrón
Bolermán Boanerges Tobar Duque
Joffre Erney Torres Cueva
Byron Javier Trujillo Solórzano
Gustavo Fabián Villarruel Guerra
José Asdrúbal Yépez Gómez
Eddiu Antonio Andrango Veintimilla
Fidel Vasili Jirón Cruz

www.ingramcontent.com/pod-product-compliance
Lightning Source LLC
Chambersburg PA
CBHW051447170526
45166CB00001B/143
9781522802051

www.ingramcontent.com/pod-product-compliance
Lightning Source LLC
Chambersburg PA
CBHW030941180526
45163CB00002B/662
* 9 7 8 0 5 9 5 2 3 4 6 5 3 *

I0488807

MEDICAL THEATRE October 10, 2011

St. John's Medical College

Johnnagar, Sarjapur Road,
BANGALORE - 560 034, INDIA

DEAN'S OFFICE
Telephone : (080) 22065030/31
Fax : 91-80-25530737
E-mail : deansjmc@sjri.res".in.

LETTER OF COMMENDATION

This is to certify and acknowledge the contributions made by Dr. Jagdish Chaturvedi in introducing an innovative and effective teaching methodology by conducting workshops for students and faculty of this institution during his endeavor as a Post Graduate Resident from the Department of Otorhinolaryngology at St John's Medical college and hospital.

The following were the events conducted by him using theatrical methodologies to learn, understand and teach medical subjects for a selected group of participants from St. John's Medical College.

April 2011 - Medical Theatre Workshop- St John's Medical College
Facilitators- Sibu Vaz, Rajesh PI (Script People Theatre)
Dr. Jagdish Chaturvedi- PG ENT (Organizer)
Topics- Meniere's disease, Mechanism of action of Penicillin.
Participants- 35 Medical Students, Duration- 1 day workshop

May 2011- Medical Theatre Workshop- St John's Medical College
Facilitators- Sibu Vaz, Rajesh PI (Script People Theatre)
Dr. Jagdish Chaturvedi- PG ENT (Organizer)
Topics- Broca's aphasia, Temporal arteritis, Parathyroid hormone, Sympathetic and parasympathetic system, Rheumatic fever and Cushing's syndrome.
Participants- 25 Nursing students + 10 Medical Students, Duration- 2 Days workshop

August 2011: Event: NATIONAL WORKSHOP ON RESEARCH METHODOLOGY Presented: "Hypothesis and framing of objectives" using an innovative methodology of Medical theatre. The 1 hour talk was presented to an audience of 500 participants using Stand up comedy and dramatic narration. This unique event was well accepted by the participants .

September 2011: Event: ANATOMY LECTURE SESSION FOR 1ST YEAR MBBS
Presented: Anatomy of the nerves of the Brachial plexus by creating and narrating stories describing the origin, course, distribution and action of the major nerves in the hand. Students were made to participate in the remaining half of the session by creating stories and tableau's representing the remaining nerves. The theoretical content was reinforced by Dr. Nachiket and Dr. Anu from anatomy. All the 60 students enjoyed and learnt with creativity in this one hour session blending into the normal curriculum.

We congratulate him for this initiative and wish every success in the future.

Dr. Prem Pais, MD
DEAN

14th September 2011

ST. JOHN'S MEDICAL COLLEGE

A STUDENT'S NOTE

Medical Theatre was a unique concept introduced by Dr. Jagdish - the friendly neighbourhood ENT PG. Attending both the workshop gave a new insight into theatre, learning techniques and more importantly combining the two. The first workshop gave a orientation into basic theatre skills the process of writing a script, and different learning techniques which were facilitated by Dr. Anil & Dr. Kasturi.

The best part of the whole program was the 2nd workshop Dr. Jagdish assigned different chapters in Harrison to be converted into scripts. Apart from giving me the "opportunity" to actually read Harrison, it enabled me to actually understand the topic and relate it in a simple & practical way during the process of converting into a script. How did all this help me. Well I never have to read those topics again and also could reproduce the entire concept during exams. Medical theatre in a structured format should be encouraged for batches to come. Hats off to Dr. Jagdis for organising both the workshop

Nitin Yashas
Batch 2007
nitinyashas@gmail.com

Medical Theatre

Dramatic improvement in health care!!

A three part dramatic experience for Theatre lovers, script writers and medical students using tools of theatre

FAQ'S

What is Medical Theatre?

The art of learning and teaching medicine by the means of dramatics and theatre.

What is the aim behind this workshop?

With an aim to explore the possibilities of ACTive learning modules are created for participants interested in the dramatic possibilities of medical science. This unique collaboration will provide an opportunity to create enriching experiences for select participants. Structured as short interventions keeping in mind the demands of a regular profession – the aim is to provide an enriching experience of learning.

Who can participate?

1. Anyone who likes to be a part of theatrical plays and wants to have fun exploring a new methodology using theatre and dramatics so that they can learn and teach medicine in an innovative and interesting way.
2. Actors , Directors and Script writers
3. Medical students (Undergraduate and post graduate)
4. Science and Art students – Media and education
5. Students from other fields (nursing/paramedical etc)

Reasons to participate in this workshop?

- To overcome long, complicated and at many times extremely cumbersome medical topics included in the curriculum.
- Students who prefer visual and auditory stimulation to learn and understand topics rather than to read the same from books.
- To avoid learning by "rote" or in other words mugging up of difficult topics.
- To reduce stress and pressure by adding variety to the studying methodology.
- To attain skills in this innovative method of learning and teaching.

Who will conduct the workshop?

This workshop will be conducted by **Dr. Jagdish Chaturvedi and The Script Peoples Theatre Group** who are a professional group of theatre artists immensely experienced in conducting such workshops over the past decade. They currently are a part of the Theatre- in – education Department in Christ University and promise an exciting, fun filled, energy packed dramatic experience for medical students and theatre lovers.

Format of the Workshop

- **Part I** – Experience the joy of theatre through structured games, activities and explorations
- **Part II** – Explore identified medical topics and bring in relationships through learnt theatrical patterns/formats.
- **Part III** – Create a theatrical presentation on created script/s.

What is the duration of this workshop?

This is a day and a half long event carefully structured by a professional group of theatre artists.

How many people can participate in the workshop?

30 + 5 members can participate.

30 Members via registration

5 members – Volunteers for the event

How to register?

Contact – Dr. Jagdish Chaturvedi – 9740076668 (Organizer)
EMAIL- hemoglobin84@yahoo.com
(Send Name of participants- registration forms and payment details will be emailed back to you_

A registration fees of 500/- (Students of St John's) and 2000 /- (outsiders) can be paid by online account transfer or by cheques or by cash payment.

Outcome of the Workshop

- To understand the basics of theatre and evolve methods to convert medical topics into theatrical representations
- To enjoy the process of collaborative creation on identified medical topics
- Create a basic theatre presentation.
- At the end of the workshop a student should be able to-

    1. Use theatrical skills to convert medical topics into stories
    2. To use this methodology to teach juniors or colleagues
    3. To understand difficult topics without having to learn by "rote" or skip them.

Date of Next Workshop

Visit www.yukti-mi.com for updates on upcoming workshops and events.

science by logic

Yukti medical innovations brings to life a unique learning experience where medical subjects are taught and understood using theatre and dramatics.

presenting

MEDICAL THEATRE

Format – workshop (1 day duration) – can be extended to 2 or 3 days as per content

Concept- Medical topics are converted into theatrical outcomes (narrations, monologues, stories, clinical scenarios, skits, poems, songs and role-plays). How this can be done in simple steps is demonstrated to the participants. The participants then convert a given topic into a theatrical outcome in a group activity. They also perform a skit on a given pre scripted topic to an invited audience. In addition to all this, at the end they make their own script on the topic of their choice. THEY LEARN BY PARTICIPATING IN THEESE ACTIVITIES SPECIFICALLY DURING THE TIME WHEN THEY CONVERT A TOPIC INTO A THEATRICAL OUTCOME.

The entire workshop is mixed with lots of theatre games to help learn basic concepts of acting along with the medical learning so that everything turns out to be a fun and enjoyable process.

Participants- Students – medical/ dental / nursing / science/ PUC/ high school.
Pharmaceutical companies – to train medical representatives
Medical device companies – to make related medical topics easier to understand
Non medical persons who need to have medical knowledge in specific areas.

Contact details – Dr. Jagdish chaturvedi - 9740076668 , hemoglobin84@yahoo.com
Mrs. Dhanya Parameshwaran , dhanya.p@gmail.com

Comments from participants of the previous workshops-

"I learnt how to convert complicated topics into simple stories for exams and viva's" – ***2ND MBBS, MAY'11***
"I learnt how theatre can be useful for a doctor with patients and students"- ***3rd MBBS, JUNE'11 Workshop***
"Learning script conversion was essential" – ***Psychiatry Post Graduate, MAY '11 Workshop***
"I feel this makes more sense than pointless mugging up of topics, at least this way we relate to what we learn"
"We don't need to learn in a boring and monotonous way, learning can be fun" ***1ST YR BSC Nursing, JUNE'11***
"This methodology must be included in the teaching curriculum" ***Nursing Teacher, St John's, JUNE'11.***
" I am much more confident with my work because of this workshop"- ***AstraZeneca, MR , JULY'11***

Session 1 – Introduction

Medical Theatre – A Dramatic improvement in Medical education. Our first experiences May and June 2011

The medical curriculum comprises a vast database of information that a medical student has to understand and apply in order to become a doctor. This on an average takes five to seven years for an undergraduate in various medical institutions across the world. For these reasons medical education has always been thought to have been a long and tedious journey taken up by those who have the perseverance and dedication to become a doctor. (1,2) Added to this, specialization in a particular field has become a necessity for doctors especially in a developing nation such as India.

This adds to stress and the pressure to finish learning the entire curriculum in a speculated period of time which forces many students to learn by rote or skip topics that they fail to understand easily. (1,2) Many resort to suicidal tendencies due to poor coping skills.(2) Dramatics in the form of plays and skits have always been popular amongst children and adults universally. (3,4) Dramatics presents a means to represent elements in multiple sensory dimensions - visual and auditory stimuli being the most common. Surveys have shown that incidences which an individual has experienced are easier to recollect and effortless to remember. (4)

In this process we hope to attempt at combining medical education with dramatics and theatre to create an innovative methodology that enables medical and non medical students to learn and understand medical concepts more effectively .

Session 2 – Theatre Games

Explore The room

Freeze in poses

The Concept – The concept involves the ability of an individual to create a theatrical outcome from a selected medical content such that during the process of creating an outcome, the content is well understood. The outcome could be a story, a script for a play, a poem, a prose, a song, a game, a narration/monologue, a group activity, a cartoon/ comic, a clinical scenario or a role play. The type of outcome depends on the content and the discretion of the individual making the outcome. The quality of the outcome is not stressed upon as the outcome only serves the purpose of revision material to that individual alone. So if the individual has created a story from a selected medical content, then that story will be useful as revision material for that individual only. It may or may not serve the same purpose for someone else, but might create inspiration amongst others to try and create something for themselves for their own content.

The understanding behind the concept- For an individual to make a theatrical presentation or outcome from the selected medical content, the process would automatically involve reading and trying to understand the medical content first. After this the individual tries to create a story or plot using the content to make a final theatrical outcome. This process involves a lot of concentration, imagination and creation where in the individual creates his own words in the story and links the contents such that it forms a sequence of events when trying to remember the topic. The outcome is special and specific to the individual's thoughts and experiences and hence allows room for excitement, curiosity and challenge to create an interesting outcome which is easy to remember and recollect during revisions. This process need not be performed for every topic in the medical curriculum, but only to those where the student finds it difficult to remember or has to learn by rote.

Session 2 – Theatre Games

Imagine a KEY
as various objects

Compete with each other

The type of intervention selected- In view of the concept involving individual participation for effective learning, an intervention in the form of a one day workshop was considered appropriate. Subsequent workshops being conducted in the same institution were structured similarly; however the content was entirely new in each of the workshops so as to enable the same participants to lean new topics in a fun and innovative manner at monthly intervals.

Number of Participants- Workshop 1- 23 participants (Medical students from different years)

Workshop 2- 23 participants (Nursing students from different years)
10 participants (Medical students from workshop 1)

Number of facilitators-

For both the workshops the facilitators included the first author of this paper for the medical contnt and conversion of medical topics into stories/scripts. The third author and team of 3 others from a professional and recognized theatre group conducted the entire workshop in between sessions taken by doctors. Two invited guest doctors as facilitators for their sessions and the second author participated mainly in the organization and management of the workshop.

Content- Workshop 1 – 4 topics-

Mechanism of action of penicillin's (Group activity -converted into story)
Syphilis (Guest doctor- Demonstration of song)
Community survey (Guest doctor- Demonstration of role play)
Meniere's Disease (pre written script- performed by participants)

Workshop 2- 6 topics-

Broca's Aphasia (Demonstration of Monologue)
Temporal Arteritis (Demonstration of theatrical case scenario)
Action of Parathyroid Hormone (Demonstration of story)
Autonomic Nervous System (Group activity by participants)
Rheumatic Fever (pre written script- performed by group A participants)
Cushing's Syndrome (pre written script- performed by group B participants)

Session 3 – Basics on scripts

Story Building Exercise

Structure of a script

The methodology - Two workshops were conducted in the span of two months between May and June 2011 in a medical college premises for medical and nursing students. The workshops were structured in a manner to enable learning and understanding of both the medical content and the basics of theatre in the same intervention. The workshop included theatrical games, group building exercises, demonstration sessions showing conversion of a few medical topics into a variety of theatrical outcomes, group activities in which students attempted to convert a topic into a theatrical outcome, sessions by two invited guest doctors who demonstrated how they have been using theatre in their teaching sessions highlighting the benefits that dramatics brings to the quality of a doctors profession and finally the workshop conclusion with a theatrical performance by the participants on pre-written scripts based out of medical content in front of an invited audience. The participants were then instructed to make a theatrical outcome of their own within a week of the workshop so as to apply their learning's into their own topics of interest. These were submitted before the issue of certificates. The structure of the two workshops is listed in Table 1.

Table 1 – Shows the schedule of the workshop for a one day and two day intervention.

ONE DAY WORKSHOP	TWO DAY WORKSHOP DAY 1	TWO DAY WORKSHOP DAY 2
Introductory session in theatre	Introductory session in theatre	Theatre games competitiveness
Theatrical skills	Script writing	Reading and rehearsing the allotted play with leaders
Script writing session	Script content and format	Technical show Suggestive changes
Stage 1 – script reading of topic	Theatre games (Bonding, competition , collaboration and understanding using play method)	Creating of the performance
Stage 2- presentation and review	Script creation	Performance in front of an invited audience
Stage 3- rehearsal	Dramatic elements of a performance	Playback theatre and closure
Stage 4- performance for an invited Audience.	Role play- Dramatic elements and use in medical field	
Circle time- de-briefing session	Learning's of the day and forward links	

Session 4 – Medical Topic conversion to a Story

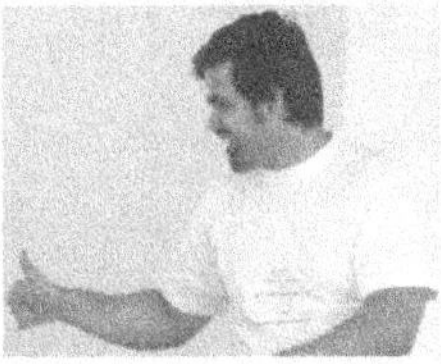

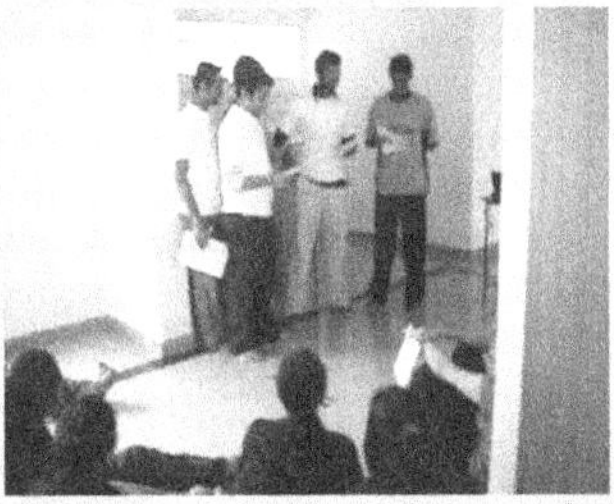

Narration of their creative story

Basic steps on conversion
Taught

Divided into groups and
asked to construct story

Assessment of the workshops-

Both the workshops included a pre-workshop assessment form and a post workshop assessment form. (Table 2 and 3 respectively) These were filled by the candidates and submitted before and after the workshops. In addition to these written assessment forms there was video documentation of the entire intervention. At the end of the workshop a one hour De- briefing session was carried out in which each participant shared their experiences and gave their individual feedback which was also video recorded. Within a week of the workshop each candidate submitted their script/story which was compiled into a book as a documentation of the outcomes created by these students and demonstrating the efficiency of the intervention.

The content created as scripts submitted by students was utilized in the subsequent workshops for the final performances. Students were also awarded the best scripts award for their efforts.

Session 5 – Theatre co-ordination games

The clapping co-ordination game

Co-ordination Exercise

Catch with your head

Table 2- Showing the pre work shop assessment questionnaire

MEDICAL THEATRE WORKSHOP
PRE WORKSHOP ASSESSMENT

Age-

Sex-

Profession/ medical student specify year-

1. Reason for participating in this workshop?
   a. Interest in Theatre and dramatics
   b. Learning problems/ to avoid learning by rote
   c. Peer pressure
   d. To attend guest facilitator sessions(Dr. Kasturi and Dr. Abraham)
   e. To miss class and get attendance too
   f. Any other reason-

2. Are you satisfied with the current teaching methods?
   a. Yes
   b. No

   Why-

3. Are you satisfied with your learning methods?
   a. Yes
   b. No
   c. Why-

4. What do you expect to gain by the end of this workshop?

5. Do you want to be an entertaining teacher in the future?
   a. Yes
   b. No

   Why-

Session 6 – Dr. Anil Abraham – Role Playing

Trust Exercise

Role Playing

Session 7 – Dr. Arvind Kasturi – Role Playing

Improvisation

Demonstration of Role play

Table 3- showing the post assessment workshop assessment questionnaire

MEDICAL THEATRE WORKSHOP

POST WORKSHOP ASSESSMENT

Age-

Sex-

Profession/ medical student specify year-

1. Did you find this workshop useful?

g. Yes very much
h. Didn't make any difference
i. No, it was a waste of my time
   Why?

2. Did you enjoy the workshop ?

c. Yes
d. No
Why-

3. Would you want to attend this workshop regularly to help understand difficult topics ?

d. Yes
e. No
f. Why-

4. Did you gain what you expected at the end of this workshop?

A. Yes
B. No
   Why?

5. Should this teaching methodology be included in the routine teaching program?

c. Yes
d. No
   Why-

Session 8 – Script reading, rehearsal and performance
"The Mystery of the Mean Ears Disease"

Rehearsal- directed by Upendra

The Performance

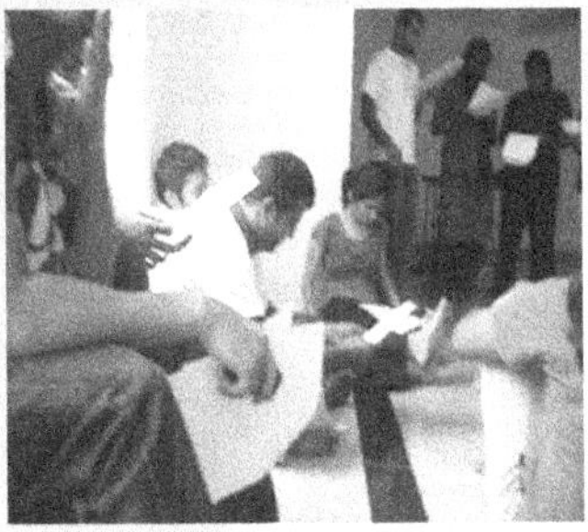

Script Reading

Workshop 1- May 2011–

The first workshop was a one day intervention conducted for 23 undergraduate medical students.

Pre workshop assessment data-

1. Reason for attending workshop-

40% participated in the workshop to attend guest facilitator sessions.
32% participated due to learning difficulties
18% participated due to peer pressure
10% attended out of curiosity

2. 63% were satisfied with the current teaching methodologies.
3. 27% were satisfied with their learning methodologies
4. 80% wanted to learn a way to avoid learning by rote
   Remaining 20 % wanted to have a good time and make friends during the workshop
5. 73% wanted to be entertaining teachers
   The rest did not want to teach in their profession

Post workshop assessment data-

1. Was the workshop useful?
   92% found the workshop useful. The rest felt they were not inclined to use the methodology as they prefer straight forward reading and lack creativity.

2. 100% participants enjoyed the workshop
3. 58% wanted to attend this workshop regularly. The rest would want to attend it occasionally.
4. 92% gained what they expected from the workshop. The remaining expected more of topics being incorporated.
5. 85% felt that this methodology must be included in their normal teaching curriculum.

Video De- briefing summary-

Positive feedback –
1. Creative thought processes were challenged
2. Learnt a new way to understand and learn
3. Learning medicine appeared easier and entertaining
4. Remembering and recollection was easier
5. Enjoyed the process and interacted with students from different batches

Negative Feedback-
1. Need for more demonstrations of how a topic is made into a story
2. More topics need to be covered
3. Some students felt overloaded and tiered
4. Some suggested a two day workshop

WORKSHOP
JUNE
GAMES
Body
Art

Changes Implemented in second workshop- In view of the negative feedback received , the subsequent workshop was converted into a two day intervention with more topics included. The performances were kept on the second day so as to avoid overloading of content on the participants. There were two performances instead of one, by essentially dividing the group into two.

Workshop 2- June 2011–

The second workshop was a two day intervention conducted for 23 undergraduate nursing students and 10 undergraduate medical students.

Pre workshop assessment data-

1. Reason for attending workshop-

62% participated due to learning difficulties
25% attended out of curiosity
10% participated in the workshop to attend guest facilitator sessions.
3% participated due to peer pressure

6. 76% were satisfied with the current teaching methodologies.
7. 24% were satisfied with their learning methodologies
8. 87% wanted to learn a way to avoid learning by rote
   Remaining 13 % wanted to have a good time and make friends during the workshop
9. 79% wanted to be entertaining teachers
   The rest did not want to teach in their profession

Post workshop assessment data-

2. Was the workshop useful?
   96% found the workshop useful. The rest felt they were not inclined to use the methodology as they prefer straight forward reading and lack creativity.

3. 100% participants enjoyed the workshop
4. 69% wanted to attend this workshop regularly. The rest would want to attend it occasionally.
5. 96% gained what they expected from the workshop. The remaining expected more of topics being incorporated.
6. 89% felt that this methodology must be included in their normal teaching curriculum.

Video De- briefing summary-

Positive feedback – 1. An innovative learning and teaching methodology
2. Makes it simpler to understand topics

Script converting Session

3. Increases self confidence and explores creativity
4. Remarkable improvement from the last workshop
5. Lot of fun and enjoyment

Negative Feedback- 1. Could be conducted for a larger student population
2. Topics must be specific to the group and the workshop must be conducted for specific years of batches.

Table 4 A, B and C summarizes all the data from the feedback forms of both the workshops.

Table 4A- Summarizing the reasons for participation assessment for both workshops

Reasons for Attending Medical Theatre	**Workshop1** (%)	**Workshop2** (%)
Attend guest facilitator sessions	40	10
Learning difficulties	32	62
Peer pressure	18	3
Curiosity	10	25

Table 4B- Summarizing the pre workshop assessment feedback for both workshops

Pre-workshop Assessment	**Workshop1** (%)	**Workshop2** (%)
Satisfied with the current teaching methodologies	63	76
Satisfied with their learning methodologies	27	24
To learn a way to avoid learning by rote	80	87
Have a good time and make friends during the workshop	20	13
Become entertaining teachers	73	79

Table 4C- Summarizing the post workshop assessment feedback for both workshops

Post-workshop Assessment	**Workshop1** (%)	**Workshop2** (%)
Usefulness of the workshop	92	96
Enjoyed participating in the workshop	100	100
Want to attend work regularly	58	69
Keep up to their expectation	92	96
Methodology included in the teaching curriculum	85	89

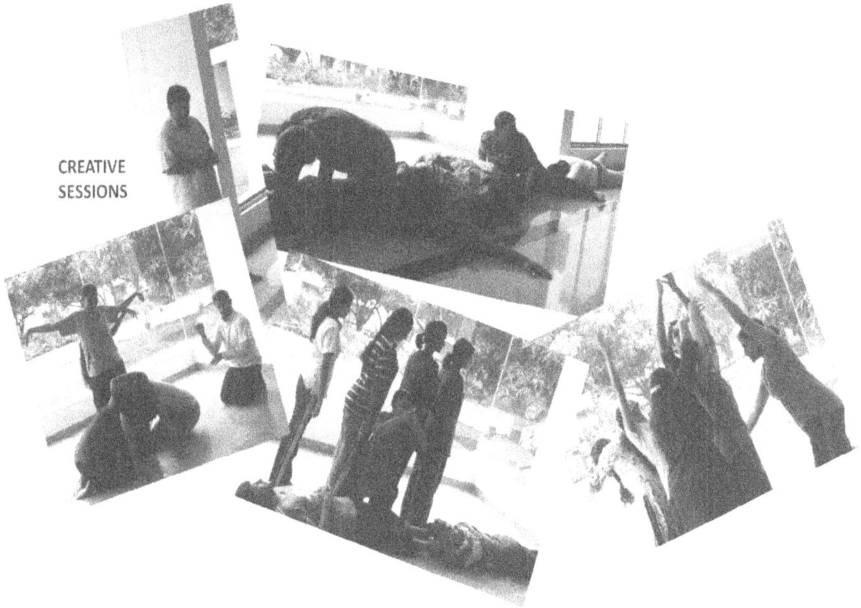

CREATIVE SESSIONS

The medical curriculum comprises a vast database of information that a medical student has to understand and apply in order to become a doctor. This on an average takes five to seven years for an undergraduate in various medical institutions across the world. During this period there are many commonly used methods that a student utilizes to understand medical subjects. Such as in addition to the teaching carried out by doctors in undergraduate classes and clinical case discussions, the student spends a significant amount of time reading the curriculum after college hours. The student might also prepare charts and drawings to remember complicated conditions and stick them up on the walls of their rooms to revise it frequently. Many often come up with interesting mnemonics to remember classifications and differential diagnosis etc. Some make notes from multiple books and read them for revision and most of the students learn by rote for all the difficult to understand medical topics.(8) This often leads to saturation of knowledge and on many occasions confusion amongst similar content. There are a large number of students who do not believe in sacrificing their time on sports and recreation and are further burdened at the time of examinations for the need of leaning large amount of content over a short duration of time. This invariably affects the future of all these students with their career in health care. For these reasons medical education has always been thought to have been a long and tedious journey taken up by those who have the perseverance and dedication to become a doctor.(1,2) Added to this, specialization in a particular field has become a requirement for doctors especially in a developing nation such as India. This adds to stress and the pressure to finish learning the entire curriculum in a speculated period of time which forces many students to learn by rote or skip topics that they fail to understand easily. (1,2) Many resort to suicidal tendencies due to poor coping skills.(2)

There are newer methodologies for teaching medical subjects that have come to existence in recent times. The most popular concept is the Problem based learning (PBL) wherein

Rheumatic Fever – A Red Riding Hood Adaptation

the student learns everything entirely with respect to the problem instead of reading various aspects at different stages of the curriculum. (5,6)This is usually carried out in groups where each student reads up a particular aspect about the disease and discusses it amongst their peers. Other methods such as Fish bowl learning , teaching from experiences using memories of being taught (7,8,9) and group discussions have also been attempted. All these however carry a significant amount of seriousness which appeals only to those who would even otherwise be interested to read on their own in order to learn the subjects. For the rest, there is no current existing methodology that gives room for fun and entertainment along with a high quality of learning. The concept of medical theatre reaches out to these group of students who can learn with as much fun and excitement as they would while playing sports or watching a movie and it also challenges their creativity and imagination to help them nurture into doctors with the ability to apply their knowledge instead of learning just for the sake of clearing examinations. Medical theatre is a method which results in *learning in the process* of a seemingly independent theatre activity.

Dramatics in the form of plays and skits have always been popular amongst children and adults universally.(4) Surveys have shown that incidences which an individual has experienced are easier to recollect and effortless to remember.(3,4) The probable reasons for this can be attributed to the stimulation of the limbic system along with the auditory visual stimuli received along with the information. On regular reading of any topic, no emotions are triggered, however when the entire process is made into a creative challenge to convert a topic into a story or enact in a play, it gives an emotion to the process and helps in remembering this as an experience.(3,4) Knowledge of theatre also helps a doctor to be more expressive and empathetic towards a patient and also trains the improvisational skills that a doctor frequently requires during clinical practice and academic teaching.

The
Performance
3
Greedy
Queens
Skit on Cushing's Disease

A classic study by Dr Alexander Luria in 1968 of his patient 'S' with five-fold synesthesia presents an extreme case of use of sensory modalities in remembering words or numbers or other elements. 'S' was synaesthetic and often long row of words became multi-dimensional images. For example he would remember a long sentence by taking a mental walk down a street from his home in Moscow, slowly placing images of houses, gates, trees and people to represent the constituting words. The technique helped him remember sentences even 15 years after the incidents (The Mind of a Mnemonic – A.Luria).

The two workshops conducted for medical and nursing students demonstrated the acceptance of this method in an overwhelming manner and emphasized the need for such a methodology to help improve learning in the medical curriculum. The feedbacks from the first workshop were invaluable in helping us improve it to an acceptable structure that benefits both students and teachers in the subsequent workshops. This methodology also promises benefit in other fields such as dentistry, pharmaceutical companies to train medical representatives with respect to the medications they promote, Medical design and equipment companies to train their team for a better medical understanding in their domain and in awareness programs for non medical population in the community where medical theatre can be utilized to create awareness among various medical elated conditions.

This article highlights the acceptance and benefits of an innovative teaching methodology for the learning and understanding of medical topics that are otherwise difficult to comprehend amongst students. This methodology however does not substitute the conventional teaching and learning methods in any manner, it only provides an option to learn certain topics in a unique manner so as to break the monotonous process of usual methods of studying and to have an enjoyable learning experience in the process.

The Magic of
PLAYBACK
THEATRE

From the workshops conducted the following were specific conclusions derived.

1. The stories and scripts converted by participants are useful as revision material for the participant who created it only. These theatrical outcomes do not promise to improve the understanding of a topic for other readers. In the same manner, if a play is enacted, the participants gain maximum benefit in understanding the subject rather than the viewers.
2. The process of learning theatrical skills and a new teaching methodology adds to the qualities of a doctor in being a better listener, empathizer and communicator.
3. These workshops help in relieving stress from the usual routine and works as a very effective group bonding exercise amongst peers and helps train the students to co-ordinate, communicate and function as a team which is very useful in future multidisciplinary approaches to various illnesses.
4. A two day intervention was much more accepted and useful when compared to a one day workshop.

This methodology needs to continue for a larger number of students with focus on more specific content and must be applied to all possible areas where this methodology can prove beneficial. Further data will be required to establish the efficacy of this methodology in the future. The current data only shares our initial experiences.

THE DE- BRIEFING SESSION

1. Dahlin M, Joneborg N, Runeson B. Stress and depression among medical students: a cross sectional study. Med Educ. 2005;39:594–604.
2. Bramness JA, Fixdal TC, Vaglum P. Effect of medical school stress on the mental health of medical students in early and late clinical curriculum. *Acta Psychiastr Scand.* 1991;**84**:340–5.
3. Alisha C. Holland and Elizabeth A. Kensinger : Emotion and Autobiographical Memory. Phys Life Rev. 2010 March; 7(1): 88–131.
4. Patrick S. R. Davidson, Craig P. McFarland, and Elizabeth L. Glisky : Effects of emotion on item and source memory in young and older adults. Cogn Affect Behav Neurosci. 2006 December; 6(4): 306–322.
5. Lewis AD, Menezes DA, McDermott HE, Hibbert LJ, Brennan SL, Ross EE, Jones LA. A comparison of course-related stressors in undergraduate problem-based learning (PBL) versus non-PBL medical programmes. BMC Med Educ. 2009 Sep 13;9:60.
6. Chang G, Cook D, Maguire T, Skakun E, Yakimets WW, Warnock GL. Problem-based learning: its role in undergraduate surgical education. Can J Surg. 1995 Feb;38(1):13-21.
7. Polan JH: Experiential Anamnesis and Group Consensus: An Innovative Method to Teach Residents to Teach. Academic Psychiatry, 34:4, July-August 2010
8. Dewey CM, Coverdale JH, Ismail NJ, et al: Residents-as teachers programs in psychiatry: a systematic review. Can J Psychiatry 2008; 53:77–84.
9. Hill AG, Yu TC, Barrow M, et al: A systematic review of resident-as-teacher programmes. Med Educ 2009; 43:1129–1140

THE GANG !!

Medical Theatre for Medical and Nursing Students

Objectives – 1. To introduce an innovative methodology and to demonstrate various possibilities in which any form of medical content can be converted into an interesting outcome.

2. To make them convert a medical topic into a script by themselves after learning few basics of theatre along with having fun and enjoyment during the process

Body – For these group of students we had chosen the structure of a workshop to be most suitable, for the reasons that since this being a new methodology, we wanted them to experience it in a structured format so that they could enjoy and benefit form the process. A shorter didactic one hour session would be too short for them to even comprehend what we were trying to do. We conducted a workshop for medical students which was one day long, and conducted another workshop for medical and nursing students which was two days long. In each of these events we had mixed up a variety of sessions for these students. From theatrical games, group exercise's, story building exercise's and basics on script writing skills to converting a topic into a story and giving a performance on a pre-written script. The students were very curious and sceptical about the entire program when they started off and soon found to have lost themselves in the process eventually having a great time and enjoying it thoroughly resulting in many strong and positive responses. The nursing students were extremely enthusiastic and energetic throughout the workshop and commended on the benefits of the entire process.

Conclusion- At the end of these workshops we concluded that this process was a worthwhile effort as it gives an experience to the learning's along with a soft skill development to the individual by encouraging creativity and interest among the students. We realised for ourselves that learning or teaching anything can be an enjoyable and an effective methodology.

Session Plan
May 2011

Institution: St John's Medical College

Subject: MBBS Syllabus

Topic: Meniere's disease, Mechanism of action of Penicillin.

Class: 1st to 4 th Year MBBS undergraduate students

Objective:

1. To introduce an innovative methodology and to demonstrate various possibilities in which any form of medical content can be converted into an interesting outcome.

2. To make them convert a medical topic into a script by themselves after learning few basics of theatre along with having fun and enjoyment during the process

Requirements: 1000 sq. feet room with empty floor space, an LCD projector, a laptop, a guitar.

No participant: 23

Opening activity: Walk and freeze

Core activity: Story building, Conversion of a topic into script, guest sessions, a theatrical Presentation.

Closure activity: De- Briefing session

Forward link: Write and submit a script on the topic of your choice.

Session Plan
June 2011

Institution: St John's Medical College

Subject: MBBS Syllabus

Topic: Broca's aphasia, Temporal arteritis, Parathyroid hormone, Sympathetic and parasympathetic system, Rheumatic fever and Cushing's syndrome.

Class: 1st to 3rd yr. B.Sc. Nursing students, 1st to 4th Year MBBS undergraduate students

Objective:

1. To introduce an innovative methodology and to demonstrate various possibilities in which any form of medical content can be converted into an interesting outcome.

2. To make them convert a medical topic into a script by themselves after learning few basics of theatre along with having fun and enjoyment during the process

Requirements: 1000 sq. feet room with empty floor space, an LCD projector, a laptop, a guitar.

No participant: 23

Opening activity: Walk and freeze

Core activity: Story building, Conversion of a topic into script, guest sessions, a theatrical Presentation.

Closure activity: De- Briefing session

Tips – Memento for guest doctors, feedback forms submit and collect before session, foot ware outside the room.

Forward link: Write and submit a script on the topic of your choice.

AstraZeneca

Medical Theatre session
Soft Skill Stories for Medical Representatives of AstraZeneca
July 2011

In the process of acquiring sponsorship for a workshop, we were introduced to AstraZeneca, a well-established pharmaceutical company. They were impressed with the whole idea of medical theatre and saw potential in it being utilised for training medical representatives. They immediately requested for a workshop to be conducted for their new scientific task force team that was being created in the field of their cardiac medication division. The intention was to help this new scientific task force team to be well versed with certain attributes that a medical representative requires to possess as a quality in his/her career. This was a three day workshop planned 3 hours every day. There was a competitive performance at the end of their third day during their symposium.

Medical Theatre for Pharmaceutical companies

Objectives – 1. To introduce an innovative methodology by which certain soft skill attributes can be learnt and understood.

Body – For these group of pharmaceutical companies we mainly concentrated on soft skills that they require when dealing with doctors and hospital staff from various institutions. Medical representatives need to have certain qualities which can help them be more efficient in their responsibilities. For this we conducted a three day intervention where we worked on certain attributes such as Influencer, Negotiator, Multi-Tasker, Alliance Builder, Partnership Manager, Strategist, Planner and New Account acquirer. Along with theatrical games and group building exercise's we brought a varied group of professionals from different backgrounds and experiences out from their comfort zones. They were divided into groups based on the attributes assigned to them and each group was facilitated to create a story plot for a short performance highlighting that particular attribute. We noticed that after a while all of them were extremely focussed to give their best performance and enjoyed the process by experiencing these attributes for themselves.

Conclusion- At the end of these workshops we concluded that this process can be utilised to emphasize the importance of certain qualities that any profession might require. Even though no elaborate medical content was communicated, even something as simple as soft skills was learnt effectively through this methodology. We learnt that it is the group that experiences the process actually decides the structure and outcome of this methodology.

Session Plan
July 2011

Institution: AstraZeneca

Subject: Attributes for medical representatives

Topic: Influencer, Negotiator, Multi-Tasker, Alliance Builder, Partnership Manager, Strategist, Planner and New Account acquirer

Class: Medical Representatives

Objective:

1. To introduce an innovative methodology by which certain soft skill attributes can be learnt and understood.

Requirements: 1000 sq. feet room with empty floor space, an LCD projector, a laptop, a guitar.

No. participant: 85

Opening activity: Walk and freeze

Core activity: Story building and theatrical Presentation.

Closure activity: De- Briefing session

Tips – props, feedback forms submit and collect before session, foot ware outside the room.

Forward link: nil

Medical Theatre

Schedule and format for AstraZeneca

We leave at 6:20 from Christ University

Sunday – 3rd July 2011

7:00 pm - Reporting time and setting the four venues for training
(two main halls with floor space for 30 participants to move freely)
7:30 to 8:30 pm – Introduction to Theatre – Mohammed Arif /Rajesh P I/Upendra/Bharath
8:30 to 9:30 pm – Understanding the themes of presentation for observation at Jayadeva – Jagdish/Sibu

Monday – 4th July 2011

6:30 pm to 7:30 pm – Observations and story line/format – Sibu/Jagdish/Rajesh/Upendra/Bharath
7:30 pm to 8:15 pm – Creating presentation format – Sibu/Jagdish/Bharath/Rajesh/Upendra
8:30 pm to 9:30 pm – Completion of presentation format - Sibu/Jagdish/Rajesh/Upendra

Wednesday – 6th July 2011

7:30 am to 8:30 am – Rehearsal of presentations – Sibu/Jagdish/Bharath/Rajesh/Upendra,
10 am – Presentation of theme 1,2 - Rajesh
11 am – Presentation of theme 3,4 - Upendra
2 pm – Presentation of theme 5,6 - Bharath
3 pm – Presentation of theme 7,8 - Jagdish

Incardia

Options for Script Framework-

1. An old lady (Remains common throughout all stories) narrates to a small boy who wants to become someone useful in life so that every human being can benefit from his work.

She narrates to him the story of "The man who wore a tie". (Remains common throughout all stories- but makes presence to the audience only after the last story is narrated)

Each story will site an example of the following attributes (These examples will be different from a doctor – Medical representative scenario)–

Influencer

Negotiator

Multi-Tasker

Alliance Builder

Partnership Manager

Strategist

Planner

New Account acquirer.

There is a problem depicted in each scenario and at the end of each story the "Man in the red tie" solves their problems. In the end, (after the 8th theatrical presentation) the boy asks his grandmother as to who the man in the red tie was? – she calmly replies – A medical Representative, as he has all these qualities , which makes him make the difference to the world because of this. If there were no Medical Representatives, A doctor would never know what type of medication should be prescribed to what kind of patients.

AstraZeneca
Incardia

2. OR , ANOTHER STYLE OF PRESENTATION ALONG THE SAME LINES COULD BE

A vikram and Betaal scene at the beginning of the first performance, where Betaal narrates these 8 stories to vikram , and at the end questions to vikram " who the man in the red tie was?, vikram smartly replies – all these qualities can be present only in a Medical Representative. Betaal says- correct answer, and flies away.

3. THE SAME FRAMEWORK can apply to a KBC situation, here the additional point that can be made is – a question at the end of each story where the answer can be the moral understood in the story or an answer like an influencer, negotiator etc. At the end of the 8th scenario the question for 1 crore can be- who was the man in the red tie? And the correct answer can be a medical representative.

Basic understanding of a medical representative (for facilitators only) –

A medical representative is an important person for a pharmaceutical company as he promotes a medicine to the doctor who prescribes these medicines to the end user- the patients.

A medical representative (MR) usually spends his entire day marketing and promoting the product he has been asked to promote. For instance if a pharmaceutical company like AstraZeneca has a wide range of products, each MR would be allotted various segments to promote at different hospitals and doctors(for eg, only cardiology, only respiratory medications etc). It's a stressful and demanding Job as the sales of a product reflects on the abilities of a MR.

On a typical day, the MR awaits at the doctors OPD (Out Patient Department) till he finishes seeing all his patients. The MR then enters the already exhausted doctor's chambers to promote the

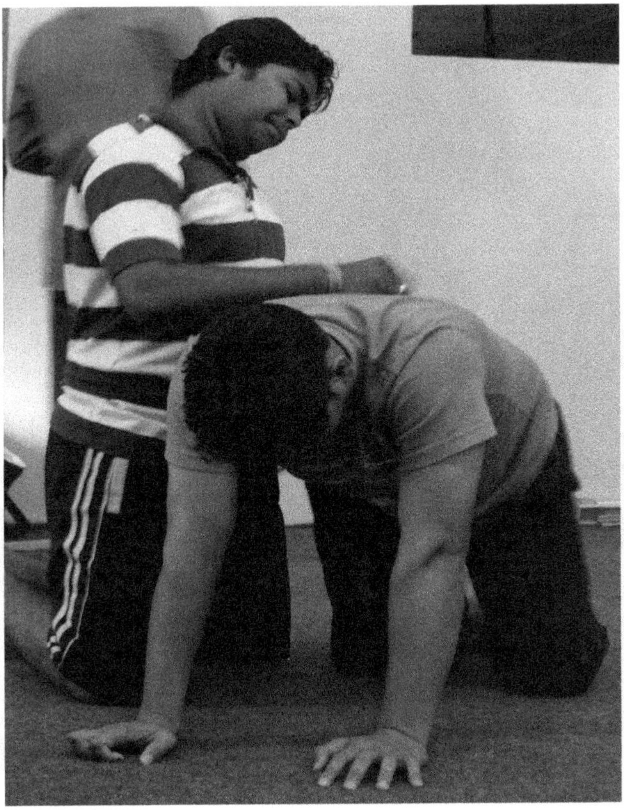

medications. This is usually done by showing charts, Placards and nowadays even on laptops etc. Many MR'S also distribute gifts, pens, pads, and also medicine samples to the doctors for trying the medications and to remember the brand names. They also organize CME's (Continuing Medical Education) , Organize and sponsor Conferences, Symposiums, Talks and host dinner/lunch parties.

In spite of all this, and the increasing competition between companies, the doctor usually prescribes a medicine at his own discretion. There could be many reasons for this – if the patient is poor, an affordable brand is prescribed. If the patient has many complaints, a medicine which has combinations of two or three drugs is used so that the patient does not have to take too many pills. Sometimes the doctor might be receiving profits or benefits by prescribing a particular brand etc etc.

So, to conclude, it's a tough occupation which is usually underrated – but the entire therapeutic team relies on these medications and their availability which an MR does efficiently by bringing to the doctor's notice the benefits of various medications.

Stories for each attribute-

1. **Influencer**: What are the influencers for the utilization for a particular medication?

   At a corporate hospital like Fortis- the Doctor at the emergency medicine or casualty usually follows strict hospital protocols for any patient that enters the casualty. If a patient enters with chest pain, there is a protocol for what the next steps are and what medications to give. Here this casualty doctor does influence the use of this medication in the casualty, whereas it is the cardiologist who has made this protocol for the duty doctor to follow in his absence or until it comes mainly influences the medications after his arrival. When the patient is in the ICU, the ICU doctor is the influencer and when patient is getting discharged the treating doctor again is the influencer. At a government set up the duty doctor, nurses and technicians

AstraZeneca
Incardia
VALFECT-II
SANJEEV

all have the ability to influence the use of a medication. On many occasions the nurses can alter charts and the technicians can promote the benefits of a medication to the doctor at a government set up. A chemist can also alter the brand by simply saying that a particular brand is unavailable at his shop and replace it with a similar brand.

Main objective- The participants must know that it is just not one doctor that influences, but many others like doctors at various stages of the flow of patient in a hospital influence the usage of medications. From Emergency room – to Cath lab- to ICU – to ward- to discharge.

Story – In the royal Mughal era, a small businessman is caught for parking his bullock cart in the no parking area. The royal soldier who identified this mistake asks for the bullock cart to be towed. The small business man begs and pleads for forgiveness but the royal soldier does not listen to his pleas and tows the bullock cart away to the royal barricade. The business man then goes to the chief sultan who was in charge of controlling crime in the Mughal city; he was a loyal warrior to the king. He asks for forgiveness and the release of his bullock cart but the sultan is stern –"your offence has no forgiveness, your bullocks will be sacrificed". Disheartened he goes to seek the advice of "the man who wore the red tie"

After meeting the "man who wore the red tie", the business man enters the royal palace. He meets the maids, the cleaners, the ministers and the royal chefs. He tells them that there shall no more be delivery of the special rice to the royal palace as his bullock cart had been towed away. If they wanted that rice to be delivered again, they would have to convince the king to release his cart. Soon the king questioned why his special rice was not served and the queens,

SANJEEV
Incardia

maids and chefs explained to him the reason. The next day the sultan was called and asked to release the bullock cart to the business man after charging him a fine instead.

Moral – it's not only the main persons who are always important, lots of others can play a role in influencing an outcome.

2. **Negotiator + Alliance builder**

Negotiator- In an institution there is a purchase committee meeting at fixed intervals usually at 2 months) durations to decide what medications should the hospital pharmacy purchase? This committee usually comprises of a group of doctors and administrative faculty. The MR keeps a track of these meetings and influences these committee members to promote his companies brand in the next forth coming meeting. At the same time applies the medication proposal to the institution for the next committee meeting so that the brand is accepted.

Alliance builder- This is to describe how he has to bring various stakeholders across departments including doctors and pharmacists on the same table. Example could be him conducting an activity across 2-3 departments i.e getting speakers from diabetology to speak to cardiologists about holistic treatment and dealing with co-morbidities.

Story- In the land of the jungle ruled by a lion king, there lived a horse that was struggling to make a living. The horse would mainly run in search throughout the day to bring a rare breed of hairy rabbits into the forest for the wild animals to hunt. This was for his own survival, as the horses would be the next targets once the deer and cows are finished. This rule was followed in the jungle and only when a particular species is eaten to fulfilment, then only the next in line would be hunted. There were very few deer's and cows left, which made the horse worry even more. He goes to the tiger and says why you consider eating these rabbits

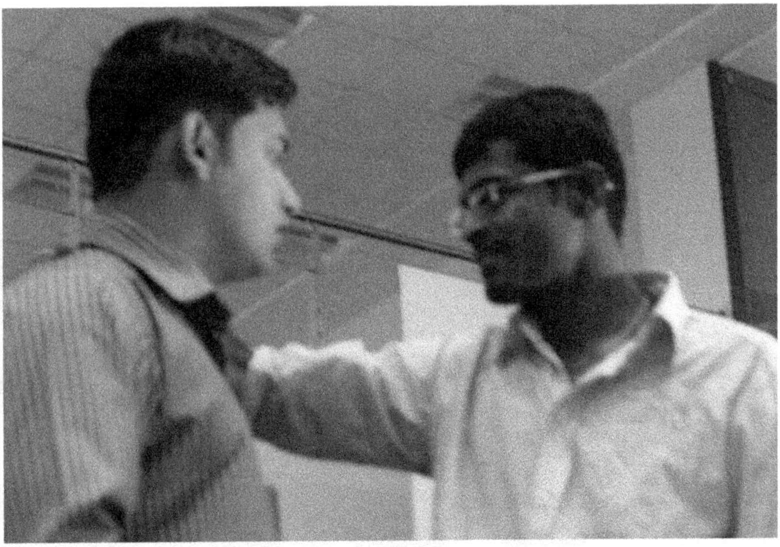

next. The tiger smirks, "rabbits? I thought we ate them all". The horse explains that these are a rare breed that has survived and there are lots of them. The tiger is not pleased – "why run behind rabbits when fine meat like your self is next in serving". The horse is dismayed and goes to the "man who wore the red tie". After receiving the advice the horse asks the tiger, who takes the decision on these rules of which animal gets eaten next. The tiger replies" the lion, the elephant and the peacock.. don't worry, they would never agree to include your hairy rabbits for hunting anyways".

The horse then organises a fitness meeting where all the animals from different domains are invited to check if they were overweight. The lion, elephant and the peacock were also present. the horse invited a doctor cheetah to give a talk who reported that 80 per cent of the animals were turning fat and are at risk for heart disease. He advised them lots of exercise and high fibre in diet. The horse then mentions to everyone that these hairy rabbits were high in fibre and tough to catch, so would be great exercise while hunting. In the next committee meeting the rabbits were decided as next on target.

3. Negotiator + planner

A MR strategizes how a medical product / medicine can be sold in large numbers. So in addition to marketing it to individual doctors, the MR tries to sell the medication to the pharmacy. The pharmacy usually requires some number of prescriptions(20 -30/month) to come from the doctor in order to procure a particular drug into the pharmacy. Therefore even though that particular brand is unavailable in the pharmacy, the MR asks the doctors to

prescribe that same brand, just so that it bounces of the pharmacy counters and the pharmacy then decides to procure these medications.

Story- The fashion industry was at its peak, when a crazy fashion designer came up with the idea of creating a furry tail out of animal fur. He was very proud of his creation and was certain this would make a fashion breakthrough. He calls his marketing team and asked them to have these tails sold at traffic signals. They all thought he had lost his mind but reluctantly agreed to do what he ordered. The sales were zero and no body wanted to buy a tail that they could attach to their bottoms. The sales men were frightened that they might get fired if they do not sell any tails. They then go to the "man who wore a red tie." Soon after this they all head out a clothes showroom all dressed as normal civilians. Each of them goes to the shopkeeper and asks for a tail that matches their clothes. Initially the shopkeeper finds it amusing, but as more and more of these "civilians" ask for such tails he begins to think it can make a good profit. After a while a sales man dressed as himself enters the showroom and asks if they were interested in purchasing the latest hit in the fashion industry- tails for clothes. The showroom executive places a big order and the sales men sell their quota for the day.

4. **Multi-tasker + partnership manager:**

Multi-tasker : This is to bring out how he plans his tasks and balances multiple priorities. For example, he would have clashing appointments with doctors, doctors requesting reps to sponsor meetings, respond to data requests from his line managers, wait for a doctor and hence miss calls and sometimes not fulfil his day's call quota.

Partnership Manager : This is to describe how he has to bring various stakeholders across departments including doctors and pharmacists on the same table. Example could be him

Being human

conducting an activity across 2-3 departments i.e getting speakers from diabetology to speak to cardiologists about holistic treatment and dealing with co-morbidities.

Story- A 35 year old happily married man was living with his lovely wife and 2 children in a 2 bedroom flat in Bangalore. He was a successful businessman with a flourishing business. He had 3 relationships with three other women apart from his wife. Typically he would spend 3 months with each of them, claiming to tour the rest of the year for business, but to his surprise he was told that all the three are arriving to Bangalore to meet him for one day, each unaware of his other existing commitments. He could manage all of them when they stayed in their own cities but had no clue how he would manage them together. He then went to the "man who wore a red tie" and took his advice.

In a days time he organized an "employees reunion party" event where all the old employees would meet for a get together. In a short notice many of his friends, colleagues and employees turned up. He told his other relationships individually to meet him in the party as he has organised it. In this party he brought along his wife and openly met and greeted his other three relationships one at a time. Each woman thinking that he is meeting an old colleague. After spending time with each of them, they all left back to their cities again.

5. **Strategist + new business acquirer** – Understanding where the prescription is getting switched/substituted and taking action accordingly; understanding what is the drug policy/protocol of the hospital; understanding who are the doctors who could give you more business & how to get new departments to give the rep business.

MPLY SANJEEV

Nurses and technicians are also useful influencers. This means that If the MR is in good terms with these nurses and technicians, he can basically monitor whether the doctor is prescribing that particular brand or not. If not, then what is the reason and how to tackle it.

Story- This is the first day of a hard working Medical representative, who tries really hard to impress the doctors to prescribe his companies medications. But when he checks at the pharmacy he realises that there have been no prescriptions for his brand medication. He continues to work hard day and night, yet no prescriptions. He seeks the advice of " the man who wore the red tie" and takes his advice.

The next day the MR be friends the nurses and technician and asks them to find out whether the doctor is prescribing his brand name. To his surprise he finds out the the doctor has been prescribing only his brand name. But why are there no sales of the medications. He then convinces the technician to pretend that he is a patient and takes his brand prescription to the chemist; the chemist says no stock for that medication and prescribes him a similar brand. That's when the MR realises that the switch was occurring at the chemists shop. He makes a deal with the chemist to give him a share for issuing the drug mentioned. The MR also finds out from the nurses and post graduate students, who is the doctor who makes the medicine protocol and specifically promotes the medications to him. Soon the sale of his brand increases.

THE TEAM !!

Medical Theatre Session
National Workshop on Research Methodology
August 2011

Following the introduction of medical theatre in the form of two workshops held at St John's Medical College, immense amount of interest spurt out from students and faculty of various departments in the hospital. One of the most eager organisations that were very enthusiastic about applying medical theatre was the St. John's College of Nursing. On special request a session was introduced into a national workshop on research methodology. This special request triggered a thought process that structured this innovative teaching methodology into a one hour session.

In order to carry out such a session in front of close to 500 persons, a team of enthusiastic nursing students were trained for a performance that would involve dramatization of a story conveying educational content. A plot was created on "Hypothesis and framing of objectives". The nursing team underwent 8 hours of practice and training distributed over 4 days and in the process learnt the subject as well as its methodology for conversion. The performance was then carried out on the day of the event along with the entire team.

The 1 hour talk was presented to an audience of 500 participants using Stand up comedy and dramatic narration. This unique presentation was well accepted by participants and teachers.

A Statistically Significant Performance

The statistics script

A HAND BAG OF ERRORS

SCENE 1

In an all India villan and robbers association conference, the pick pocketing associations chief in charge raises some concerns to the president of the association.

Chief- sir, it is with deep regret that I bring to your notice our poor success over the last few months. We are helpless to find out what has been going wrong?

President- why? What happened? Your team is our pride and we respect you for having been the most successful pickpocketing team in the entire underworld and the underuniverse. What is this that has caused your downfall?

Chief- well sir, there have been significant failures from the hand bag robbing division mainly, as we were unable to rob these handbags successfully.

President- you were outdone by ladies? Is that what you are trying to say? ladies got the better of you?

Nurses in Action !!!

Chief- I have a few real life scenarios .

Robber 1- sir, I was robbing the hand bag of a lady in Jayanagar, as soon as I got hold of the bag and was about to run... suddenly from somewhere she sprayed something into my eyes and it burnt for hours.

President- That must have been pepper spray you idiot.. Havent you read that you must always wear glasses before robbing?? Do you know the three advantages of wearing glasses ? and the soda buddi classification?

Robber 1- No Sir.

President- Read up and come.. I ll ask you tomorrow.

Robber 2 - Sir, When I tried to rob another lady from jayanagar area, I took the hand bag and ran..

President- Very Good

Robber 2- But it started ringing loudly like an alarm until people around me identified me as a robber and beat me to pulp.

President- Did that purse have lots of colours on it? (robber nods

The Underworld and Under-universe

vigorously) You should be ashamed... A simple alarm you didnt know how to diffuse.. You must always avoid hand bags with too many colours. what is the coloured bag sign?... dont know?? read up and come.

Robber 3- Sir , I robbed a hand bag from a lady in jayanagar successfully.

President- Shahbhash!! my boy shahbhash!! see learn from him...

Robber 3 - But when I ran for a few minutes I was arrested by the cops..

President- what?? how >> ..

Robber 3- Sir ,, the Hand bag had a GPS enabled detecting device...

President- Technology.....What is happening to the world we are living in? .. how has this revolt against us started?... we must stop it...Now its in Jayanagar.. Tomorrw it will be in Koramangala.. Then in mekri circle...What will we do.. what will our kids eat.. what will our children become..all our future is at stake.. we must act soon..

Chief- Sir..What if it starts in other areas like bank roberry division and

ATIONAL CONFERENCE
ON
SEARCH METHODOLOGY
2011

Gambling departments.

President- We must act fast.. we need to fix these errors

Robber 4- errors? Sir.. I think I know who can help us..

President and chief- Who? Who?

Robber 4- I was once robbing a group of doctors who were attending a Biostatistics workshop.. It wasnt difficult as all were either asleep or too confused to respond.. and the person speaking was going on claiming that he can identify errors... I still remember that day when I almost found it difficult to escape as I my eyes too were closing.. But I was determined and got out of there.

President- he can fix errors? bring him here immediately.

SCENE 2

Entry of a man in formal dress enters being escorted to the president by two thieves on either side who have already stolen his wallet. The statistitian is terrified...

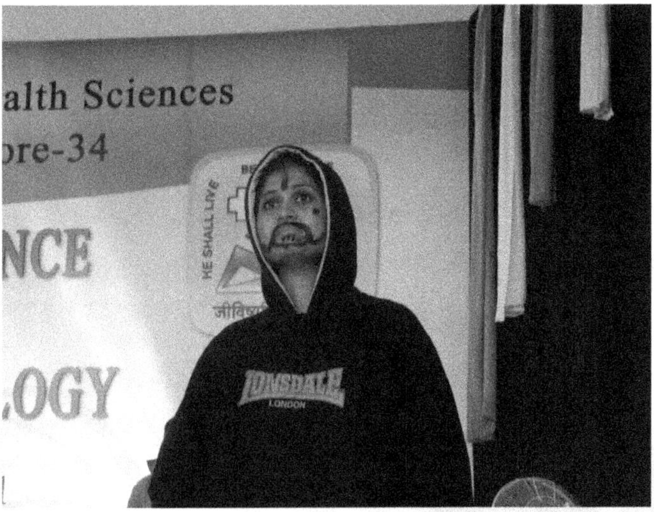
alth Sciences
ore-34
NCE
OGY
HE SHALL LIVE
LONSDALE
LONDON

ademy of Health Sciences
rsing, B lore-34
BECAUSE OF ME
HE SHALL LIVE
NCE

President- Welcome O error detector. I have heard great things about you. Well you must be knowing that there is a very small difference between me and you.. you find error and we find terror.. ha ha ah aha ah ahhha hahah ah a.. whats your name?

robber 4- Sir.. his name is Sunder Prasad Shivashankar.. They call him SPSS for short..

president - Did I ask you??? did I ???..... (to SPSS).. Why are you so scared?

SPSS- is my hypothesis true if I state that the purpose for you to bring me here is the same thought that has 90 percent prevelance in every mind when they think about someone being picked up by a group of thugs and that the likelihood of you robbing my valuables has a 95 percent confidence interval of 97 - 99.

President- what did he say?.. did any one understand that?

robber 4- sir.. when I attended his session during my roberry at the workshop, I learnt a few basics that he was teaching. may be I can try to interpret..

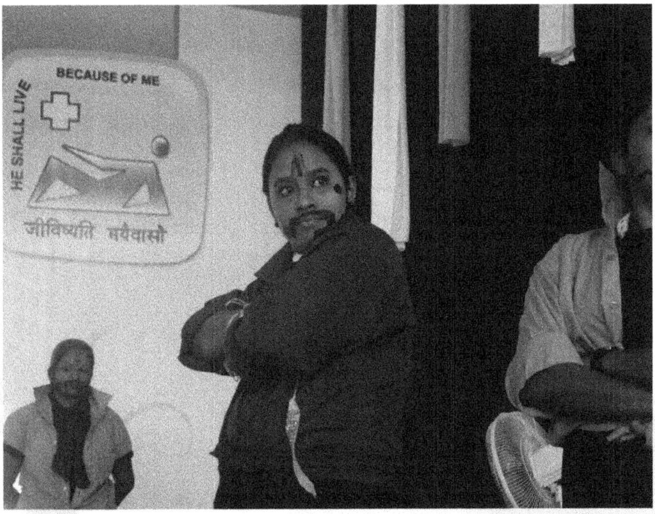
BECAUSE OF ME
HE SHALL LIVE

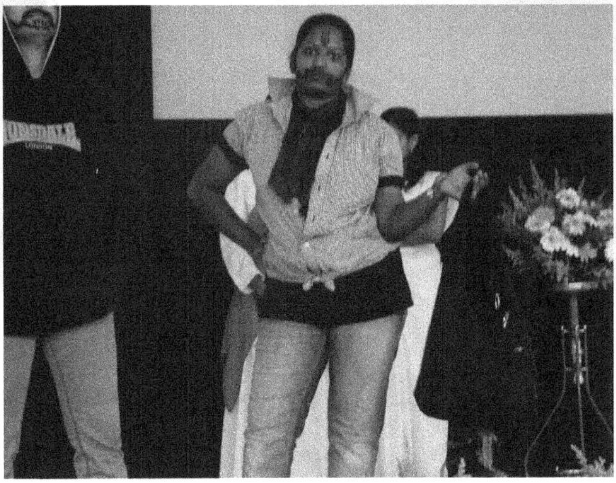

President - Ok... But what did he just say?

Robber 4 - sir .. he said that he belives we are going to steal from him.

President- are you sure thats what he said, cos he spoke for a pretty long time.. anyways.. dont be scared.. we want your help. we want you to solve our problems.

SPSS- oh.. well.. thats some relief..do you want me to teach you statistics?

President- no no.. we want you to help us be better robbers.. we have some problems we are facing

SPSS- what problems? you need to be specific..

President-we need to find out why some women have started to get brave enough to fight back with us and spoil our hard earned efforts in making our living.

SPSS- See you are begining to get very vague again. you need to tell me very specifically what you want me to find out. Your objectives need to be framed properly otherwise the results will get effected.. for example

do you want to estimate the incidence of women that are brave enough to revolt against robbers in bangalore?

or do you want to estimate the fear levels in girls who carry handvags?

or do you want to estimate the fear levels in girls with handbags who shop in groups with friends or in those who shop alone?

or do you want to estimate the ammount of money in the handbags of thin girls, normal sized girls and fat girls.

or do you want to estimate the incidence of which girl will attack the robber back, the one who was robbed by scaring or the one who was robbed by being friendly.

or do you want to see the strength of association between size of hand bag with ammount of money in the hand bag?

President- bring me my gun or immediately translate all that he just said?

Robber 4- sir.. all that he is asking is for us to clearly tell him what exactly we need to find out about these women.

St.John's Nat
College
NAT
RESE
9th

St.John's Nat
College
NAT

President - Now that you put it this way, maybe we can try and find out whether robbing by being friendly is more useful.. I know a gang who used to do that.. but never heard from them again... maybe they did really well.. anyways..

well we need to know only two things..

1. which is a better method? robbing girls by being scay or by robbing girls after making friends with them?

2. and what is the average ammount of money they carry in hand bags.

SPSS- OK .. so you want me to estimate the ammount of money in the handbags of these girls and also want me to estimate the incidence of which girl will attack the robber back, the one who was robbed by scaring or the one who was robbed by being friendly.

president- well that made sense and thats what I want to know. How long would you take?

SPSS- maybe one week, but I will need your people to help me out.

President- Anything my friend, they are all yours.

SPSS- Here is what we need to do.. this group here goes to jayanagar and snatches the hand bags of 25 gils of age group 18 to 25 years by using the conventional scaring methodology. You two will act

Medical Theatre

- It was a new experience.
- It is an effective way of teaching a complex concept like statistics.
- It was a great experience working with you sir.

Lovely David

1st year MSc. Nursing

as news reporters and interview these girls after the robbery has taken place and will ask them these 5 questions.

That group goes to jayanagar and snatches the hand bag of 25 girls of age group of 18 to 25 years by being friendly with them first and then robbing their hand bags. You two again will act as reporters and interview these girls for feed back.

The filled feedback forms will be brought to me and you .. over there.. you will count how much money was in each of the hand bags and submit it to me. After this I shall run it on my computer and shall tell you the results.

President- Well , that doesnt sound too difficult.. lets begin..

1 Week later all the data is collected from a sample of the universe and submitted to SPSS.

SPSS- Hmm.. I have run the results on my computer and an interesting finding has surfaced.

pesident- what ? what? tell us.

SPSS- well the incidence of hand bags being robbed by comparing the two methods, robbing with fear and robbing with being friends shows that robbing with being friends has a higher number of

Name – Madonna D. C. Ferns

* It was an amazing experience working with you sir. the practice sessions were fun and innovative and absolutely spontaneous.

It definitely made statistics classes to me, especially the parts I was finding hard to understand.

You were GREAT and if given a chance to take part in a medical theatre again, I'd gladly participate.

All the best!

Madonna Ferns

MSc (N) 1st yr.

robberies. And it is statistically significant with a p value of < 0.05. Interpreting that robbing a hand bag sfter being friendly is a much beter methodology.

President- wow... so that what we have been doing wrongly. robbing by being friendly is much better. are you sure? ae you confident?

SPSS- well the 95 percent confidence interval for robbing by being friendly is 97-99.

president- what does that mean? what language is this.

robber 4- sire. he means that if we rob the hand bags of hundred girls , 97 to 99 will be robbed successfuly by the being friendly method.

President- ok.. thats good news.. and what about the ammount of money in the purses?

SPSS- Well, this was a quantitative data and the mean ammount of rupees in the stolen hand bags was roughly 1500 rupees . And the maodel was

President- model?/? where?? where is she

Chief and robbes- model? where? we want her?

SPSS- The statistical model is mean + and - standard deviation of 500.

Hi Dr. Jagadish,

Working with you in the Medical Theatre was an awesome experience. I think, you have the ability to put the most complicated concept into a simplified way through your ability to organise Medical theatre. We will miss you. I'm sure you can steal the show regardless of the number of specta-tors. ~~we~~ I wish you good luck in your future endeavours.

Lotsa wishes & prayers

[signature]

(Raja Uma Veni)

President- Areey.. what is this that you are saying again.. I am not able to understand anything.. tell him to talk nomally or ill shoot his brain into peices.

Robber 4 - sire.. he means that the ammount in the purse is about 1000 to 2000 rupees.

President- ar you confident?

SPSS- well the CI is 92- 95.

President takes out his gun

Robber 4- sir, sir -- he means that if we rob a hundred purses then 92 to 95 times tha ammount in the purse will be 1500 + or - 500.

president- Hmm.. we ar happy with you O SPSS. You may go now and thank you for helping us. I hope there ae no erros in you study.. or else you will see a different side of us

SPSS-- Well when you apply these results into the universe again, if thee is an error then thats the standard reeor

President- take him away before I do something to him.

Name: Monica Hendricks

* It was a very novel experience
* It made statistics really interesting
* It was fun-filled easy learning
* Comprehension of concepts was even more clear
* It expressed serious the topics in a lighter manner.

Monica
4 BSc Nursing 2011

Now, having known these new results, the robbers apply this to their target population. within on week the entire robbery crew is missing. only the chief is left behind.

Chief- sire.. sire.. that SPSS fooled us.. he made a mockery out of us..

Psident- what? what happened?

Chief- sir.. all our robbers have gone missing..

President- but how? who took them?

Chief- No one took them.. they ran away them selves.. they started being friendly with girls and the girls liked them back. They didnt have the heart to steal from them so they ran away with them... we are doomed... our entire department has fallen down to nothing. and there was no money in the purses.. all wee filled with cedit and debit cards. he lied to us and played with our emotions.

Prsident- The SPSS will pay for this.. he told us something that was not true as true.. this is a grave error on his part. Kill him...

soon the SPSS is found and hacked to death. SPSS then goes to heaven where he meets god.

God- SPSS ? YOU ? WHAT ARE YOU DOING HERE.. I THOUGHT YOU HAD ANOHER 20 YEARS BEFORE YOU WERE SCHEDULED TO DIE?

Name : Ms. Nishanthini. R.

– Medical Theatre on Research and statistics was very interesting.

– It made the topic simple and easily understandable.

– It was an excellent experience ~~for~~ to participate in medical theatre.

–

Nishanthini
(MSc. Nsg)

SPSS- O holy one.. I made a type 1 error. The result of something false came as statistically significant and thats why I have been given this fate.

GOD- BUT.. YOU ARE ONE OF THE VERY FEW WHO KNOWS STATISTICS. IF YOU DO NOT TEACH THIS TO MORE AND MORE PEOPLE, THE WORLD WILL GET MORE AND MORE CHAOTIC.. THIS CANNOT HAPPEN. ILL SEND YOU DOWN AGAIN AT THE TIME WHEN YOU WERE RUNNING THE TEST ESULTS. THIS TIME USE YOUR BRAIN...

soon SPSS was sent down to earth as he realised that it was the time when the data was submitted to him and he was running the results on his computer

SPSS- Hmm.. I have run the results on my computer and an interesting finding has surfaced.

pesident- what ? what? tell us.

SPSS- well, the results show that robbing by being scary is better than robbing by being fiendly but the results are not statistically significant. May be this is because we have a smal sample size. Further studies will be required to get moe conclusive data.

President- what does this mean

Medical theatre workshop was an amazing experience. Thanks to Dr Jagadish, we had a whole new insight into theatre skills. A combination of theatre & medicine was something ~~so~~ so unique. A lot of fun & interaction with new people. Both the sessions ~~through~~ I attended were innovative & we were made to write our own scripts because of which I've never had to read again. Thanks to Mr Upendra who kept the sessions alive with a lot of games. Not only did this help us improve our medical knowledge but also our theatre skills. On the whole a wonderful experience!

(LAVIENA MALLELA)
laviena_sizz@yahoo.in

Robber 4- sire.. what SPSS is trying to say is that our method for robbing by scaring these gils is better that robbing by being friendly but it is not significant by the soft wae. he says w need to coninue this on larger population

President- well if you know so much then why do we need spss. I appoint you as our new statistitian and ask you to take this study forwads. lets get rid of SPSS. set him free.

SPSS- phew... this type 2 error saved my life. Here something true came as statistically insignificant but atleast it doesnt kill any one. thank you god. thank you for the second chance.

Hospital Soft Skills and Continuous Quality Improvement in NABH
Vasan Eye Hospital, Jayanagar
August 2011

National Accreted Board of Hospitals (NABH) maintains a rigorous set of rules and regulations that any hospital needs to possess in order to reach their standards. These rules were converted into theatrical outcomes in the form of a story narration and tableau games. This methodology turned out to be a very effective tool in communicating NABH content using medical theatre. The 85 participants enjoyed the experience and learnt through it effectively.

a. **Procurement of medication pertaining to Patient needs**- Doctor prescribes a medication and it should be available.
b. **Meeting laws and regulations** – patient privacy if 2 or 3 are being seen at same time, patient files being accessible to strangers, patient confidentiality. Female pts should have a nurse or attender during examination.
c. **Monitoring risk Managements**- risks related to electrical shocks, fire, patients fall down from beds if no rails present, radiation related. Laser beams, infections (sanitizing solutions with foot pedals at various places).
d. **Utilizations of space and man power**- no seating available for pts and family members. Less equipment and more patients. Man power for investigations and scanning.
e. **Employee satisfaction**- working hrs (8 hrs – every 2 hrs there should be a break), salary levels and upgrades. Doctors may shout at sisters and ward workers.
f. **Adverse events and near misses**- wrong operation. On wrong side, or wrong medications. Allergies or reactions. Out dated eye dilatation medications can cause blindness.

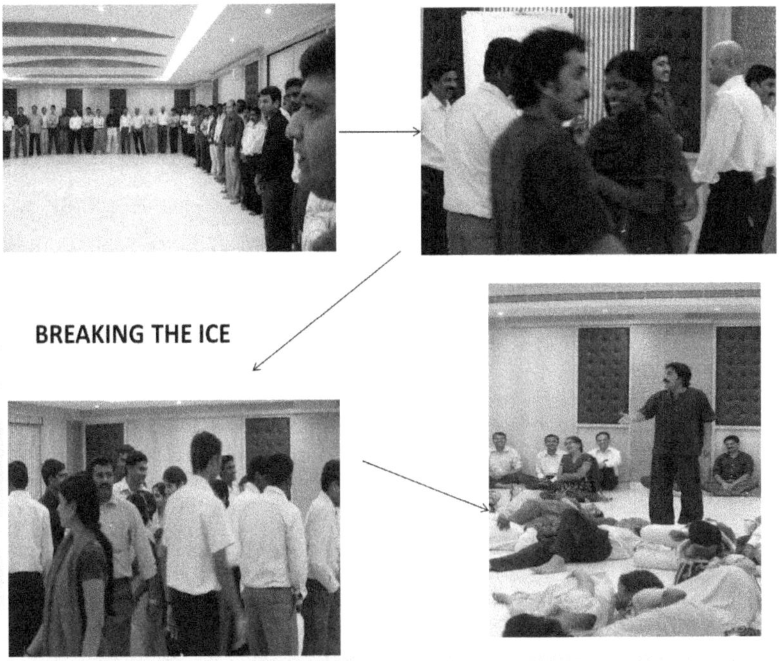

g. **Data collation and support evaluation of the improvements** – summaries, data/ audit, quality of documentation. 1 year data should be collected and kept. Data must be reviewed and analyzed.

Medical Theatre for Hospital Administration and staff

Objectives – 1. To introduce an innovative methodology by which certain NABH guidelines can be learnt and understood effectively.

Body – For these group of hospitals we mainly concentrated on NABH content, specifically on Continuous Quality Improvement (CQI). The National Accreted Board of Hospitals gives a set of guidelines that a hospital needs to follow in order to receive accreditation of NABH. We conducted a half day intervention for a group of hospital professionals including doctors, nurses, customer care, housekeeping and administrative faculty. Between theatre games and timely breaks, the content of CQI was dealt with by facilitating the participants to create Tableau's or still images using each other in groups. The rest of the participants tried to interpret which of the CQI content was depicted in these tableaus. They also experienced a story narration and created a short theatrical presentation on certain soft skills such as punctuality, attitude, business mindedness and dedication. The group took some time to get comfortable as the hierarchy was intimidating, however as soon as the events started to get more and more interesting the group united together to experience a fun filled yet educational process.

Conclusion- At the end of these workshops we concluded that this process can be utilised to create awareness on NABH guidelines amongst hospital professionals from all backgrounds. The effectiveness of this methodology was demonstrated by reaching out to non- medical and administrative members creating importance of NABH in an effective manner. It turned out to be a useful methodology when it comes to learning and understanding a set of rules and regulations.

Yukti overlooking at Zandig

A Theatrical Explanation !!

Session Plan
August 2011

Institution: Vasan Eye Hospital

Subject: NABH

Topic: Continuous Quality Improvement + soft skills

Class: Doctors, Customer care, OT Staff, Housekeeping.

Objective:

1. To introduce an innovative methodology by which certain NABH guidelines can be learnt and understood effectively.

Requirements: 1000 sq. feet room with empty floor space, an LCD projector, a laptop, a guitar.

No. participant: 85

Opening activity: Walk and shake

Core activity: Content Tableau , story narration and soft skills theatrical Presentation.

Closure activity: De- Briefing session

Tips – props, feedback forms submit and collect before session, foot ware outside the room.

Forward link: nil

TABLEAU – MY GRANDMOTHER'S HOME

MEDICAL THEATRE WORKSHOP - VASAN EYE HOSPITAL, JAYANAGAR

23/08/2011- Tuesday, 1:30 pm to 7:00 pm

- **1:45 – 2:00 – Introduction – Sibu, Jagdish**
- **2:00 – 3:00 – Theatre games – Upendra**
- **3:00 – 3:30 – Tableaus – Arif**
  Introduction to Tableaus

✓ Maiya's hotel - *Composition*
✓ Animals in the zoo – *Body position*
✓ My grandmother's house – *Story picture*
✓ Tree of life - *Symbolic*
✓ Vasan Eye Hospital – *My truth*

- **3:30 – 4:15 – Practical Tableaus – Arif & Jagdish**
  For CQI- Continuous quality Improvement

1. Procurement of medication pertaining to Patient needs
2. Meeting laws and regulations
3. Monitoring risk Managements
4. Utilizations of space and man power
5. Employment satisfaction
6. Adverse events and near misses
7. Data collation and support evaluation of the improvements
8. Patient satisfaction including waiting time

- **4:15 – 4:30 – Tea**
- **4:30 – 5:15 – Movement - Upendra**
- **5:15 – 5:45 – Story – Jagdish**
- **5:45 – 6:15 – Create story presentation - Arif, Rajesh Jagdish, Bharat**

a) **Importance of Communication** - A corporate office boss has an important meeting for a huge deal with another MNC in a few days time. Due to his short temperament and anger outbursts, all his employees are very terrified of him. He has fired many people from their jobs due this behaviour. His secretary has made all necessary arrangements for his talk with the MNC. She just receives information that the meeting has been preponed to today and her boss is expecting it to be a few days away. She is afraid to tell him but fears he might get angry at her for this. She tells a few of her friends to pass the message along. They try but due to his anger but do not gather courage enough to tell the boss. Finally 1 hour before the talk the secretary tells this to her boss. He is angry, furious and anxious as to why the secretary did not keep him prepared. He is furious and storms into the toilet. The MNC team begins to gather outside his office for a talk. He comes out in anger and takes his papers and rushes out to make his presentation for the meeting. The secretary is shocked but hesitates to tell her boss that he forgot to wear his pants... The deal is broken as the boss

TABLEAU – SPIRITUAL TREE

has been a laughing stock and the secretary is fired from her job. *Moral – Communicate however bad the situation is, it will be worse if it is not communicated.*

b) **Attitude** - A well-established restaurant called "fast food center" that has been doing moderately well now has a new manager who has taken over the business. He makes a new rule that all the waiters will address the customers by addressing them as brothers or sisters, mothers or fathers. The new rule also includes apologizing to the customer at the time a mistake has been made and replacing the dish free of cost. No tips will be accepted by any waiter. Hearing to this all the waiters are unhappy and make fun of these strange rules and regulations. A few of them hesitantly follow the rules but are ridiculed by their peers. Slowly all the waiters stop following any rules and continue to be their rude and irritable selves, always unhappy with their work and making their living out of tips. The manager notices this behaviour and decides to sell off the hotel to someone else as the staff is very un co-operative. He opens a restaurant called" home meals" in the Next Street and employs people who would to listen to him. Slowly people stop coming into fast food center and everyone in the hotel faces a huge loss. On enquiring a few customers they found out that home meals was a very homely and hospitable restaurant where all the waiters were friendly and more understanding. People loved that approach and it is now one of the leading hotels in India. *Moral - No matter whose mistake, a positive and understanding attitude makes a huge difference.*

c) **Concept of Business** - In a royal kingdom the king decided to set up an advisory board for common men to share their problems.
He paid the advisors well but for some reason not many common men found their satisfaction from the solutions given by the royal advisory board. As it turns out, one of the kings best ministers has his own advisory board in the village which was successfully solving all the problems of many common men in the king dom. The king disguised himself as a farmer and went to the minister's advisory with a problem. To his surprise he found he was very well looked after, was given shelter, was well listened to and was given a very wise advising for his problem. The next day the king calls his minister and asks for the secret behind his success. The minister smartly explained that since his advisory runs on funds given by a few zamindars, he and his fellow advisors work to their best potential to ensure adequate returns so that more and more funding comes to his advisory. Also he uses the advisors houses, land and food as resources to help the needy , hence making a very effective problem solving team. The king realized that since his advisors were getting a fixed pay, they never tried too hard to do well and eventually led to the downfall of the royal advisory. *Moral - even if you are employed by a company, to understand its business potential and help in generating revenues if very important for the sustainability of the company.*

d) **Punctuality and Discipline** - The king of the jungle wanted a wooden throne to be made for the upcoming festival in the jungle kingdom. He called for the rabbits that were the best in business. He asked how long it would take for the throne to be made. The rabbit replied at

THE PERFORMANCE – CONCEPT OF BUSINESS

the fastest it would take a week. A monkey from the crowd shouted that he knows someone who could make it in three days. The lion asked for that person be brought in front of him. The monkey brings the turtle in front of the king. The rabbit chuckles that a slow turtle can never make an entire throne faster than the rabbits. He mocks the king for listening to the monkey. The lion is annoyed but asks the rabbit to go ahead with his team to make the throne. He asks the turtle to make the throne as a back up. The rabbit gets his team together. He gathers his team together. It takes a day to get everyone as all have returned from a vacation. He sends a few to get the wood. And the rest to get the hammer and nails. He also calls for the polisher who would do the final finishing for the throne. His rabbit team takes their own time, laughing and cracking jokes on the turtle. Soon a week is over and the throne is almost done. It takes a couple of days extra but the throne looks fabulous. The rabbit takes it to the lion to watch his breath-Holding look. To his shock the king is sitting on a huge and magnificent throne. The rabbit is speechless.

The king explains how the turtles made the throne in just two days. The rabbit asks the turtle how they did it. The turtle simply replied that they work on time every day with punctuality and determination. The wood is stored in their backyard throughout the Year and basic wood pieces are cut out since months together. When an order is placed they only need to modify the shape and polish it.

Moral- Punctuality and discipline can make even the slowest, the most efficient and powerful.

- **6:15 – 6:30 – Showcase Performances**
- **6:30 – 7:00 – Debrief – Sibu**

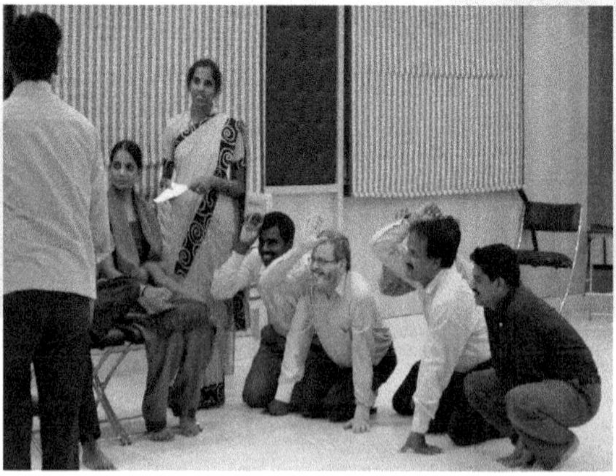

A still from the performance on – Punctuality and Discipline

ST. JOHN'S MEDICAL COLLEGE

BANGALORE - 560 034

TESTIMONIAL

I have had the pleasure of interacting with Dr. Jagdish Chaturvedi on a number of occasions, both academically as well as in extra-curricular activities. Dr. Jagdish has been actively involved in organizing 'medical theatre' workshops in St. John's Medical College over the past year or so. The idea behind this concept is to employ techniques used in theater to medical education. As a member of the medical education department I became interested in the concept and wanted to try out this technique in the setting of a conventional didactic lecture session. The topic that I had been assigned to take was 'An overview of the major nerves of the upper limb', which had to be covered in an hour for sixty first year undergraduate medical students.

A PowerPoint presentation with the learning objectives and the important points to be covered was prepared before the session. Once the objectives of the class were defined, Dr. Jagdish helped to write out a script based on important points that would have been covered during a lecture. A story was created for each of three major nerves, the median, ulnar and radial nerves. Dr. Anu, one of our postgraduate students, who was interested in the concept, enacted the script which was narrated by Dr. Jagdish. Once the script was enacted for each nerve, itwas then linkedto the PowerPoint presentation which covered the important points for that particular nerve.

Once the concept became clear to the students, they were divided into five groups of twelve students each. Two groupsof students wrote a script for two other nerves, musculocutaneous and axillary and followed the same process described above. The three other groups created a tableau for three nerves, the long thoracic, thoracodorsal and spinal accessory. They had to freeze in a particular frame that depicted the important points about that particular nerve and the rest of the class had to guess the nerve that was represented.

While no formal assessment was conducted, it was amply evident that the students enjoyed the session and actively participated in the learning process. The success of this new concept was almost entirely due to the enthusiasm and skill of Dr. Jagdish and possibly paves the way for its more widespread use in didactic lecture sessions.

Dr. Nachiket Shankar
Associate Professor
Department of Anatomy
St. John's Medical College
Bangalore

Medical theatre on Anatomy of the Nerves of the Brachial plexus September 2011

As the popularity of the concept of medical theatre grew around St. John's Medical College and Hospital, it was brought to the notice of their Medical Education Cell. It was discussed with regards to incorporation of this methodology into the current curriculum. A debate on whether this type of theatrical process would be possible to execute in a one hour session without compromising on the amount of content. Leading from this the Department of Anatomy put their first step forwards in experimenting a session with this methodology.

A one hour session was created for all the nerves from the brachial plexus by using stories in one half of the session and asking the students to make stories on the remaining nerves along with tableau formations for some nerves. The session concluded in time with all the theatrical methodologies applied, with a phenomenal positive response from the students along with the display of the standard content in the form of a power point presentation, all in one hour. This has proved the application of this methodology to be effective in a one hour session within the existing curriculum.

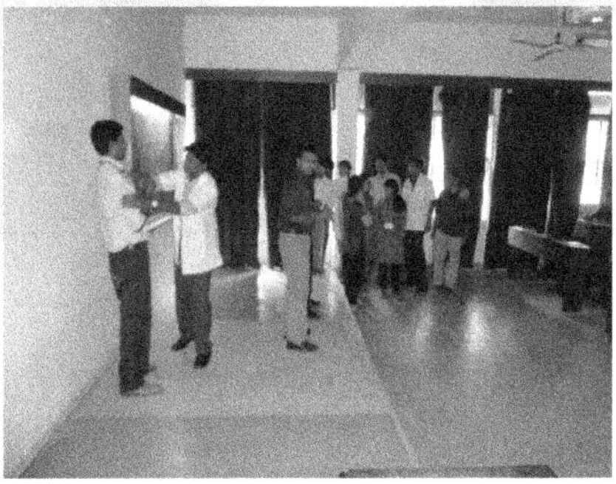

Medical Theatre for undergraduate students in a one hour session on a specific topic

Objectives – 1. To introduce an innovative methodology by which a specific topic can be learnt, taught and understood in a one hour session within the normal curriculum.

Body – The concept of medical theatre explores participation rather than watching or listening to a particular session. It was therefore challenging to identify how a specific session can be conducted using medical theatre. The structure employed was narration of created stories on specific topics followed by their respective power point presentations. This was followed by students being asked to create stories on the spot along with creating tableaus on topics that have not yet been covered.

A one hour session was therefore created for all the nerves from the brachial plexus by using stories in one half of the session and asking the students to make stories on the remaining nerves along with tableau formations for some nerves. The session concluded in time with all the theatrical methodologies applied, with a phenomenal positive response from the students along with the display of the standard content in the form of a power point presentation, all in one hour. This has proved the application of this methodology to be effective in a one hour session within the existing curriculum.

Conclusion- At the end of this session it can be concluded that this methodology can be as effective and informative even though implemented in a short duration. However the preparation and process for creating the content and structure takes significant effort prior to the session. This event also emphasises the importance of creating the structure based on the audience and content concluding that the participants who are going to experience the event, are the main deciding factors for the format and outcome of the event.

Session Plan
September 2011

Institution: St John's Medical College

Subject: Anatomy of the upper limb

Topic: Brachial Plexus

Class: 1st year Undergraduate students

Objective:

1. To introduce an innovative methodology by which a specific topic can be learnt, taught and understood in a one hour session within the normal curriculum

Requirements: 1000 sq. feet room with empty floor space, an LCD projector, a laptop,.

No. participant: 60

Opening activity: stand up comedy

Core activity: Content Tableau , story narration and story creation by students.

Closure activity: De- Briefing session

Tips – task chits

.

Forward link: nil

Medical theatre for the teaching faulty of College of Nursing September 2011

With the growing success and popularity of Medical Theatre, it was about time when an official request was sent by the Principal of Nursing College requesting for a workshop for the entire teaching faculty from the college of nursing. This request was embraced willingly and a workshop was scheduled for the faculty of nursing which comprised of college teachers and clinical instructors.

The Challenging aspect of this workshop was the seniority of the participants who are already a trained group of teachers. To introduce to them an unconventional and new methodology would have to be prepared carefully. In order to conduct this workshop without compromising on quality, The script Peoples Theatre Group involvement was vital. The second year Msc. Nursing students took time out of their daily schedule and organised the entire event for us.

Medical theatre for the teaching faulty of College of Nursing September 2011

Objectives –

1. To introduce an innovative methodology to a group of professional trainers from the college of Nursing.

2. To demonstrate four aspects of Medical theatre-
   a. Creating a script
   b. Creating a performance
   c. Importance of Games
   d. Theory and Practical's on Theatre-in-education.

3. To assign the creation of a script on any topic of their choice such that after its creation it is performed on a selected date along with the performances of the others.

Body – The concept of medical theatre explores participation rather than watching or listening to a particular session. It was therefore challenging to identify how a session for teachers could be employed. Moreover , since the expected participation was for about 60 nursing trainers, it was difficult to decide on an appropriate structure for this methodology. After multiple brain storming sessions with the Script People's Theatre group it was decided that the entire group of nursing faculty would be divided into 4 groups. Each group will enter a particular

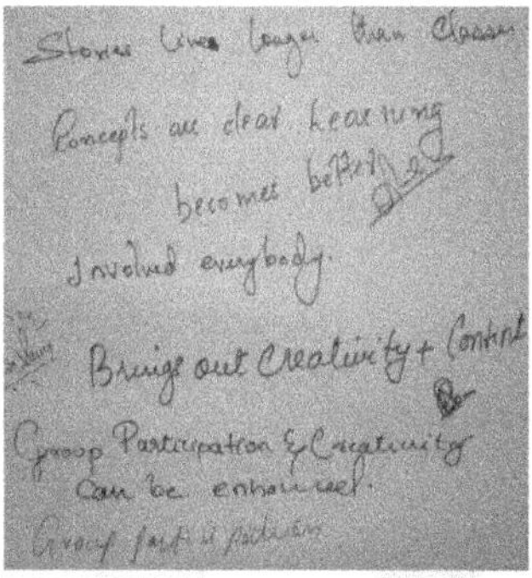

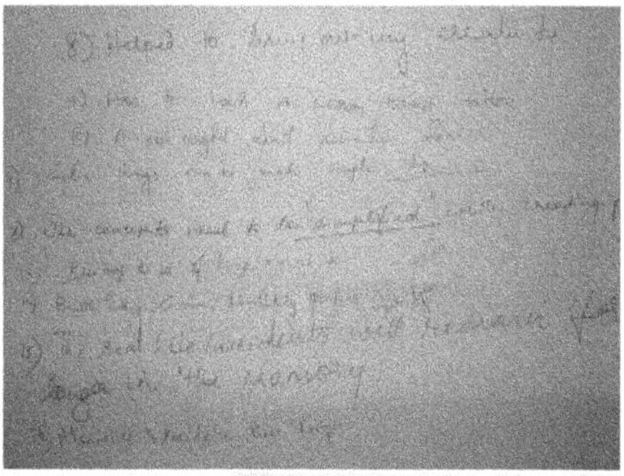

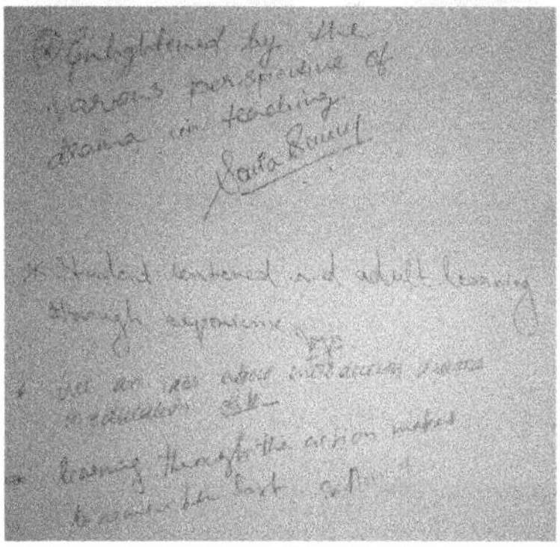

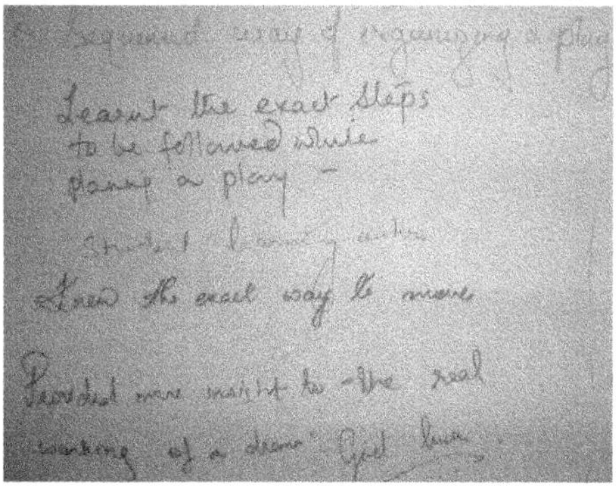

room which will be specific to a particular session. Therefore each group enters 4 rooms consecutively for 45 minutes each. The rooms had facilitators for their own sessions.

One room was for creating a script – here the participants underwent an entertaining session in which various medical topics are presented in an interesting and creative format.

The next room was for creating a performance – where the participants learnt some basics on how a performance is created from a script. The script chosen for performance was on mechanism of action of penicillin's.

In another room there was a session on theatrical games – where the participants played a variety of games and understood it's importance in the field of education.

The fourth room was kept for some theory and practical's – where the participants were shown a snippet of a professional performance after which they were taught some basics about theatre. Each group visited these rooms in different permutations and combinations

To start with, the participants assembled in their main auditorium where a common introductory session was taken. Following which they were divided into 4 groups. Each group went into different rooms to experience an enriching session of that speciality. At the end of being through the 4 rooms there was a common closing session where feedback was taken and the workshop was concluded by the narration of a short story on Florence Nightingale. Individual session feedback was taken on chart papers stuck in the respective rooms. The forward link that was given to the participants in pairs involved them to develop a story on any topic of their choice within 3 days of the workshop and to direct it into a play 2 weeks after the workshop, converting the outcome of the program into a full day theatre festival.

Very innovative method.

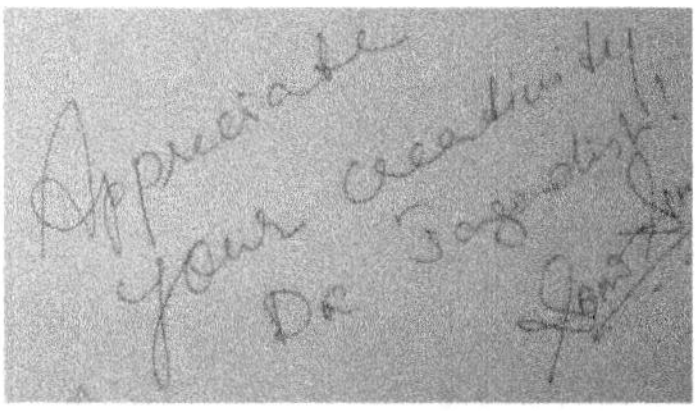
Appreciate your creativity Dr Jagadish!

Well organized
helped me to understand
thanks

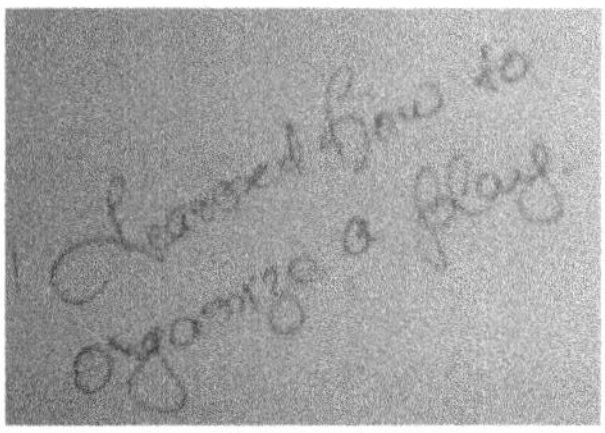
Learned how to organize a play.

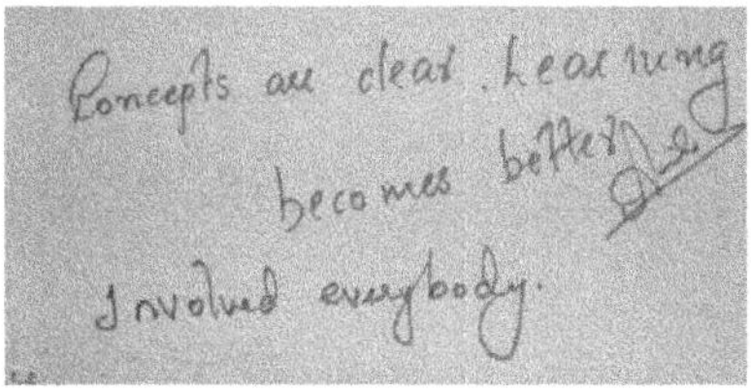
Concepts are clear. Learning becomes better
Involved everybody.

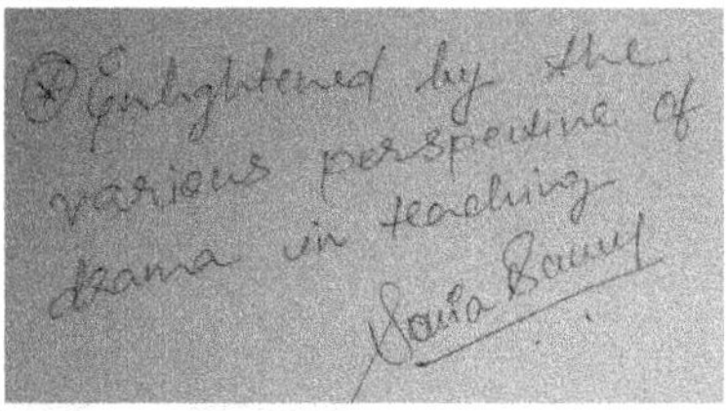
Enlightened by the various perspective of drama in teaching

Conclusion- At the end of this half a day ordeal it can be concluded that this methodology can be as effective and informative even though implemented in a short duration. The teachers who participated definitely felt that they are more confident in implementing creativity and emotions into their sessions. This kind of workshop was only an informative trailer for medical theatre, there must be longer duration of interventions for a proper and structured training and implementation of medical theatre.

Medical Theatre Workshop for St. Johns Nursing faculty

16th September 2011 (Friday)

Pre – event activity

Write a four paragraph report on – *My Understanding of Medical Theatre* based on the four page report provided and e-mail to hemoglobin84@yahoo.com.

Event

Sl. No.	Time	Batch 1	Batch 2	Batch 3	Batch 4
1.	9:00 a.m. – 9:15 a.m.	Introduction – Jagdish, Sibu & Batch Leaders			
2.	9:30 a.m. – 10:15 a.m.	Jagdish (*Creating a script*)	Upendra (*Games*)	Rajesh (*Creating a Performance*)	Arif (*Practicals & Theory*)
3.	10:15 a.m. – 11:00 a.m.	Upendra (*Games*)	Jagdish (*Creating a script*)	Arif (*Practicals & Theory*)	Rajesh (*Creating a Performance*)
4.	11:00 a.m. – 11:15 a.m.	Break			
5.	11:15 a.m. – 12:00 p.m.	Rajesh (*Creating a Performance*)	Arif (*Practicals & Theory*)	Upendra (*Games*)	Jagdish (*Creating a script*)
6.	12:00 p.m. – 12:45 p.m.	Arif (*Practicals & Theory*)	Rajesh (*Creating a Performance*)	Jagdish (*Creating a script*)	Upendra (*Games*)
7.	12:45 p.m. – 1:00 p.m.	Closure Story – Jagdish			

Post – event activity

Showcase a 5-7 minute performance on any Nursing topic on Gandhi Jayanti (2nd October).

Centre for Education Beyond Curriculum and Theatre- in Education

Facilitator training

June 23,24, 25(Thu, Fri, Sat) - 2011

Self Evaluation Feedback

Review of Day -1

Day 1 covered the basic introduction to theatre, some interesting insights into introducing music in education and conducting a model TIE session. The learning's from the day made me realise how structured an otherwise unstructured appearing field of theatre actually is.

We began with introductions from the attendees and were asked to write our expectations from the course and what we expect ourselves to be after 10 years. This introspection was an excellent way for me to look into myself and think where I would want to be in the future. It was a unique method to retrieve everyone's expectations.

What followed was an exciting chain walk, when we were taken as a group, blindfolded. People were left wondering as to where they were going and where this was heading towards. I personally felt that we were being taken into a hall where we would be asked to perform before an audience. At the end of the activity I realised the importance of group work, concentration, trust and coordination. The session was fun filled and an enjoyable beginning to the course.

We moved onto depicting various objects and people in the real world. This prompted us to think on our feet and be very creative. I believe this is a critical aspect of any theatre course where the participants need to apply themselves creatively for showcasing the real world.

Integrating one form of music into a completely different lyric was as challenging and interesting as the tests before. As mentioned by Ms. Chitra, music spreads across all languages and cultures and is a good way of application of theatre. The Tableau activity split us into different groups. The groups were told to enact situations in a railway station, festival and even a rehabilitation centre. While the

time limit of 5 minutes made us act quickly and perform, the question was if other groups could guess the situation. The activity was great and was very enjoyable. I got the knowledge of three dimensional usage of stage while performing the same, positions of the stage, and three levels which we should be used while enacting. This helps create a beautiful outcome.

Ms. Meeta spoke about the journey of theatre experience. She spoke about the origins of theatre in the ice age, types of theatre like community, street, political theatre etc. and difference between other performing arts and theatre which gave me good insights into the differences with other art forms. Overall I also realised that the journey of a theatrical career is not always very smooth and on many occasions is quite a struggle.

Wrapping up the day was a group counting from 1-20 as told by Mr.Arif. This was also an interesting activity where without any prior sequencing, the group members must shout out numbers consecutively without clashing or saying the same number together with anyone else in no particular order of participants.. This can really liven up a theatre class. Dr. Kennedy's speech about innovation was also good. He said innovation always talks about the next level. We always need to think using a critical mind.

All in all, day 1 was way above expectations and I really enjoyed the various activities and sessions throughout the day. The learning from the day would stay with me for a long time to come. I was eagerly looking forwards to what the other two days had in store for me.

Review of Day - 2

After an exciting day, I was looking forward to day 2 with sessions like models of education, dance in education and experiences across stage in store. The day proved more fun filled than I imagined.

We kicked off with models of education by Mr.Sibu and Mr.Rajesh. It was about how to teach in effective ways. We had a discussion on cone of learning experience developed by Edgar Dale. The cone of experience is given below in the pictorial representation:

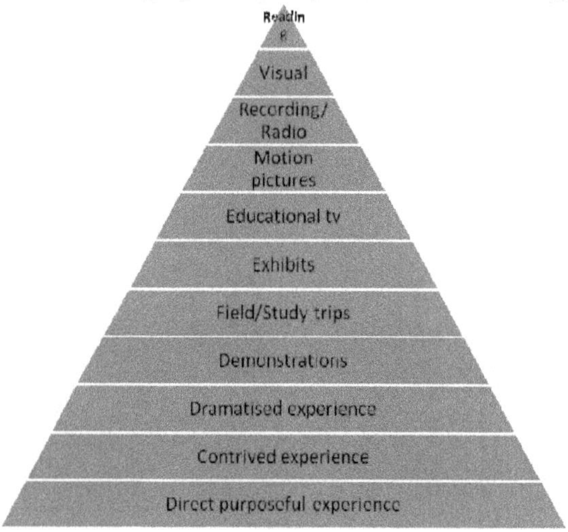

The discussion then centred around usage of different methods in teaching, how to assess the children and how to develop concepts using different methods. The cone of experience starts with basic, intermediate and advance form of teaching or developing the concepts.

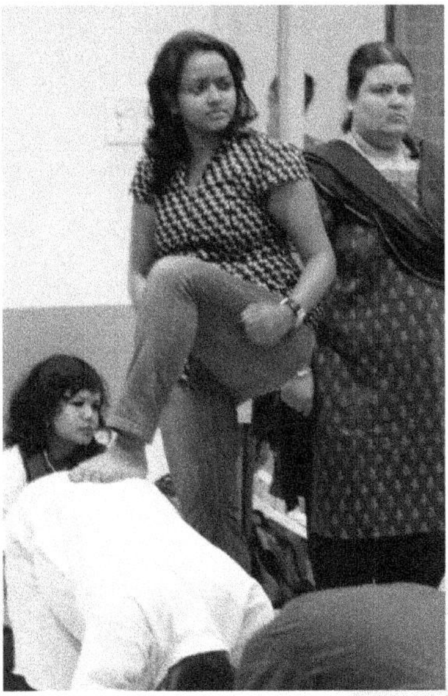

This was followed by the importance of the golden words in theatre. Body, words and mind are the golden words in theatre.

The theatrical star which consists of students or participants, body, words, staff, management and the mind are the critical elements to the success of theatre. We discussed the importance of each of these elements and how they play a role in our goal of excellent theatre.

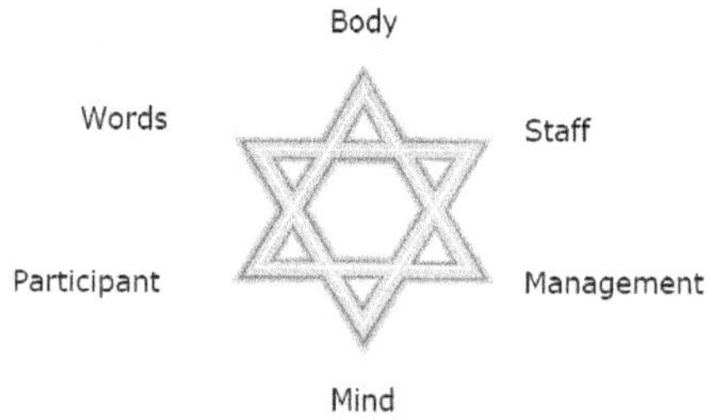

We were then put into different groups. We were asked to categorize the activities performed to date into understand, great & comprehending. This was an excellent way of recollecting and understanding which activities are the most effective and appropriate given the theatrical situation.

Mr.Prabin introduced us to dance in education. It was a good learning experience for me even though dance is one of the last things I ever intended to do. He started with the warm up activity, bending toes. Followed by running backward, forward and freeze as the rhythm of the delightful music freezes. He provided a ball and asked us to stand in a circle. Then, we had to throw the ball to any one of the people standing in the circle by shouting out their name. This according to me is a great game to help each other get introduced to each other. It helped in people's trouble of overcoming their inhibition of introducing themselves.

Some of the activities we got exposed to were bending toes, running, paint brush with hand, sculpture, walk in a maze, walking according to speed, star fish, rhythm etc.

Afternoon session started with Mr.Sibu, which was about stage dynamics and how to use stage while performing. Different positions of the stage like center, upper right, upper back, center right, center left, etc. was discussed.

UR	UB	UL
CR	C	CL
DR	DF	DL

And then we discussed lesson plan for theatre. The lesson plan consists of objective, special requirements, opening activity, core activity, closure activity and forward links.

This was also the key take away for me in the second day, to develop concepts using theatre. The fact that we need to start with an opening activity to warm up and then we can develop the core activity to meet the objectives and then complete the class with a closure activity by giving them to think and find out the forward link. This completed my understanding of making a good lesson plan and using various activities as pertinent to the situation.

Review of Day - 3

We started off with a micro teaching demonstration. I was a part of the Arif Group where I created an activity where in I could make the participants understand the concept of memory and how one remembers things. I asked everyone in the class to write one memorable occasion from their lives. I then called out three people and made one of them the limbic system, the other the amygdala and the third as the hippocampus. I instructed the limbic system to score the occasions read out by the participants on a scale of 1 to 10 on the basis of love, happiness or sadness in those lines. Similar instructions were given to the amygdala and the scoring was made on reward and fear. The hippocampus was asked to simply total the score. If <3 then the participant would be asked to sit down back in the class, if between 3 to 6 then the participant was asked to stand near the door and if more than 6 then the participant was asked to enter another room by opening the door. This represents in a simple manner as to how the brain works. This activity helped me relate to how the core activity can be used given an objective.

The next session in the day was about cleaver Heron and the fishes in the pond. This was to understand a created script. Different groups enacted on different story. It was a great learning experience and helped me understand the implementation of stage positions, different levels, how to occupy stage while doing a play etc.

In the afternoon, we got an interesting challenge. The session was about watching and working on the play. The story given was from Akbar's life – Akbar and the nine gems. The part given to us was called 'Swayamvara'. We were asked to change the situation of the story but character and contact should not be changed.

We devised the situation as a nursery annual day program. The situation was that nursery children are performing on Akbar and 9 gems. Here I got an idea of writing script, delivery, dialogue etc.

The day ended with most desired qualities of facilitators and the session was done by Mr. Sibin.

As mentioned in the class, most desired qualities of facilitators in ascending order are:

- Creating curiosity,
- Animating group (involving them in different way) with
music, body and voice
- Being encouraging
- Providing possible straws
- Planning and organisation
- Creating and enhancing wanting within the group
- Being genuine
- Listening skill
- At being non-interrupted
- Sense of humour

The Training

The Facilitator

The Performance

The Hero

The Car

The Tableau

OTHER DRAMATIC EXPERIENCES

Theatre in education workshop for MBA Students

I am thankful for the opportunity provided to me by script peoples theatre for facilitating 3 complete workshops for 40 MBA students out of 240 in each workshop. The journey of testing my skills as a facilitator was unbelievable and the amount of experience gathered during the entire event by conducting theatrical games and encasing an entire performance for these MBA students from Christ university all by myself was simply priceless.

These workshops involved –

1. Introducing myself as a facilitator to 40 MBA students assigned to my group
2. Conducting warm up games – Exploring the room and freezing
3. Conducting Tableau sessions and Visual / Kinetics and auditory sessions
4. Script reading session
5. Direction and encasing a performance in front of 200 other MBA students

Achievement's –

1. Best Poster Award for the first workshop
2. Best Play award "The white elephant" and the best poster award in the second workshop.
3. The performance in the third workshop was performed for the opening ceremony for 'DARPAN" which is an inter- collegiate theatre festival organized by Christ University.

CHRIST
UNIVERSITY
The Apricot Jar

OTHER DRAMATIC EXPERIENCES

The Apricot Jar – A Public Performance

I am extremely grateful to Script Peoples theatre for giving me an opportunity to perform in their yearly production called Vision 2020 – The Apricot Jar. This experience was the most useful and productive time in theatre that I have spent after a very long time. The journey of this year's vision 2020 started about a month before the performance and involved an early morning practice session 3 times a week. Not only did this help me wake up every morning but also helped me accquire a whole lot of theatrical knowledge with fun and creativity. I express my deepest gratitude to Mr. Upendra for all the theatre training bestowed so generously.

My role in the play was short, but I was allowed to play multiple characters at frequent intervals. This was new for me and I enjoyed the variety.

I portrayed an auctioneer in the Arabic tale where I am auctioning the protagonist Ali's belongings. This small scene helped me to realise that its not the auctioneer that is important, it's the auction scene that is more important. I also played the apricot merchant, the plastic toy seller and the audience in a court scene.

Apart from this enjoyable experience of performing in one of the best auditoriums in front of more than a thousand people, I also made some great friends whom I will always remember.

OTHER DRAMATIC EXPERIENCES

Teacher Training Program for B. Ed Students

I am once again thankful to script people's theatre for allowing me to explore my role as a facilitator at a professional level. In this intensive three days training program for teachers, I was asked to conduct a 90 minute session on "Teaching a topic in an interesting manner" for three different groups over three days.

My session involved exploring the mind of a student when a teacher takes a class and how a student might misinterpret simple information if not conveyed in an effective manner. Further in the session I demonstrate how a complicated topic can be made into an emotional narration, a lucrative story or an exciting fun filled game. Towards the latter half of my session, I allotted specific content to four groups and helped them create an interesting method for these topics.

It was a great experience being invited to conduct sessions at a professional platform. This has helped me be more mature, more structured and more effective.

COURSE MATERIAL FOR MEDICAL THEATRE

CONTENTS

Introduction to conversion of a medical topic into a story/script

IMPORTANT STEPS

STEP 1 - Select a topic of your interest

(Try to take small sections of 2 to 3 paragraphs instead of the entire chapters in the beginning, this will help create small stories in a short duration of time)

STEP 2 - Select 10 to 15 sentences or points from your selected topic of interest.

(This is the most difficult step and it is your decision to take those selected sentences. Try to only take the most important ones. Omit the sentences that are supporting the same point which you have already selected. This is to help you form a schematic and you can include as many number of details later)

STEP 3 - Formation of schematics

(look at each of the sentences that you have selected and replace it with as few words that you can use in a simple manner such that the intention of the sentence does not change, do this for all sentences)

STEP 4- Formation of a plot

(Now by looking into the schematics only, try to see what kind of situation fits to incorporate the entire sequence –

- A sequence similar to a movie – adapt the story to the movie
- A sequence similar to a real life incident- politics/controversies
- A sequence similar to serials – spoofs can be adapted
- A monologue can be considered if no story is formable.
- If there is room for rhyming words- try to make it a poem
- If it's too clinical- try to keep the features in the clinical situation but build a story around it.
- A stand - up comedy routine can be created

STEP 5 – Outcome

Based on the schematics and plot, the outcome can be

A story

A dramatic skit script

A role play script

A monologue

A poem/ rap/ song

A stand up comedy routine

An outcome in the mind is volatile, if converted into a material like a script, the revision is faster as you would be reading what you created with an experience.

It may be a bit more of an effort, but it is challenging, interesting and ensures a longer memory of the subject.

It would be sensible though to try these methods only for areas where you otherwise find it difficult to understand or remember. There is no point in making easy things easy, you can learn those as they are.

Topic 1 – Conversion of Broca's aphasia into a theatrical outcome

Broca's Aphasia – Theory from a standard text book

STEP 1- SELECTING AND READING THE TOPIC

Speech is nonfluent, labored, interrupted by many word-finding pauses, and usually dysarthric. It is impoverished in function words but enriched in meaning-appropriate nouns and verbs. Abnormal word order and the inappropriate deployment of *bound morphemes* (word endings used to denote tenses, possessives, or plurals) lead to a characteristic agrammatism. Speech is telegraphic and pithy but quite informative. In the following passage, a patient with Broca's aphasia describes his medical history: "I see . . . the dotor, dotor sent me . . . Bosson. Go to hospital. Dotor . . . kept me beside. Two, tee days, doctor send me home."

Output may be reduced to a grunt or single word ("yes" or "no"), which is emitted with different intonations in an attempt to express approval or disapproval. In addition to fluency, naming and repetition are also impaired. Comprehension of spoken language is intact, except for syntactically difficult sentences with passive voice structure or embedded clauses. Reading comprehension is also preserved, with the occasional exception of a specific inability to read small grammatical words such as conjunctions and pronouns. The last two features indicate that Broca's aphasia is not just an "expressive" or "motor" disorder and that it may also involve a comprehension deficit for function words and syntax. Patients with Broca's aphasia can be tearful, easily frustrated, and profoundly depressed. Insight into their condition is preserved, in contrast to Wernicke's aphasia. Even when spontaneous speech is severely dysarthric, the patient may be able to display a relatively normal articulation of words when singing. This dissociation has been used to develop specific therapeutic approaches (melodic intonation therapy) for Broca's aphasia. Additional neurologic deficits usually include right facial weakness, hemiparesis or hemiplegia, and a buccofacial apraxia characterized by an inability to carry out motor commands involving oropharyngeal and facial musculature (e.g., patients are unable to demonstrate how to blow out a match or suck through a straw). Visual fields are intact. The cause is most often infarction of Broca's area (the inferior frontal convolution; "B" in Fig. 27-1) and surrounding anterior perisylvian and insular cortex, due to occlusion of the superior division of the middle cerebral artery (Chap. 364). Mass lesions including tumor, intracerebral hemorrhage, or abscess may also be responsible. Small lesions confined to the posterior part of Broca's area may lead to a nonaphasic and often reversible deficit of speech articulation, usually accompanied by mild right facial weakness. When the cause of Broca's aphasia is stroke, recovery of language function generally peaks within 2–6 months, after which time further progress is limited.

STEP 2 – SELECTING IMPORTANT SENTENCES

Broca's Aphasia

Speech is nonfluent, labored, interrupted by many word-finding pauses, and usually dysarthric. **(slow and interrupted speech)**

It is impoverished in function words but enriched in meaning-appropriate nouns and verbs. Abnormal word order and the inappropriate deployment of *bound morphemes* (word endings used to denote tenses, possessives, or plurals) lead to a characteristic agrammatism. **Speech is telegraphic and pithy but quite informative.**

In the following passage, a patient with Broca's aphasia describes his medical history: "I see . . . the dotor, dotor sent me . . . Bosson. Go to hospital. Dotor . . . kept me beside. Two, tee days, doctor send me home."

Output may be reduced to a grunt or single word ("yes" or "no"), which is emitted with different intonations in an attempt to express approval or disapproval. In addition to fluency, naming and repetition are also impaired. **Comprehension of spoken language is intact**, except for syntactically difficult sentences with passive voice structure or embedded clauses.

Reading comprehension is also preserved, with the occasional exception of a specific inability to read small grammatical words such as conjunctions and pronouns.

The last two features indicate that **Broca's aphasia is not just an "expressive" or "motor" disorder and that it may also involve a comprehension deficit for function words and syntax**.

Patients with **Broca's aphasia can be tearful, easily frustrated, and profoundly depressed**. Insight into their condition is preserved, in contrast to Wernicke's aphasia.

Even when spontaneous speech is severely dysarthric, **the patient may be able to display a relatively normal articulation of words when singing.** This dissociation has been used to develop specific therapeutic approaches **(melodic intonation therapy)** for Broca's aphasia.

Additional neurologic deficits usually include right **facial weakness, hemiparesis** or hemiplegia, and a buccofacial apraxia characterized by an inability to carry out motor commands involving oropharyngeal and facial musculature (e.g., patients are unable to demonstrate how to blow out a match or suck through a straw).

Visual fields are intact. The cause is most often infarction of Broca's area (the inferior frontal convolution; "B" in Fig. 27-1) and surrounding anterior perisylvian and insular cortex, due to occlusion of the superior division of the middle cerebral artery (Chap. 364). Mass lesions including tumor, intracerebral hemorrhage, or abscess may also be responsible. Small lesions confined to the posterior part of Broca's area may lead to a nonaphasic and often reversible deficit of speech articulation, usually accompanied by mild right facial weakness. **When the cause of Broca's**

aphasia is stroke, recovery of language function generally peaks within 2–6 months, after which time further progress is limited.

STEP 3 – FORMATION OF SCHEMATICS

Broca's aphasia-

- Broca's area is Brodmann's area 44 located on the Posterior inferior frontal gyrus.
- Slow and interrupted speech,
- Speech is telegraphic and pithy but quite informative.
- Comprehension of spoken language is intact,
- Reading comprehension is also preserved.
- Broca's aphasia can be tearful, easily frustrated, and profoundly depressed.
- The patient may be able to display a relatively normal articulation of words when singing (melodic intonation therapy).
- Additional neurologic deficits usually include right facial weakness, hemiparesis.
- Visual fields are intact.
- When the cause of Broca's aphasia is stroke, recovery of language function generally peaks within 2–6 months, after which time further progress is limited.

STEP 4 – FORMATION OF A PLOT

Since this topic has room for a theatrical demonstration showing the way a Broca's Aphasia patient talks and keeping in mind I will be teaching medical and nursing students , I selected my plot as – sharing an incidence from my life/ monologue.

STEP 5 - OUTCOME

Stages of sharing an incidence-

The distraction- *It is best if the audience is unprepared for what you are about to say, and the appearance of a spontaneous narration of an incidence, being led as a continuation of another topic works most effectively.*

In this session – "Hi, before we start off with this script conversion session I'd like to emphasise on something we all do but never notice...(To sibu)- Few minutes extra just before I start.

(To audience) see, there will be a few script reading sessions where you might have to read the script. Script reading is an art, its words on a piece of paper, but by the way you say it gives life and meaning to the sentence. For a small example – suppose one has to read a sentence like "hey, what way is it to go to Indiranagar... what? What... (The other person from within an auto)- Go straight, turn right at the signal – (me) - oh ok ... thanks.

This is many times read like this (in a manner that a child reads from a book- demonstration in front of audience).

The drift - There might be many reasons not to read it like a real life situation, as many might have reading issues, vision issues, language issues, stammering issues.

Stammering is the most troublesome...in fact when I was in school and used to act in plays.. I had this classmate who was mainly a very good classical singer... But when it came to telling dialogues he used to say it in a very peculiar manner... Something like this (demonstration)

The incident- I used to imitate him occasionally because I had never heard any one talk like that before, and soon I became popular because I could imitate him. This lead to lots of other people teasing him and then one day he just left school and I never met him after that.

The sad part is when I was told he was actually suffering from Broca's aphasia. Initially I was like... What bullshit because he was an amazing singer... But apparently I realised that guys with Broca's can sometimes sing without difficulties and it is also used in a type of therapy.

Anyways let's get back to the point.. Before that I hope you guys know what Broca's aphasia is right? – A motor aphasia where in there is slow and interruptive speech, but comprehension is normal. It is due to a disorder in the area 44 of the brain in the inferior frontal lobe according to Brodmann's area. Usually has associated facial palsy or hemiparesis but this guy didn't have all that.

But what's interesting is that reading and vision is normal. And they get depressed, frustrated and tearful very easily due to their daily difficulties. To think of it as kids sometimes we are very mean to our friends without knowing what pain we are causing to them.

The conclusion- Anyways. Let's start our script writing session, I'm sure Sibu must be cursing me for drifting off... Wait a minute... I think we already finished our first topic..."

Topic 2 – Conversion of Temporal Arteritis into a story / script

STEP 1- SELECTION OF TOPIC AND READING IT

Temporal (giant cell) arteritis is an inflammatory disorder of arteries that frequently involves the extracranial carotid circulation. It is a common disorder of the elderly; its annual incidence is 77 per 100,000 individuals ages 50 and older. The average age of onset is 70 years, and women account for 65% of cases. About half of patients with untreated temporal arteritis develop blindness due to involvement of the ophthalmic artery and its branches; indeed, the ischemic optic neuropathy induced by giant cell arteritis is the major cause of rapidly developing bilateral blindness in patients >60 years. Because treatment with glucocorticoids is effective in preventing this complication, prompt recognition of the disorder is important.

Typical presenting symptoms include headache, polymyalgia rheumatica (Chap. 319), jaw claudication, fever, and weight loss. Headache is the dominant symptom and often appears in association with malaise and muscle aches. Head pain may be unilateral or bilateral and is located temporally in 50% of patients but may involve any and all aspects of the cranium. Pain usually appears gradually over a few hours before peak intensity is reached; occasionally, it is explosive in onset. The quality of pain is only seldom throbbing; it is almost invariably described as dull and boring, with superimposed episodic stabbing pains similar to the sharp pains that appear in migraine. Most patients can recognize that the origin of their head pain is superficial, external to the skull, rather than originating deep within the cranium (the pain site for migraineurs). Scalp tenderness is present, often to a marked degree; brushing the hair or resting the head on a pillow may be impossible because of pain. Headache is usually worse at night and often aggravated by exposure to cold. Additional findings may include reddened, tender nodules or red streaking of the skin overlying the temporal arteries, and tenderness of the temporal or, less commonly, the occipital arteries.

The erythrocyte sedimentation rate (ESR) is often, though not always, elevated; a normal ESR does not exclude giant cell arteritis. A temporal artery biopsy followed by treatment with prednisone 80 mg daily for the first 4–6 weeks should be initiated when clinical suspicion is high. The prevalence of migraine among the elderly is substantial, considerably higher than that of giant cell arteritis. Migraineurs often report amelioration of their headaches with prednisone; thus, caution must be used when interpreting the therapeutic response.

STEP 2 – SELECTING IMPORTANT SENTENCES FROM THE TOPIC

Temporal Arteritis

Temporal (giant cell) arteritis is an inflammatory disorder of arteries that frequently involves the extracranial carotid circulation.

It is a common disorder of the elderly; its annual incidence is 77 per 100,000 individuals ages 50 and older. **The average age of onset is 70 years**, and women account for 65% of cases. **About half of patients with untreated temporal arteritis develop blindness due to involvement of the ophthalmic artery and its branches**; indeed, the ischemic optic neuropathy induced by giant cell arteritis is the major cause of rapidly developing bilateral blindness in patients >60 years. **Because treatment with glucocorticoids is effective in preventing this complication, prompt recognition of the disorder is important.**

Typical presenting symptoms include **headache, polymyalgia rheumatica** (Chap. 319), **jaw claudication, fever, and weight loss.**

Headache is the dominant symptom and often appears in association with malaise and muscle aches. **Head pain may be unilateral or bilateral and is located temporally in 50%** of patients but may involve any and all aspects of the cranium.

Pain usually appears gradually over a few hours before peak intensity is reached; occasionally, it is explosive in onset. The quality of pain is only seldom throbbing; it is almost invariably described as **dull and boring**, with **superimposed episodic stabbing pains similar to the sharp pains that appear in migraine.**

Most patients can recognize that the origin of their **head pain is superficial, external to the skull**, rather than originating **deep within the cranium (the pain site for migraineurs)**.

Scalp tenderness is present, often to a marked degree; brushing the hair or resting the head on a pillow may be impossible because of pain. **Headache is usually worse at night and often aggravated by exposure to cold.**

Additional findings may include **reddened, tender nodules or red streaking of the skin overlying the temporal arteries,** and tenderness of the temporal or, less commonly, the occipital arteries.

The erythrocyte sedimentation rate (ESR) is often, though not always, elevated; a normal ESR does not exclude giant cell arteritis.

A temporal artery biopsy followed by treatment with prednisone 80 mg daily for the first 4–6 weeks should be initiated when clinical suspicion is high.

The prevalence of migraine among the elderly is substantial, considerably higher than that of giant cell arteritis. Migraineurs often report amelioration of their headaches with prednisone; thus, caution must be used when interpreting the therapeutic response.

STEP 3- FORMATION OF SCHEMATICS

- Temporal (giant cell) arteritis is an **inflammatory disorder** of arteries
- It is a common disorder of the **elderly**
- The average age of onset is **70 years**
- **C/F- headache, polymyalgia rheumatica, jaw claudication, fever, and weight loss.**
- About half of patients with **untreated temporal arteritis develop blindness** due to involvement of the ophthalmic artery and its branches
- **Head pain may be unilateral or bilateral** and is located temporally in 50%
- **Headache is dull and boring with episodic stabbing pain**
- **Scalp tenderness is present**
- Headache is usually **worse at night** and often **aggravated by exposure to cold.**
- **Reddened, tender nodules or red streaking of the skin overlying the temporal arteries**
- **ESR usually elevated – not always though**
- Treatment is **glucocorticoids – 80 mg prednisolone for 2 to 6 weeks.**

STEP 4 - FORMATION OF A PLOT

In this disease I know there is an important "not to do "here. That is if a temporal arteritis is not suspected and treated wrongly, then the patient will be blind soon. So I felt it is best to take the students through this journey so that they in the process make the mistake of getting a patient blind in a virtual scenario so that they are cautious in the future. I do not think making this into a story would make a great impact, but yes I can have a dramatic skit in which I make the doctor treat this patient wrongly and get into legal issues due to blinding the patient. And yes by this way the agony a patient goes through can also be emphasised.

However to keep the audience involved and thinking, I have chosen to present the outcome as a scenario.

STEP 4 – OUTCOME IN THE FORM OF A CLINICAL SCENARIO

Phase of distraction - I am an old lady, miserable at this age, no one to look after me and no one to care. I have hypertension, heart disease, diabetes mellitus and osteoarthritis. Have been taking treatment for all this and somehow carrying out my daily activities, waiting for a day when I will die. I am forgetful and irritable. I have had two surgeries already, one for liver stones and another for kidney stones. I have suffered from so many diseases that I probably know more medicine than you do. I am ok with all this but please free me from this very bothersome illness that I have. It's this headache and it's unbearable. Please treat me for this.

Phase of interaction – (to audience) – This patient has come to you; you can ask any Questions to this patient and treat her. May be you could start from taking a history.

History- Well... I have this head ache... It's more on the left side... It's not very severe but it is there... all the time... Dull and boring type... I can't even touch that area or comb my hair and it worsens when I sleep. Last week I had cold and the pain became unbearable. Sometimes I feel all of a sudden that someone is stabbing a knife into my head. I also have fever... joint pains... one one day one one joint.. And have lost a lot of weight. I don't know what to do... I just want to die in peace.

(to audience) – ask them to elaborate on history and ask for clinical examination findings

On Examination- clinical features that correlate with the history like fever, scalp tenderness, nasal congestion , weakness, joint tenderness etc

(To audience)- Ask what investigations they would want- answer all results in correlation to the comorbidities like high sugars, normal counts, raised ESR, High BP etc

Then ask the audience to treat this lady and what is their probable diagnosis.

This session can then take any one of the following course

1. **The course of suspense -** here most likely the audience is going to guess migraine or sinusitis most commonly and when they say that, don't let them know that they are wrong. Ask them to treat the patient then, with whatever they have diagnosed and promise them that you will let them know the outcome.

   The expected responses would be antibiotics, pain killers, steam inhalation, nose drops, vision check-up etc. etc. and they would probably want to review the patient in 3 to 5 days.

That time lay down the news to them (This has to be done in style) – Tell them that the patient has come back with the headache showing no improvement but has lost vision completely in the left eye. You have done something or maybe not but have caused this patient to become blind (enjoy the shocked silence for a few minutes and let the audience feel the moment)

Here probably someone will guess it is temporal arteritis and do not forget to acknowledge that student for coming up with the right answer (very very important step)

Then explain to them how the blind ness is caused and what happens if steroids are not given on time. Also emphasise the consequences that would occur if a migraine is misdiagnosed as temporal arteritis and steroids are given as steroids worsen migraine.

2. **The course of solution**- Maybe one smart allec from the class will tell it is temporal arteritis... still don't give up.. Ask for what treatment they would want to start... Maybe the student might not be aware of steroid as a treatment and you may still have a chance of continuing this session as planned. If the student has answered steroids then acknowledge the answer (ask the audience to clap) and tell them what you were trying and that it's good presence of mind to diagnose this answer. Here the class ends short but at least it was interesting.

Note – If your teaching session is longer , then this is a wonderful way to teach all the possible causes of a head ache in an interesting manner, keeping the audience gripped at all times, for eg. When they say migraine – explain to them how it is unlikely by telling the features of migraine, if someone guesses cluster headache, tension headache or somatic headache, tell them the features of each and show them that the current patient doesn't have these symptoms. By this way the students learn the entire differential diagnosis and their features in one interesting challenge to diagnose this case. The success of the class is based on your acting skills, improvisation, how you mislead them into keeping them guessing and how you interact. All these require theatrical skills which this workshop aims to provide.

Topic 3 – conversion of PTH and PTHrP Hormone Action into a script/ story

STEP 1 –SELECT TOPIC AND READ IT

Both PTH and PTHrP bind to and activate the PTH/PTHrP receptor. The 500-amino-acid PTH/PTHrP receptor (also known as the PTH-1 receptor, PTH1R) belongs to a subfamily of GPCRs that includes those for glucagon, secretin, and vasoactive intestinal peptide. The extracellular regions are involved in hormone binding, and the intracellular domains, after hormone activation, bind G protein subunits to transduce hormone signaling into cellular responses through stimulation of second messengers. A second receptor that binds PTH, originally termed the *PTH-2 receptor* (PTH2R), is expressed in brain, pancreas, and several other tissues. PTH1R responds equivalently to PTH and PTHrP, whereas PTH2R responds only to PTH. The endogenous ligand of this receptor is now believed to be a peptide distinct from PTH, a 39-amino-acid hypothalamic peptide (*tubular infundibular peptide*, TIP-39). PTH1R and PTH2R can be traced backward in evolutionary time to fish. Zebrafish PTH1R and PTH2R exhibit the same selective responses to PTH and PTHrP as do human PTH1R and PTH2R. The evolutionary conservation of structure and function suggests important biologic roles for these receptors, even in fish (which lack discrete parathyroid glands).

Studies using cloned PTH1R confirm that it can be coupled to more than one G protein and second-messenger kinase pathway, apparently explaining the multiplicity of pathways stimulated by PTH. Stimulation of protein kinases (A and C) and calcium transport channels is associated with a variety of hormone-specific tissue responses. These responses include inhibition of phosphate and

bicarbonate transport, stimulation of calcium transport, and activation of renal 1 -hydroxylase in the kidney. The responses in bone include effects on collagen synthesis; increased alkaline phosphatase, ornithine decarboxylase, citrate decarboxylase, and glucose-6-phosphate dehydrogenase activities; DNA, protein, and phospholipid synthesis; and calcium and phosphate transport. Ultimately, these biochemical events lead to an integrated hormonal response in bone turnover and calcium homeostasis. PTH also activates Na^+/Ca^{2+} exchanges in renal distal tubular sites and stimulates translocation of preformed calcium transport channels, moving them from the interior to the apical surface to mediate increased tubular uptake of calcium. PTH-dependent stimulation of phosphate excretion (blocking reabsorption—the opposite effect from actions on calcium in the kidney) involves the sodium-dependent phosphate cotransporter, NPT-2, lowering its apical membrane content (and therefore function). Similar shifts may be involved in other renal tubular transport effects of PTH.

PTHrP exerts important developmental influences on fetal bone development and in adult physiology. A homozygous knockout of the PTHrP gene (or the gene for the PTH receptor) in mice causes a lethal deformity in which animals are born with severe skeletal deformities resembling chondrodysplasia (Fig. 347-2). The heterozygous PTHrP gene knockout animals display reduced mineral density consistent with osteoporosis. Experiments with these mouse models point to a hitherto unappreciated role of PTHrP as a paracrine/autocrine factor that modulates bone metabolism in adults as well as during bone development.

STEP 2 – SELECT IMPORTANT POINTS FROM THE TOPIC

PTH and PTHrP Hormone Action

Both PTH and PTHrP bind to and activate the PTH/PTHrP receptor. – **Binding causes activation**

The extracellular regions are involved in hormone binding, and the intracellular domains, after hormone activation, bind G protein subunits to transduce hormone signaling into cellular responses through stimulation of second messengers. (**outside cell- binds , inside cell – stimulates second messenger)**

A second receptor that binds PTH, originally termed the *PTH-2 receptor* (PTH2R), is expressed in brain, pancreas, and several other tissues.(**new binder- brain, pancreas etc)**

PTH1R responds equivalently to PTH and PTHrP, whereas PTH2R responds only to PTH. (**one responds to both , another responds to only one)**

Studies using cloned PTH1R confirm that it can be coupled to more than one G protein and second-messenger kinase pathway, apparently explaining the multiplicity of pathways stimulated by PTH. Stimulation of protein kinases (A and C) and calcium transport channels is associated with a variety of hormone-specific tissue responses. These responses include inhibition of phosphate and

bicarbonate transport, stimulation of calcium transport, and activation of renal 1 -hydroxylase in the kidney.

(Stimulation of calcium transport channels - transport inhibited phosphate/bicarb , stimulates calcium transport, activates enzyme in kidney)

The responses in bone include effects on collagen synthesis; increased alkaline phosphatase, ornithine decarboxylase, citrate decarboxylase, and glucose-6-phosphate dehydrogenase activities; DNA, protein, and phospholipid synthesis; and calcium and phosphate transport. **(Calcium transport increased)**

Ultimately, these biochemical events lead to an integrated hormonal response in bone turnover and calcium homeostasis. PTH also activates Na^+/Ca^{2+} exchanges in renal distal tubular sites and stimulates translocation of preformed calcium transport channels, moving them from the interior to the apical surface to mediate increased tubular uptake of calcium. **(Ionic exchange at renal sites and causes uptake of calcium)**

PTH-dependent stimulation of phosphate excretion (blocking reabsorption—the opposite effect from actions on calcium in the kidney) involves the sodium-dependent phosphate co-transporter, NPT-2, lowering its apical membrane content (and therefore function). Similar shifts may be involved in other renal tubular transport effects of PTH.**(phosphate is lost)**

PTHrP exerts important developmental influences on fetal bone development and in adult physiology. A homozygous knockout of the PTHrP gene (or the gene for the PTH receptor) in mice causes a lethal deformity in which animals are born with severe skeletal deformities resembling chondrodysplasia (Fig. 347-2). The heterozygous PTHrP gene knockout animals display reduced mineral density consistent with osteoporosis. Experiments with these mouse models point to a hitherto unappreciated role of PTHrP as a paracrine/autocrine factor that modulates bone metabolism in adults as well as during bone development.

STEP 3 – FORMATION OF SCHEMATICS

1. A binder
2. Binding to receptor- causes activation
3. Outside- binding , internally- stimulation of a second messenger
4. A new receptor that binds only in brain and pancreas. This new receptor doesn't bind to receptor protein.
5. Causes stimulation of transport channels – increased calcium transport, reduced phosphate transport and kidney enzyme activation.
6. Increased re-uptake of calcium from kidney and bones – increased calcium levels
7. Excess phosphate loss
8. Absence of this receptor at birth – bony deformities.

Modified Schematics – Deleting sentences that support same points

1. A binder
2. Binding to receptor- causes activation
3. Causes stimulation of transport channels – increased calcium transport, reduced phosphate transport and kidney enzyme activation.
4. Increased re-uptake of calcium from kidney and bones – increased calcium levels
5. Excess phosphate loss
6. Absence of this receptor at birth – bony deformities.

Formation of objective -

I want to remember that parathyroid hormone causes high calcium in blood due to reabsorption from bones and kidneys and low phosphate levels in blood due to renal excretion.

STEP 4 – FORMATION OF A PLOT

Since this is a description of a hormone and its effects, the most appropriate method would be to make it into a story.

STEP 5 – THE OUTCOME- STORY

Poison to Hell - PTH

A thin, weak husband of a pretty girl (arranged marriage).

Depressed that his wife keeps saying how she likes strong men and threatens to leave him for a body builder neighbour.

He decides to end his life, looks at the internet on howtoendlifeforfree.com to get the cheapest available options.

He finds a watch that can end life once the time for death is set. He orders it immediately.

He got a cool hairband with it for free too.

On reading the instruction manual he learnt that there was a small bottle with a pink fluid in it that came with the watch. It was labelled PTH – Poison To Hell. Two drops need to be poured on the watch and the poison would absorb at the given time.

So he decided that he would end his life at 9 pm after his favourite daily soap. Anyways his wife doesn't come back from work till morning due to her recently started night shifts. Its hard work at the bakery nowadays.

As soon as the clock ticked 9 pm , he waited for his death, till 9 05 nothing happened, he then added more of the solution , still no death, he added more and more and still no death.(excess of parathyroid hormone emphasised) He felt cheated, and decided to sleep and plan for another method the next day. As soon as he was about to lie down on his bed he notices a huge muscular guy in the mirror and then screamed on the top of his voice when he realised it was him.

This made him thrilled with joy as he understood that it was the excess of the solution that caused him to become like this. He was as hard as stone with stone in his head and stone in his blood. (calcium in blood)He started to imagine himself as a super hero as the strongest man alive. But what would he call himself... well he thought CALCIMAN sounded appropriate as he is as strong as stone. The first thing he planned was to show his neighbour that he is now not weak any more.

Next morning he walks to his neighbours house, he sees his wife coming out of his house, she is shocked to see this new transformation of her husband and asked how did this happen and was it really him. He proudly said that he is now strong and she could finally be happy to be a wife of a strong man. The neighbour sees this and laughs. "Hey moron, you think this makes you strong, you can't even kill a fly." This sparks the anger in this newly reformed husband and challenges the neighbour to a power contest.

On mutual decision and lots of common sense they decide to stop a truck with their bare hands. This was appropriate as these days the transport service of these stone carrying trucks has increased. (Increased calcium transport)The Husband insisted that he go first so that he could prove to his wife his strength. The neighbour stands beside and watches Calciman try stopping a truck with his hands. In two minutes there is a sound of an approaching truck and in 4 minutes there is a sound of a scream and in 10 minutes there is a sound of a rushing ambulance. Our calciman is hurt in every bone of his body, twisted in odd shapes beyond recognition, being taken in the ambulance to a hospital. (Weak bones due to resorption – osteomalacia and osteoporosis seen in hyperparathyroidism)

The husbands wife is in tears- how could this happen, he looked so strong?

Neighbour replies- I became strong the same way, its only external, the bones are as thin as match sticks, loosing minerals in urination,(phosphate lost in urine) it's just the stone in the blood.. What use is that.. I don't think there is any point worrying about your husband any more honey.. Come... lets go home.

Topic 4- Group activity – script conversion exercise – Autonomic Nervous System The ten pointers

Autonomic Nervous System

1. The autonomic nervous system (ANS) regulates blood pressure (BP), heart rate, sleep, and bladder and bowel function. It operates automatically.

2. The ANS comprises of the Sympathetic (nerves from the spinal cord) and parasympathetic (nerves from the brain or cranial nerves) nerve systems.

3. Both these systems have ganglions (consolidation of these nerves). There are nerves that carry transmitters from the source to the ganglion called preganglionic fibres and there are those that carry transmitters from the ganglion to the organ called the post ganglionic fibres.

4. A neurochemical called acetylcholine is the transmitter for preganglionic and post ganglionic fibres of the parasympathetic system.

5. Acetylecholine is transmitter for pre ganglionic fibres of sympathetic system. Nor epinephrine is the transmitter for post ganglionic fibres of sympathetic system. Except for skin glands which is innervated by acetyl choline.

6. The effect these systems have on the body.

	Sympathetic	Parasympathetic
Heart rate	Increased	Decreased
Blood pressure	Increased	Mildly decreased
Bladder	Increased sphincter tone	Voiding (decreased tone)
Bowel motility	Decreased motility	Increased
Lung	Bronchodilation	Bronchoconstriction
Sweat glands	Sweating	—
Pupils	Dilation	Constriction
Adrenal glands	Catecholamine release	—
Sexual function	Ejaculation, orgasm	Erection
Lacrimal glands	—	Tearing
Parotid glands	—	Salivation

GROUP ACTIVITY – DRAW A STORY OUTLINE ON THE TEN POINTERS PROVIDED

OUTCOME EXAMPLE

Autonomous Nervous System- ANS- Anti Ragging Nagging Story

The scene opens with the entry of a typical hard working average man who walks towards his desk at his office to finish the days accountancy for a small company.

Sharp on time as usual, with a box of lunch neatly packed from home he reaches his desk and keeps his suitcase on the table. Just as he is about to sit......................screeeech..the chair is suddenly pulled behind and he falls on the floor. Two colleagues appear from behind laughing along with the rest of the staff.

Bully at work 1- Hey ....Shitty pants... looks like your chair is a looser like you too...

Bully at work 2- Ha ha ha.. saw the way he fell... hilarious... he has been falling for this so many times that he is starting to get good at it..

Bully at work 1- Join the circus shitty pants... you are funny...here, what's your fat wify given Mr. loose motions for food today (opening his lunch box)... hmmm smells good... dude check out the gravy with these rotis....looks like some one is not gonna get to eat these...

Hard working avg man- Give me my lunch box back

Bully at work 2- oh attitude.....(in a teasing tone) looks like some one wants to rush to the toilet once again...

Hard working avg man- no no please don't

Bully at work 1- oh yes .. oh yes.. at the count of 1, 2, 3

Both bullies together make the hissing sound- pisssssssssssssssssssssssssssssssssssssssssssssss

Hard working avg man—no... no..(trying to control his urge to pee)... I cant.. (rushes to the toilet at once) (Voiding of urine due to decreased tone- parasympathetic activity)

All laugh as he runs

Bully at work 2- Ha ha ha... this is funnier than the last time when we urged him to shit and he did it right there... ha ha ha... chal lets get back to work. (Increased Bowel motility- parasympathetic activity)

Narrator- (Speaks as the hardworking avg man walks back into the stage weeping softly to get back to his desk and continue his work)

Hi all.. I welcome you to the usual morning in the life of Mr. AC Nicholas.. Or as we call him Chole for short. (Acetyl Choline- parasympathetic neurotransmitter)*He was a normal child at birth born to a good business family. He was put into a reputed school, was a favourite amongst teachers for his top grades. But he faced ragging in school in 3rd 7th 9th and 10th standards.* (parasympathetic nerves are derived from the 3rd, 7th, 9th and 10th cranial nerves*). Today he is one of the best charted accountants in this company. Why must he be treated like this?*

Maybe because he had only one small problem.... He was a little too soft for a man if you know what I mean. Always careful, meticulous, slow in everything, never getting angry and as is clearly evident ... crying for everything. Not to forget his poor control on his bowel and bladder habits which has been an embarrassment for him since his school days.

AC Nicholas- For how long am I going to face this every day? . I haven't eaten lunch in two weeks. All I do is salivate in hunger and cry in dismayal and again salivate and again cry. (Increased salivation and lacrimation- parasympathetic activity) How can I put an end to this. I must end my life. But how?

Narrator- *Determined to end his life he tries to strangulate himself. He strangulates till he chokes and triggers an asthma .I forgot to mention he has this as well. He takes out his inhaler and relives himself falling unconscious on the desk.* (Bronchoconstriction- parasympathetic activity)

Suddenly he is noticed by other colleagues and they begin to panic.

Colleague 1- Look.. Chole is out cold.. call the ambulance

Bully at work 1- Nah... don't worry.. its normal.. that's how he works..

Colleague- Its not funny..what if he is dead then you guys will be responsible for this

Bully at work 2- (checking the pulse)—relax.. his slow and lousy pulse is felt...just like him (Low heart rate- parasympathetic activity)

A paramedics personnel with a BP apparatus ,torch and a stethoscope measures his BP..

Paramedics- Wow—that's the lowest I have ever recorded..and his pupils are constricted ..I really doubt if he is alive (Low Blood pressure, constricted pupils- parasympathetic activity)

Bully at work 1- chillax.. ill tell you how to get him up and running.. at the count of 1, 2, 3....

(In chorus) pisssssssssssssssssssssssssssssssssssssss

Narrator- *Nicholas immediately gets up and runs up embarrassed towards the toilet.*

Feeling bad for him the boss's hot secretary – noriene (norepinephrine – sympathetic neurotransmitter) *starts to melt her heart for our miserable Nicholas.*

As he is sitting on his table trying to get back to work she makes an attempt to talk with him. But there is one small issue; Nicholas has never spoken to a woman before. Not like he isn't attracted to them, he even has an errecsion mobile phone for the same reason thinking that one day he will have a girlfriend who will call him on this phone (Erection – parasympathetic action)

Hot Secretary/ norine- hi.. are you alright there? Seems like the guys have been very rough on you.

Nicholas- Oh hi.. err.. he.. err... hi..I I I I am fine really... I just play around with them so that we have some fun at work..

Hot secretary/noriene- oh really.. well that's good to know. I thought you were those sensitive guys.

Nicholas- no .. oh no.. you are way off.. (pretending to be brave) ive been a part of dangerous gangs .. these kids are nothing

Norine- gangs? (ganglions)

Nicholas- yeah—the Sea clearing gang.. have you heard of them? (cilliary ganglion)

Norine- oh—you know the don of terror .. edinger westphal, who used to hide all the loot under the sea?(parasympathetic fibres arise from the edinger westphal nucleus)

Nicholas- yeah.. I was his best pupil. Short and adjustable i.e I could accommodate any amount of work required. (causes pupillary constriction and accommodation)

Norine- wow.. im starting to get impressed. Whom else have you worked with?

Nicholas- well the usual's,, I was the sub head for the man bully gang, (sub mandibular ganglion) then I was a part of the terrible plate throwing gang (pterygopaltine ganglion) and finally our local ooty gang (otic ganglion) also I have had special training from las vegal on how to slow the heart and stab the stomach till its empty. (vagal innervation to heart and stomach)

Norine- well, then you do a pretty good job acting sissy in front of these bullies. May be you should teach them a lesson some time.

Nicholas- Well, definitely ... let the right time come.. by the way what made you come and talk to me.. I mean I feel great as no one has ever sympathised with me before? (Sympathetic stimulation)

Norine- oh ive been bullied too in my school days throughout from ist standard to 2nd year of graduation (The sympathetic nerves arise from first thoracic vertebra to second lumbar vertebra).. I know exactly how it feels.

Nicholas- just out of curiosity if I might ask.. how did you tackle your bullies?

Norine- well its simple.. the best revenge is living happy..I continued to be happy even when they bullied me and they eventually gave up and stopped. Hey look its your bullies.. lets have some fun..

Nicholas- em,.. er.. norine..

Norine- Hey guys... Nicholas thinks you guys are fat ladies who are fit for nothing and do girly gossip all day.....(giggles)

Bullies- in chorus- What?

Nicholas- What?

Norine- come on.. show them who you are.

Narrator- Bullies charge towards Nicholas who is gulping like never before.

Bully at work 1- Oh so Mr. Piss pants is trying to act smart in front of the lady enh?

Bully at work 2 – Looks like he is missing the toilet again .. aren't you Mr. wet pants?

Narrator- *Nicholas was extremely scared, terrified and petrified... his pupils dilated , heart beat racing and his blood pressure touching the roof* .(Pupillary dilatation, increased heart rate and increased BP – Sympathetic action) *All he remembered is what norine told him about remaining normal. So he summed up courage to say*

Nicholas- Yes ladies...I am Mr. wet pants and shitty pants.. I love being this way..Firstly I get all the attention, hot girls come voluntarily to talk to me and my wife takes good care of me when I get home. Not to forget I am not overloaded with work because every one pities me and all that work is given to you fellows. So its time I thank you for doing my work.

Bully at work 1- Well think about what you said in the loo Mr. loose stools....

Narrator- *Even after saying these bold words to the bullies Nicholas was terrified as to how he would control his bladder and bowel. He starts to sweat and breath heavily in anticipation of what happens.* (Sweating and bronchodilatation are sympathetic activities) *He only concentrates on what Norine said.*

Bully at work 2- In chorus- 1, 2, 3 pissssssssssssssssssssssssssssssssssssssssssssssssss

Nicholas- (even though has an irresistible urge to pass urine he clenches his fist and tries to control his bladder) Hmmm.. Its time for me to stop acting ladies... I used to pee in the pants just to make you guys happy.. not effecting me.. (Bladder sphincter tone increased- sympathetic activity)

Bully at work 1- Well lets try to make him shit.. make constipating noises

Narrator- *After being able to resist urination Nicholas developed confidence in this method and laughed at the efforts of these bullies and sat down on his desk to finish his work as nothing effected him.* (Decreased bowel motility- sympathetic activity) *The bullies continued to make noises around him for two or three hours and then saw no point to it. The following days the bullies stopped their mean ways and developed respect for Nicholas, who just thanked norine for giving the sympathy he required and how it changed his life.*

Now Nichole and norine are not just at the erricson phone talking level... they are well ahead of these and beyond..living happily ever after (Ejaculation and orgasm is a sympathetic activity)

Topic 5 – Performance activity on pathogenesis of rheumatic fever

10 Pointer on rheumatic fever

1. Acute rheumatic fever (ARF) is a multisystem disease resulting from an autoimmune reaction to infection with group A streptococci.(A type of circular bacteria that are seen in chains)

2. Pathogenesis-
    a. Organism factor- ARF is exclusively caused by infection of the upper respiratory tract (cold and cough with fever) with group A streptococci.
    b. HOST FACTORS-Particular HLA class II alleles appear to be strongly associated with susceptibility. *Associations have also been described with high levels of circulating mannose-binding lectin and polymorphisms of transforming growth factor 1 gene and immunoglobulin genes. High-level expression of a particular alloantigen present on B cells, D8-17, has been found in patients with a history of ARF in many populations, with intermediate-level expression in first-degree family members, suggesting that this may be a marker of inherited susceptibility.*
    c. Immune Response- When a susceptible host encounters a group A streptococcus, an autoimmune reaction results, which leads to damage to human tissues as a result of cross-reactivity between epitopes on the organism and the host.
    (Fig. 315-2).

3. ARF is mainly a disease of children aged 5–14 years. Initial episodes become less common in olderadolescents and young adults and are rare in persons aged >30 years.

4. clinical features- History of infection(cough and cold) is present. Usually within a period of 3 weeks of a latent phase the following can occur.

    a. The most common clinical presentation of ARF is polyarthritis and fever.
    b. Polyarthritis (multiple fleeting joint pain) is present in 60–75% of cases
    c. Carditis(infection of heart) in 50–60%. Valvular damage is the hallmark of rheumatic carditis. Mitral valve is m/c valve effected
    d. The prevalence of chorea(abnormal involuntary movements of hands,legs,tongue) in ARF varies substantially between populations, ranging from <2% to 30%.

e. Erythema marginatum(red rashes) and subcutaneous nodules(lumps beneath skin) are now rare, being found in <5% of cases.

5. Treatment- Antibiotics, pain killers. Steroids in heart involvement.

   Penicillin is the drug of choice and can be given orally (as penicillin, 500 mg PO twice daily for 10 days) or as a single dose of 1.2 million units IM benzathine penicillin G. Erythromycin, 250 mg bid, may be used for patients with penicillin allergy

6. primary prevention for ARF remains primary prophylaxis, i.e., the timely and complete treatment of group A streptococcal sore throat with antibiotics. If commenced within 9 days of sore throat onset, a course of 10 days of penicillin V (500 mg bid PO in adults) or a single IM injection of 1.2 million units of benzathine penicillin G will prevent almost all cases of ARF that would otherwise have developed.

Acute Rheumatic Fever – A Little Red –Riding hood Adaptation

Plot – Little Red Ridinghood (heart) is going to meet her grandmother.

Her brother Immaneul is a hunter, who hunts down foxes in the forest. He uses B-power arrows to kill them. (immune system)

Her mother is Mrs Jointy rodes, who asks her to get fruits for grandma from the forest. (bones)

Step-to-Foxus is an evil fox, which always steps from behind in groups to scare people. (streptococcus)

Grandmother

Play

Narrator: There once lived a little 5 year old girl who stayed with her mother, brother and grandmother in a house at the edge of the forest. She always wore a red cloak and hood that her grandmother had made for her. And so everyone, even her mother Mrs Jointy, called her The Little Red Riding Hood. Her brother who was 14, followed his fathers footsteps and aimed to become the best fox hunter in the village.

ARF is mainly a disease of children aged 5–14 years. Initial episodes become less common in older adolescents and young adults and are rare in persons aged >30 years.

Ever since her father had died, Mrs Jointy had taken care of everything, and villagers called her the back bone of the house. But it was Little Red Riding Hood, who was everyone's favorite, always bubbly, thus known as the 'heart of the house'. Immanuel was a brave little fellow who loved to hunt and he was always chasing down evil foxes with his powerful Berry coated -arrows which he called B arrows, whenever they attacked his house. He was the 'protector of the house' and the strongest line of defence.

One day Little Red Riding Hood's mother said to her:

Mother: Dear little red riding hood, I just heard your Grandmother is not feeling well. She has been screaming at the lousy foxes all week long to keep them at bay and her throat has become sore. Will you get her some fruits from the forest to help her recover?

Streptococcus causes sore throat and URTI

Red Riding Hood: Yes, Mother.

mother: But be careful while going through the forest dear as you may fall prey to the evil foxes. And don't play along the way and cut yourself. Grandmother is waiting for you.

Red Riding Hood: Don't worry Mother, I ll fox the foxes and take the fruits to grandmother.

Narrator: Little Red Riding Hood took the basket and was on her way. She had smartly kept prepared something apt for the situation- A dress of a fox. Something her father had given to

her when after he hunted the most dangerous fox and gave it to his daughter to use it wisely while crossing the forrest. Little Red Riding Hood had gone but a short way into the forest when a great fox, Step-to-Fox stepped out from behind, as usual from a tree along with a group of other foxes that accompany him.

Step-to-Fox: Good Morning, fellow fox. I am afraid we haven't met you before, are you a traveller? Where are you going on this fine day?

Red Riding Hood: err. Hmm. I am coming from the near by village to meet a friend here. Carrying these fruits as a present to their family.

Step-to-Fox: Why not bring them some wild flowers as well? There are so many to pick from here in the forest.

Red Riding Hood: I cannot stop. I have to travel back as well and it might get very late.

Step-to-Fox: Picking flowers is not playing around. Besides, a bunch of pretty flowers will cheer your friends up and make them feel better.

Red Riding Hood: Perhaps you are right.

Narrator: Little Red Riding Hood put down her basket and began to pick wild flowers. Meanwhile, as Step-to-Fox was about to leave he noticed a part of the red hood realised it was red riding hood. He gave a wicked smile and sped to Little Red Riding Hood's home.

Step-to fox: Oh little red hood tries to play smart with me? I will show her a lesson of her life time.. ha ha ha

When he came to the cottage, he crept up to the door and knocked.

Grand Mother: Who is it?

Step-to-Fox: It is I, Little Red Riding Hood. I have brought you fruit and flowers.

Grand Mother: Come in dear. The door is open.

Narrator: As soon as the Step-to-Fox was inside, Grandmother started sneezing because of the wild-flowers in the basket. Immediately, Step-to-Fox used the opportunity to grab her by the t throat and twist it till she fell unconscious. He then hid her in the toilet and picked up the night cap and blanket.

Organism factor- ARF is exclusively caused by infection of the upper respiratory tract (cold and cough with fever) with group A streptococci.

He put on the night cap and wrapped himself in the blanket. Then he crawled into bed to wait for Little Red Riding Hood.

Narrator: In the mean while – Immunuel in the forrest is on the hunt for foxes. As a command recived from his mother he was on the lookout for step to fox so that he can put an end to this menace. He had seen enough of his family living in his fear. Suddenly he spots a fox piking flowers. He takes his B arrow and targets right to the heart of the fox and hits his target

A girlish scream follows- aaaaaaahhhh

Immunuel- That's a strange scream for a fox. Let me check

As he goes close he sees its red riding hood and immediately breaks down into tears.

Immunuel- Dear sister- what were you doing dressed like this, i thought it was step-to fox .

Immune Response- When a susceptible host encounters a group A streptococcus, an autoimmune reaction results, which leads to damage to human tissues as a result of cross-reactivity between epitopes on the organism and the host.

Red riding hood- I am hit bad brother. Please deliver this to grand ma. I don't think ill be able to make it.

Immunuel- No sister. I have the antidote to this berry coated arrow. Here take this and you will be fine.

Red riding hood drinks the medicine and soon feels better. She sends her brother home and continues towards her grand mothers house .

Three hours later where step to fox is waiting impatiently for red riding hood, hears a knock at the door.

3 weeks of a latent phase

Step-to-Fox: (like grandmother) Who is it?

Red Riding Hood: It is I, Little Red Riding Hood. I have brought you fruit and a bunch of pretty wild flowers to cheer you, Grandma.

Step-to-Fox: Come in. The door is open.

Narrator: Little Red Riding Hood went in and stood for a moment in the doorway.

Red Riding Hood: Oh, Grandmother! You look so blue!

Step-to-Fox: Its the fever, my dear. Come closer.

Narrator: Little Red Riding Hood walked closer to the bed.

Red Riding Hood: Oh, Grandmother! You smell like wild grapes!

Step-to-Fox: It must be the fruits you brought, my dear. Come closer.

streptococci.(A type of circular bacteria that are seen in grape like clusters)

Narrator: Little Red Riding Hood walked even closer to the bed. Meanwhile, Immanuel had also come to meet his grandmother to chase out the foxes which gave her a soar throat. But when he peeped through the window, he just saw grandma's blanket and Little Red Riding Hood, and thought all was fine.

Red Riding Hood: Oh, Grandmother! What big teeth you have!

Step-to-Fox: The better to eat you with!

Narrator: The Step-to-Fox grabbed Little Red Riding Hood and twisted her neck to make her un conscious.

Carditis(infection of heart) in 50–60%. Valvular damage is the hallmark of rheumatic carditis. Mitral valve is m/c valve effected

Meanwhile, Mrs Jointy rushed back home to see if Little Red Ridinghood was back home. Step-to-Fox pounced on her knees, making them red-hot and pushed her down causing terrible pain in her joints. He then made her unconscious as well. Then, feeling satisfied, the Step-to-Fox lay down and fell asleep. He muttered planning tonight he and his friends will enjoy a great feast

Polyarthritis (multiple fleeting joint pain) is present in 60–75% of cases. Erythema marginatum(red rashes) and subcutaneous nodules(lumps beneath skin) are now rare, being found in <5% of cases.

Hearing Mrs jointy rodes creak when she was trying to fight with Step-to-Fox, Immanuel heard the stereo break in the house. Quickly realising something was wrong, he got ready his pencil sharp arrow and aimed at the snoring Step-to-Fox.

Treatment- Antibiotics, pain killers. Steroids in heart involvement.

Penicillin is the drug of choice and can be given orally (as penicillin, 500 mg PO twice daily for 10

days) or as a single dose of 1.2 million units IM benzathine penicillin G. Erythromycin, 250 mg bid,

may be used for patients with penicillin allergy

Immaneul: I have you at last, you devil!

Narrator: The Immaneul fired 500 hi-profiled arrows through the central axis of Step-to-Fox within 10 minutes to get better of him. Then the Immaneul took out the remaining medicine and gave it to his sister, mother and grandmother. Out arose Mrs jointy, Little Red Riding Hood, and her grandmother, safe and whole. Grandmother gave Little Red Riding Hood a hug. Then Little Red Riding Hood, Grandmother, Mrs jointy and Immaneul sat down to feast on fruit, cakes, and honey. They all lived happily from that day on, never to be troubled by the wicked Step-to-Fox again.

primary prevention for ARF remains primary prophylaxis, i.e., the timely and complete treatment of group A streptococcal sore throat with antibiotics. If commenced within 9 days of sore throat onset, a course of 10 days of penicillin V (500 mg bid PO in adults) or a single IM injection of 1.2 million units of benzathine penicillin G will prevent almost all cases of ARF that wold otherwise have developed.

Initial episodes become less common in olderadolescents and young adults and are rare in persons aged >30 years.

TOPICS FROM THE LAST WORKSHOP HELD IN MAY 2011

1. Mechanism of action of penicillin

10 Pointer on MOA of Penicillin's

1. Staphylococci bacteria are seen in grape like clusters. They are gram positive cocci.
2. Penicillins are a group of antibiotics that act on gram positive bacteria.
3. Penicillins are notoriously known for allergic and anaphylactic reactions and they must be used after a test dose.
4. Peptidoglycan is a heteropolymer that provides rigid mechanical stability to the bacterial cell wall by virtue of its highly cross-linked structure.
5. The peptidoglycan is composed of glycan chains, which are linear strands of two alternating amino sugars (N-acetylglucosamine and N-acetylmuramic acid) cross-linked by peptide chains.)(The cross-link is completed by a transpeptidation reaction that occurs outside the cell membrane.
6. Penicillins consist of a thiazolidine ring (A) connected to a b-lactam ring (B), to which is attached a side chain.
7. The b-lactam antibiotics inhibit this last step in peptidoglycan synthesis, presumably by acylating the transpeptidase via cleavage of the—CO—N—bond of the b-lactam ring.Although inhibition of the transpeptidase is demonstrably important, there are additional targets for the actions of penicillins and cephalosporins; these collectively are termed penicillinbinding proteins (PBPs). The transpeptidase responsible for peptidoglycan synthesis is one of these PBPs.
8. A sensitive strain may acquire resistance by mutations that decrease the affinity of PBPs for the antibiotic. Because b-lactam antibiotics inhibit many different PBPs, their affinity for several PBPs must decrease to confer resistance.
9. Bacteria also can destroy b-lactam antibiotics enzymatically via the actions of b-lactamases.
10. Bacterial resistance to b-lactam antibiotics also results from the inability of the drug to penetrate to its site of action

Script- Mechanism of action of penicillins (Antibiotic) and mechanism of resistance developed.

Scene 1- Ravi and richa were a happily married couple living in their house 1 year since their marriage. Neither had any major complaints about each other , except one small unbearable issue which richa couldn't tolerate was the frequent visits made by ravi's friends.

Ravi- I think we can expect 7 people for dinner today.

Richa- what? Who's coming today? C'mon its one Sunday a week I get to relax ... why don't we go to their houses.

Ravi- They are my child hood friends darling, who are coming after a long time. Please . this one last time.

Richa- you always say that.. why don't we all go out for dinner

Ravi- its too expensive dear...plus we are watching the match together right?.. and they want to taste the wonderful food you cook....please

Richa- you always do this to me... they have been coming since last 5 Sundays...why do you do this to me.. how can you do this to me..every time you say the same reasons.... why did you do this to me?... don't do this to me

Ravi—relax.. I am sorry.. its this last time .. please..they are a group and you know they are always together like a bunch of grapes.(Streptococci bacteria are seen in grape like clusters) next time ill take care of this

Richa – (in the kitchen mumbling to her self)- It must be this peptin pinto once again. Hope he was never ravi's childhood friend and I wouldn't be facing all this today. I must think of a way to put an end to this. He is the main strength to his group of friends and If I take care of him I think ill spend my future Sundays more peacefully.(Peptidoglycan is a heteropolymer that provides rigid mechanical stability to the bacterial cell wall by virtue of its highly cross-linked structure .)

And what about his two sugar buddies cosa and mura. Their links together are the main strength of peptin. I must break this association also, but how.. these three have been linked even before they knew ravi. (The peptidoglycan is composed of glycan chains, which are linear strands of two alternating amino sugars (N-acetylglucosamine and N-acetylmuramic acid) cross-linked by peptide chains.)(The cross-link is completed by a transpeptidation reaction that occurs outside the cell membrane)

(Suddenly- as if an idea struck her) that's it.. why didn't I think of this before.. Its perfect. *(she makes a phone call)*

Scene 2- All friends have arrived at home and are about to sit for dinner. Peptin and his close friends cosa and mura are busy talking to ravi. In the mean time the door bell rings.

*Richa opens the door... bow wowoow bow woow .. a nice hairy golden retriever with a belt around the neck connected to a side chain walks inside home. (*Penicillins consist of a thiazolidine ring (A) connected to a b-lactam ring (B), to which is attached a side chain*)*

Peptin- Oh my god.. where did this come from.. keep him away from me .. I m scared of dogs.

Ravi- whats going on.. why didn't you discuss this with me..

Richa- what could I DO HONEY...Swathi just called me and asked me to look after beta on Sundays as she has to go for salsa classes. She is my best friend and you know how we co-operate for each others friends right?

(The b-lactam antibiotics inhibit this last step in peptidoglycan synthesis, presumably by acylating the transpeptidase via cleavage of the—CO—N—bond of the b-lactam ring.

Although inhibition of the transpeptidase is demonstrably important, there are additional targets for the actions of penicillins and cephalosporins; these collectively are termed penicillinbinding proteins (PBPs). The transpeptidase responsible for peptidoglycan synthesis is one of these PBPs.)

Cosa and mura start sneezing profusely

Richa- what happened ?

Peptin- these two brothers have allergy to pet dander. I better get them out as it can trigger an asthma.

Richa- ooops.. sorry for the trouble.

Peptin- ravi.. I think its not a good idea to send cosa and mura alone with all this sneezing. Ill drop them home and we'll catch up some time later.

Other friends- in chorus- well if peptin is leaving then we'll also carry on.. will come some other time...

All leave. Richa being extremely happy with her successful plan decides to relax the rest of the Sunday with her husband and borrowed pet dog beta.

1 week later-

Ravi- I think we can expect 7 people for dinner today.

Richa- what? Who's coming today? C'mon its one Sunday a week I get to relax ... is it the same peptin and group.. but beta is coming today too and they are allergic

Ravi- Its all taken care of honey. They have started to use anti allergic inhalers and are now no more allergic to pet dander.

(A sensitive strain may acquire resistance by mutations that decrease the affinity of PBPs for the antibiotic. Because b-lactam antibiotics inhibit many different PBPs, their affinity for several PBPs must decrease to confer resistance.)

Also they are carrying water squirters so that beta doesn't come close to them. (Bacteria also can destroy b-lactam antibiotics enzymatically via the actions of b-lactamases.)

I was also thinking to put a nail outside the house door so that beta can be chained there as we have the dinner.

(Bacterial resistance to b-lactam antibiotics also results from the inability of the drug to penetrate

to its site of action .)

See I told you I wont disappoint you, all this was peptins idea only. now you can cook a good meal for them again.

Richa storms to the kitchen once again.. mumbling to her

2. MENIERES DISEASE

10 pointer on Meniere's disease

1. Meniere's disease is a disorder characterized by spontaneous attacks of vertigo (with associated fluctuating sensorineural hearing loss, tinnitus, and aural fullness.

2. These predisposing factors somehow cause overproduction or malabsorption of endolymph resulting in endolymphatic hypertension.

3. Pathology- Periodic ruptures of the membranous labyrinth, which result in leakage of the potassium-rich endolymph into the perilymph, are thought to be responsible for producing Meniere's attacks. Healing of the rupture then allows restitution of the normal chemical composition of the endolymph and perilymph, resulting in termination of the attack and improvement in vestibular and auditory function.
4. Clinical feature-
    a. recurrent attacks of spontaneous vertigo, with associated nausea and vomiting.
    b. lateralized low-frequency hearing loss, tinnitus and aural fullness.

A typical attack of vertigo has three phases, each defined by the direction of the spontaneous nystagmus. The first is the so-called irritative phase, in which the nystagmus, usually horizontal or horizontal-torsional, beats towards the affected ear; this phase usually lasts less than an hour.

The second is the so-called paretic phase, in which the nystagmus beats away from the affected ear; this phase lasts several hours, sometimes even a day or two. The third is the so-called recovery phase, in which the nystagmus again beats towards the affected side; this phase lasts about as long as the second phase.

5. Diagnosis- two or more spontaneous attacks of vertigo (each lasting 20 minutes or longer), hearing loss documented by pure-tone audiogram on at least one occasion, tinnitus or aural fullness on the affected side, and other possible causes excluded.

6. A fluctuating low-frequency sensorineural hearing loss may be seen in the early stages of the disease in some patients and is very suggestive of the diagnosis

7. Electrocochleography is of some use in diagnosing Meniere's disease. It is most sensitive and specific for Meniere's disease when tone-burst and click stimuli are used, and when the responses are recorded transtympanically at the promontory. Giving patients 4 g of oral sodium chloride for three days prior to the electrocochleogram may increase the sensitivity of the test.

8. Rx- Strict sodium restriction (so that urinary sodium is less than 50 mmoL/day) and diuretics, Endolymphatic sac surgery .

9. Selective vestibular neurectomy , surgical labyrinthectomy can stop the vertigo attacks at the expense of losing any remaining hearing on that side. Those whose hearing is below the reach of standard hearing aids may be helped by cochlear implantation.

10. Intratympanic injections of aminoglycosides . As little as one injection of gentamicin may be effective in controlling vertigo in most patients, with little or no hearing loss complicating the procedure.

The Mystery of the Mean ears disease

Scene 1- Earlock homes is sitting on his desk, closing the file of his last solved mystery as his best friend from india wrongcase bakshi joins him in his room.

Wrongcase- Mr. Homes- I hope you are finding your visit to India exciting, eventhough I understand the nature of cases here must be very different from what you see in England.

Homes- well, Im simply lovin it out here Mr. Bakshi. The culture, the petty issues and the type of criminal offences done here are very fascinating to solve. Quite an experience if I may say.

Wrong case- well then I have just thing for you. Our next case is from a small middle class family who claim that someone has been stealing water from their home and they really want to put an end to this menace.

Homes- stealing water? Why would on do that?

Wrongcase- well Mr. Homes, In India it is not new to have scarcity of water and electricity so I am not very surprised that water is being stolen.

Homes- All right.. lets get down and cracking on this one immediately then.

Scene 2- At the location of the crime which is the drawing room of the house. Homes and bakshi walk in and introduce themselves. In the room there stand Mr. Kann singh, his wife kumsun and their son Behra singh. A little away from them is a little poorly dresses house maid Kaantha bai.

Wrong case- I would like any one of you to clearly explain the problem that we are facing here.

Kann singh- Sir, As you know there is a major shortage of water in our locality. We all need to get ready in the morning to get to work. Me and my wife are both earning members and my son goes to school. Since the last week there is not a drop of water in our house only, every one else around us in the neighbourhood gets supplies in the mornings atleast. We don't get anything.

Homes- At what time do the neighbours get water?

Kann singh- 6 to 8 am in the morning

Homes –hmm.. and what time do you get up in the morning

Kann singh- I get up by 7 and my wife gets up earlier at around 6 30.

Homes- who all stay here?

Kann singh- Me, My wife and My son. Kantha bai stays in the innermost room of our house (inner ear) where my father used to stay.

Homes – where is he now?

Kann singh- oh – he is very old and debilitated as he cant walk so he spends 6 months with me and 6 months with my brother. Currently he is there, but must be looked after well as he lies down all the time.

Wrong case- do you suspect anyone?

Kann singh- whom to suspect sir? God knows from where the water is being stolen and why would anyone do this to us

Kum sun- I suspect some one sir.

Homes- who?

Kum sun- Kantha bai.. who else?

Homes- What makes you suspect her mrs. Singh?

Kumsun- Sir.. she has been working here since 2 years..I took her because she was poor and needed treatment for her ear disease. But she would never do any work correctly?

Homes- What do you mean Mrs. Singh , I need you to be very clear.

Kum sun- well.. when I tell her to do some work she does it, but if I add one small extra task like to hand me over a box from the shelve or to clean the corners of the house she would suddenly complain that her head is spinning (Episodic spontaneous vertigo) with vomiting and eyes would move very rapidly.(Nystagmus) And not only that she pretends not to hear when some work is told to her.

Kann singh- that's true.. anything that I tell her she never hears and avoids any work given to her by me but while watching TV and listening to music she seems to hear very well. Its only when my wife scolds her she hears everything perfectly.(fluctuating sensorineural hearing loss to low frequency sounds)

Kum sum- sometimes she leaves work and just walks to the phone and picks it up even though it never rang. If I question why she is wasting time and piking up calls she argues that she heard the phone ring.(Tinnitus) and keeps saying there is something stuck in her ear (aural fullness)

Homes- Are you trying to imply that since you have been bothering her with these issues she is taking a revenge on you by stealing your water?

Kum sum- Its not only that sir, some other behaviours that have no explanation raise doubts. I think just two days ago I remember clearly as she was wearing this same saree I had asked her to cook paneer butter masala at dinner for us. Usually we cook ourselves but I came late from work. The moment she started cooking she suddenly fell to the floor dropping all the food on the floor.(Tumarkin's drop attack) I was so angry because I had to cook all over again.

Homes- Kantha bai.. do you have anything to say in your defence

Kantha bai—arrey sir.. madam is just putting all blame on me.. what can I do.. I am suffering from this disease since so long.. I cant hear from one ear (unilateral hearing loss), I fall down suddenly as if some one is pushing me on the floor (pushing sensation is characteristic) I am not doing this on purpose. If you don't believe me see my reports.(hands it over to wrong case)

Wrong case- she is right homes. Her condition is genuine. This electrocochleography done in tone burst and click stimuli show a positive response. (Electrocochleography is of some use in diagnosing Meniere's disease. It is most sensitive and specific for Meniere's disease when tone-burst and click stimuli are used, and when the responses are recorded transtympanically at the promontory.)

Homes- hmm- two or more spontaneous attacks of vertigo (each lasting 20 minutes or longer), hearing loss documented by pure-tone audiogram on at least one occasion, tinnitus or aural fullness on the affected side all point to her ear disease. It is very unlikely that she is faking it.

(two or more spontaneous attacks of vertigo (each lasting 20 minutes or longer), hearing loss documented by pure-tone audiogram on at least one occasion, tinnitus or aural fullness on the affected side)

Nevertheless Mrs singh, all you are telling me is just petty complaints about how your maid doesn't do her work. And the reports support her on this point. There is no strong evidence to say she is the culprit.

Kumsun- Sir.. wait a minute.. there is one evidence.. I have seen at times in the afternoon that water is seeping out of her room. She begins to act very irritable at that time. When I ask her where did this water come from she simply says there is no water and it is just my imagination.but she continues to enter the room again and again as if she is doing something fishy (The first is the so-called irritative phase, in which the nystagmus, usually horizontal or horizontal-torsional, beats towards the affected ear; this phase usually lasts less than an hour.)

Homes- didn't you call some one else to see the water seeping below.

Kumsun- I did.. but by the time I call my husband or son there would be no water from the room and at that time kantha bai would be in other rooms pretending to say nothing ever happened. (The second is the so-called paretic phase, in which the nystagmus beats away from the affected ear; this phase lasts several hours, sometimes even a day or two.)

Then after my husband and son go away after scolding me for calling them unnecessarily she again steps into her room many times.

(The third is the so-called recovery phase, in which the nystagmus again beats towards the affected side; this phase lasts about as long as the second phase.)

I cant exactly point to where she is doing something but I know for sure something is suspicious about her behaviour. She has almost convinced me that I am crazy.

Homes- Relax mrs. Kumsun- Hmm so you have given some interesting inputs. Kantha bai.. why would you repeatedly enter your room like that?

Kantha bai- arrey sir.. madam has gone crazy.. sometimes she says she sees water here, or there but nothing will be there. Its because she hasn't seen water for a week. I go in my room to clean it up. May be I don't realise it when she calls me as I cannot hear.

Wrongcase- haven't you checked kantha bai's room.?

Kann singh- we have.. there is nothing there but a bed and her suitcase.

Homes- well give me a minute.. I think I just might have caught the culprit. Wrong case, I suggest we arrest Kantha bai for stealing water and creating an opportunity for poor hygiene in this house hold.

Wrongcase- but how? And where did she keep the water and what supporting proof.

Homes- Let me explain- Firstly- remember mrs. Singh telling that when she had asked kantha bai to cook a curry she suddenly fell due to her ear disease and dropped all the curry on the floor.

Wrong case- I remember.

Homes- Since she was cooking she cant possibly escape the curry from falling on her .

Wrong case- right

Homes- then how do you explain the stainless dress she is wearing now if she had no access to water?

Wrongcase- that's brilliant. But where is she keeping the water?

Homes- simple.. in her room..

Wrong case- In her room?

Homes- yes. That room is where an ols debilitated bedridden father used to stay. He must have had to be on a water bed to prevent bed sores.

Kann singh- yes.. we did buy a water bed.

Homes- well, that's where it will be then. Now the sequence-

For some reason she stored water. For which she would get up at 6 in the morning.. collect the water and fill it in the water bed. Since the water has been stored there over time and is not being utilised it has been accumulating and stretching to the capacity of the bed. (overproduction or malabsorption of endolymph (there is a fluid filled canal inside the innermost part of the ear which bulges due to over production of fluid and decreased excretion of the fluid), resulting in endolymphatic hypertension.)

This would cause water to leak out into the room and seep out from underneath the door when mrs. Singh would notice it.

(Periodic ruptures of the membranous labyrinth, which result in leakage of the potassium-rich endolymph into the perilymph, are thought to be responsible for producing Meniere's attacks)

This would irritate kantha bai and she would repeatedly enter the room to ensure she cleans up the spilled water and repairs the leak

(Healing of the rupture then allows restitution of the normal chemical composition of the endolymph and perilymph, resulting in termination of the attack and improvement in vestibular and auditory function.)

By the time mr kann and his son would come all evidence would be gone. That IS why to divert attention she went into other rooms to play innocent.

Now lets ask kantha bai why she did it?

Kantha bai- sir.. all that you have said is correct. I also suffer from a condition known as OCD where I have to keep my room, my self and my clothes extremely clean. If I don't then I face very severe distress. with such scarcity of water I feared there will never be enough water for me so I started storing water. (obscessive compulsive disorder is a condition due to decreased serotonin levels in the brain which leads to uncontrolled repeated, distressing thoughts most commonly relating to cleanliness and bathing which is followed by compulsive acts where there Is ritualistic cleaning and bathing habits)

Sorry madam and sir for doing this but I am forced to do so because I am slave to the thoughts I get as a part of my illness.

Plus the salt restriction, medications and surgeries that I have gone through made me very vulnerable and has worsened my ocd. (Strict sodium restriction and diuretics can be effective in reducing the frequency of vertigo attacks. Endolymphatic sac surgery may result in a reduction in the frequency, duration and intensity of vertigo attacks. Selective vestibular neurectomy is effective in stopping vertigo attacks in most patients, although hearing loss and facial paralysis may occur as potential complications. A surgical labyrinthectomy can stop the vertigo attacks at the expense of losing any remaining hearing on that side. Those whose hearing is below the reach of standard hearing aids may be helped by cochlear implantation. Intratympanic injections of aminoglycosides have become popular)

Wrong case- well done once again mr. homes. Its no surprise that you are the best.

PROJECT ASSIGNMENT – Make your own script

STEP 1- Read through your topic

STEP 2 – Underline the important points that you would like to include in your script

STEP 3 – Write down the schematics below

STEP 4- Form a plot

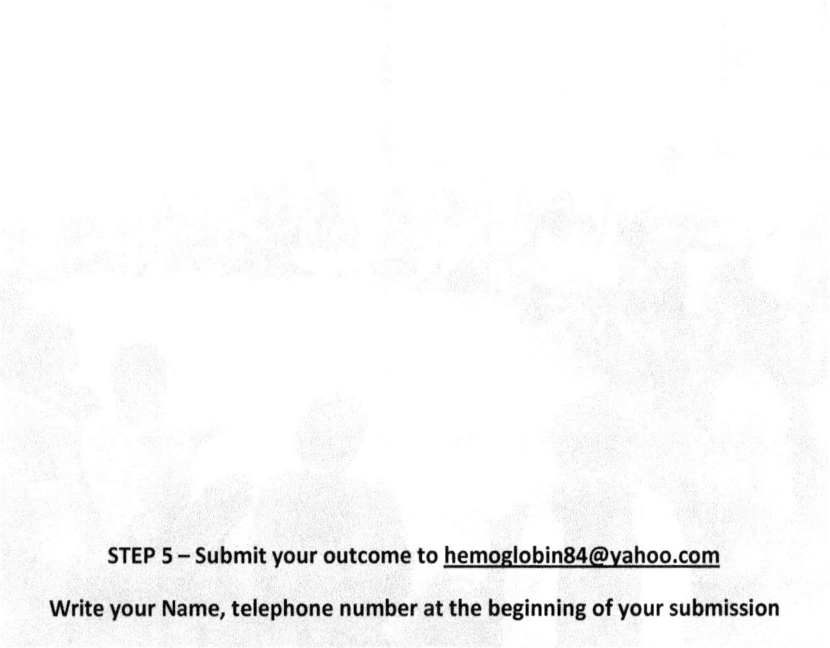

STEP 5 – Submit your outcome to hemoglobin84@yahoo.com

Write your Name, telephone number at the beginning of your submission

Write a small paragraph on your experience after writing your script- this should be at the end of your script.

Your script will be read and feedback comments would be given on the second day of the workshop.

The certificates would be issued after submission of the scripts only.

THE INSCRIPTED

IMPORTANT INFORMATION FOR READERS

These scripts/stories are created by medical students who participated in the first Medical theatre workshop held in the month of May 2011.

These scripts do not claim to guarantee an easy understanding of the topic for the reader; it however intends to make an interesting reading where the efforts of these students and their attempt at a new methodology can be highlighted.

The learning process and understanding develops during the formation of the script when the student creates something new out of existing material available in books. The output in the form of these stories/poems/monologues and scripts are just to enable revision for these participants and to share with each other their imagination.

It is therefore important to look at these scripts in the light of imagination and creativity and the variety each individual brings to every topic. The intention is to encourage students to try this methodology for topics they find difficult to remember and understand.

"Your scripts and stories will work for you and you alone, because it is your creation from your imagination, so don't try to learn from the creation of others, get inspired and create your own" – Words by a wise man

CONTENTS

Script 1

Alan in the Acid Land

Medical Theatre Script on Phenyketonuria
(Covers – Biochemistry and Clinical Features; Ref: Harrisons, 17th Ed.)

Characters

1. Alan - The Phenyl Specialist
2. Amino1 - An Amino Acid Association Member
3. Amino2 - Amino Acid Association Member
4. Amino3 - Amino Acid Association Member
5. Amino4 - Amino Acid Association Member

N.B The parts of script in italics are related to stage settings, order of appearance, movements etc.
The part in **Bold Font** *indicates the technical Medical Data, for better understanding the topic, which can be omitted during performances of the script.*

(The Amino Acid suppliers association of the body has got together for their annual grand reunion and a Ball Dinner party at the Hotel Body Park.)

Amino1: Welcome friends to the Big Party. Inviting you all to have a bash tonight. The bar will be open and I welcome all of you to help each other by making the best acid cocktails and protein rich barbeques for all your friends gathered here. (*Raise the toast*) For all the Amino Acids!

Amino2: Talking about cocktails, will Alan come tonight with his phenyl special?? Oh Man, I still cannot forget the last time we had his deadly mix. It was almost like living a 'Hangover' movie.

Amino3: Oh, I wish he doesn't turn up today. For otherwise it is goanna be a Mess... for today the bouncer Mr. Hydroxylase, the only man who can put a control on Alan's phenyl madness is on a leave too. If Mr. Hydroxylase was around he would have forced Alan to make some of that yummy Tyrosine Punch, instead of that nasty phenyl drink.

(HYDROXYLASE CONVERTS PHENYLALANINE TO TYROSINE.)

Amino4: I still cannot forget that last party night. If you even tell the name of that phenyl drink I will puke that moment!!

Alan: Hey..........Hold the puke bro.......!!!!!!Guess what!!! Daddy is here!!!!!! Tonight is goanna be one awesome party night! Well, I didn't see Mr. Hydroxylase at the door and so Boys I think it is going to be just neat 'Phenyl on the rocks' tonight (Laughs)!! Get ready to be blown out of your minds!!

(Deficiency OF HYDROXYLASE LEADS TO PHENYALANINE ACCUMULATION)

All four together: Oh No !!

Alan: Oh Yes !!! Come on boys... No more running to the kitchen to make that Nor Adr, Seretonin Fry or any such crap ! Here enjoy the world's best Amino Acid Cock tail from your own Phenyl – Alan !!!!! Now everybody bottoms up!!

(ACCUMULATION OF PHENYLALANINE INHIBITS TRANSPORT OF OTHER AMINO ACIDS AND DECREASE FORMATION OF NOR ADRENALINE AND SERTONIN)

Narrator: Few hours passed and the Aminos were smashed with Alan's Phenyl drink.

Amino1: Dude Alan, this is what I love about your drink. The party just started like any other one... a normal birth... But see the way it is progressing.... 2 hours down, but still we haven't achieved discussing any milestones on the agenda yet.

(NORMAL AT BIRTH, BUT FAILS TO HIT normal DEVELOPMENTAL MILESTONES)

Amino2: Can you believe ? I am already hallucinating!! (laughs) The bloody red painted PARTY HALL OF ours is now looking sort of light pinkish to me??? (laughs)Pink, does that turn anybody on???

(HYPOPIGMENATION)

Amino3: Oh, Shut up !!! The thing turns me on is this mousy odor in this room and from all of you. I'm loving it baby!!

(MOUSY ODOUR - SKIN)

Amino4: Oh.. That is scary & disgusting... and you love it???? Guess what; even my piss smells the same dude!!! Do you want me to unleash that on your tiny head???? And then say – if you still love it!!!

(MOUSY ODOUR – urine, MICROCEPHALOPATHY)

Amino1: Oh, Cut the crap guys!!! I think I'm the only one stable here!!!

Alan: Really, you think so??? Now spell for me BRITISH Constitution??? Or walk with your eyes closed on this line???

Amino1: Oh, That's all ! Watch I will spell it and walk on the line simultaneously... Here we go!!
B-R-ri- tee—sshh – Con- S- Tit-ut- rion (he walks and falls down)
Oh Man !!! I'm so blown !!!

(MICROCEPHALOPATHY, DECREASED CEREBRAL FUNCTION)

Amino2: Oh Allan, here look at me... I'm feeling so nice....I'm goanna dance .... (dances)Hey .... I can feel shivers all over..... See look at my hands and feet....Oh my !! They are .... (He behaves as if got fits)... all going so wired !!! I think this is going to our live the movie experience for Hangover Part 2

(HYPER ACTIVITY SEIZURES)

Alan: See Boys... that is what an awesome party means. ... Enjoy being in the 'Wolf Pack' !!! Time for one more round bottoms up!!! Here we go... for the crazy & wild night once again!!!

COMMENTS FROM THE TEAM

This script is a full blown entertainer. The concept of a party wherein the effects of a drink are compared to the features of an enzyme deficiency is a very smart association. The sentences used in this script are short and crisp with a nice and well defined structure. The context is simplified and the characterization is well maintained with humor. Every paragraph describes a particular aspect of the disease or its process.

What is worth noting here in this script is that there is no elaborate story or plot as such but it is still fun to read. That is very difficult to create as the entire strength of the script relies on the dialogues and the humor within it. Very little is dependent on the story. So this highlights the fact that even if no substantial plot can be created, a topic can be made into an interesting sequence of events such that revision of this topic becomes enjoyable. Smart writing and full marks to the author.

Script 2

Ann & Arbor; Shaadi Mubarak!

Medical Theatre Script on Ann Arbor staging of Hodgkin's Lymphoma & Prognostic Factors
(Ref: Harrisons, 17th Ed.)

Characters

1. Hodg King - The Emperor
2. Minister - Middle Aged Man
3. Ann - Marriage Consultant
4. Arbor - Ex-boyfriend of Ann, Marriage Consultant
5. Guard 1 - Middle Aged Man
6. Guard 2 - -do

N.B The parts of script in italics are related to stage settings, order of appearance, movements etc. The part in **Bold Font** *indicates the technical Medical Data, for better understanding the topic, which can be omitted during performances of the script.*

Guard1: For the attention of all men and women in the country of Bodypuri, Our emperor, His Highness Hodg King hereby announces Swayamvara for all his grown up tender lymph daughters – our beloved princesses, who have settled on both sides of Diaphra river.

(TENDER, ENLARGED LYMPH NODES IN HODGKINS)

Guard 2: His Excellency invites the kings and princes from all around the world to come and compete, to win the hands of his highnesses beautiful daughters! Ladies and Gentlemen, Come join us in the celebrations at the palace!

Narration: Meanwhile at the palace Ann and Arbor the two hired marriage consultants from Shaadi Mubarak were waiting to meet the king. Seems like the petty fights between both of them will never end !!

Ann: Arbor, Stay out of it! I know to manage this. I can do this all alone. I don't need my ex-boy friend to be with me anymore. And remember this is my idea, my business, my passion, just leave me alone.

Arbor: Yeah, Right, MY 'Ex-'Girl Friend. The idea might be yours, but remember without my expertise your business is nothing. Have you ever heard a girl being successful in business??

Ann: Well, Kiran Mazhumdar, Tanea Smith are they not ladies?

Arbor: Ah... Exceptions....but ...

Minister: Silence you both. Stop this cat fight. His Majesty is expected to be here anytime.

Guard: Please rise up as the Emperor, his Majesty Hodg king arrives

(Everybody stands up, bows to the king. King takes the seat and indicates everybody to be seated)

King: Dear Friends, I'm pleased by your good work. Minister is everything going on well ?

Minister: Yes My Lord. I'm personally ensuring that all the works are done in time. My Lord, Meet this lady Ms. Ann and the Gentleman Mr. Arbor from Shaadi Mubarak

King: Oh welcome you both ! What news you have got for me.

Ann: Your Highness, 'I' have gathered information about all the kings and princes attending the Swayamvar from tomorrow

Arbor: Of Course, under my guidance *(naughty smile)*

(Ann grins at Arbor)

Ann: Your highness, shall we proceed and stage them in front of you?

King: Carry On, Ann

Arbor: *(snatches the paper from Ann)* Your Highness the Kings and princes attending the Swayamvar are classified into 4 groups based on their demands.

Ann: Group 1, includes the kings who are not with particular demands. They are ready to marry any 'one' of your daughters, either settle on this side or that side of the Diaphra River. Few of them would just like to be your allies by managing any one of your lymphatic kingdom, otherwise.

(STAGE 1 – SINGLE LYMPH NODE REGION INVOLVEMENT OR LYMPHOID STRUCTURE)

Arbor: My Lord, But in Group 2, we have the kings who want to marry at least two or three of your daughters. But they are very particular that as cannot keep crossing the diphra river to meet their wives, they want to be married to the princesses settled on the same side of the river only. Either this side or that side.

(STAGE 2 – TWO OR MORE LYPH NODE REGIONS ON THE SAME SIDE OF DIAPHRAGM)

King: Hmmm... interesting.... Go Ahead !!!

Ann: Ah, and here we have the group 3, they not only want to marry more than one of your lymph daughters but they are particular in having lymph kingdoms or wives on both sides of the river.

(STAGE 3 – INVOLVEMENT OF LYMPH NODE REGIONS OR LYMPHOID STRUCTURES ON BOTH SIDES OF DIAPHRAGM)

Arbor: And finally, the stage 4..... (*Subtle Laugh*) My Lord, they want also to have more than one of your daughters probably, but the main concern is, they want something beyond designated extra nodal fertile 'E' areas of your lymphatic kingdom as dowry. And some are asking for your favorite fertile liver area or the rich bony marrows as dowry.

Minister: What?? Liver and Marrow areas as Dowry????

Ann: Yes! My Lord, so is the demand!

(STAGE 4- INVOLVEMENT OF EXTRANODAL SITES, BEYOND DESIGNATED AS 'E' OR INVOLVEMENT OF LIVER OR BONE MARROW)

King: Crooks, eyeing my favorite liver and bony marrows. I will teach them a lesson. Well, Ann and Arbor, did you meet my daughters? Have you both classified them as well?

Arbor: Your highness, they are so tender but beautiful .We could classify the extremely beautiful ladies only based on their anxiety levels about the Swayamvara happening tomorrow.

King: Hmmm... So how do you classify them?

Ann: Your highness, the princesses in Class A category – shows no Anxiety symptoms what so ever. However in class B- they showed few symptoms like unexplained weight loss, recurrent fever, night sweats etc.

(A-NO SYMPTOMS, B-SYMPTOMS PRESENT(WEIGHT LOSS, FEVER, NIGHT SWEATS), E-LOCALISED, SOLITARY SWELLING OF EXTRALYMPHATIC TISSUE, EXCLUDING LIVER AND BONE MARROW)

Minister: Your Highness, I cannot still get over the Class 4 kings. They are being rude with their demands.

King: I understand, here is my decision. All the kings above the age of 45 can leave the swayamvara, I don't my daughters to be married to some uncles. Then, all of them in the so called category 4 of Ann-Arbor Classification may be thrown out immediately. Among the princes, all those who look pale and anaemic **(Haemoglobin <10.5),** along with all those who have infections **(WBC>15,000)** may also be asked to leave.

(POOR PROGNOSTIC INDICATORS: AGE 45+, MALES, ANN-ARBOR STAGE 4, HB,10.5, WBC>15,000)

Ann and Arbor, thank you for your efforts. Hope you make this Swayamvara Ceremony a grand success on my behalf

Ann & Arbor: At your service, your highness! Shaadi Mubarak !!!!

COMMENTS FROM THE TEAM

The associations are simply beautiful in this script. In the sense by just trying to remember what happened in this story some points such as" ann arbour classification is used for hodgkins lymphoma and that this

classification is based on lymphnodes being present in relation to the diaphragm ... etc etc" can be remembered with ease. The story line is strong and there is a basic plot. An adaptation of the main characters to that of the film 'Band Baaja Baarat " was very neatly done. Smart usage of names for characters such as Hodge King and diaphra river, gives cues to answers while remembering or revising this topic.

The stand out here is the conversion of a classification into a story which is quite a challenging task and was demonstrated almost effortlessly here. This story enables the revision of the entire classification along with the prognostic factors. Humor quotient was maintained and the ackward conflict and competitive spirit between ann and arbour was well depicted. An ideal example of utilization of story to remember a difficult to otherwise remember topic like a classification. Brilliant work.

Script 3

CARDIA: The wonderful machine

Myocardial Ischemia and Injury

This is the story of the wonderful huge tree machine called CARDIA. The cardia used to be present in the centre of the city and used to be the lifeline of the city. It had 4 subunits which though separated worked together in harmony.It used to give people whatever they wished for and demanded.

But sometimes this machine used to conk off. When the fuel supply to it reduced due to inflation or the demands of selfish people increased or both, the machine could conk off. The worst part of it when it conked off was that it could produce a kind of tremor that was felt by the entire city. It used to find itself in the news and use when the political tension used to increase or after the return of the game show kbc(kaun banega cardiomyopathy), the machine could not handle the fame it was garnering.

The tremors most of the time used to affect an area behind the machine. This used to put pressure or heaviness on the shops situated in those areas. But this tremor could also radiate upwards in the city from where it is situated as it was the same source of electricity supply of this tree and other saplings were distributed in these areas.. whenever this happened peoples neck turned towards the cardia, their teeth would shiver in fear and their jaws would open in awe .

If the machine used to slowly and gradually conk off, people in the city used to funnily exercise more. Some used to get emotional while others started belting food. Perhaps they knew that Mr.Nitroglycerine could fix the machine in a few minutes and everything would be allright.

But what worried the people was the cardia conking off in the night. This used to make people get up from sleep . even mr. nitroglycerine would not be able to fix it. The machine used to become so hot that when people went near it, they would start sweating. Sometime it could metamorphose into a gruesome creature that would make people nauseus and vomit

COMMENTS FROM THE TEAM

A short and simple description of the mechanics of a heart and signifying its importance as the lifeline of a city. Quite straight forward with elements of subtle comedy. "Kaun Banega Cardiomyopathy " is the most unique KBC association that I have come across , very creative indeed.

This script is different in its presentation with the entire description in the form of a narration, demonstrates yet another way to convey a message. The style of narration represented here is casual, just as when talking amongst friends and this makes it all the more easy to read. Simplicity is the best policy. Good work.

Myocardial Ischemia and Injury (Theory)

Myocardial ischemia occurs when the oxygen supply to the heart is not sufficient to meet metabolic needs. This mismatch can result from a decrease in oxygen supply, a rise in demand, or both. The most common underlying cause of myocardial ischemia is obstruction of coronary arteries by atherosclerosis; in the presence of such obstruction, transient ischemic episodes are usually precipitated by an increase in oxygen demand as a result of physical exertion. However, ischemia can also result from psychological stress, fever, or large meals or from compromised oxygen delivery due to anemia, hypoxia, or hypotension. Ventricular hypertrophy due to valvular heart disease, hypertrophic cardiomyopathy, or hypertension can predispose the myocardium to ischemia because of impaired penetration of blood flow from epicardial coronary arteries to the endocardium.

Angina Pectoris

The chest discomfort of myocardial ischemia is a visceral discomfort that is usually described as a heaviness, pressure, or squeezing (Table 13-2). Other common adjectives for anginal pain are burning and aching. Some patients deny any "pain" but may admit to dyspnea or a vague sense of anxiety. The word "sharp" is sometimes used by patients to describe intensity rather than quality.

The location of angina pectoris is usually retrosternal; most patients do not localize the pain to any small area. The discomfort may radiate to the neck, jaw, teeth, arms, or shoulders, reflecting the common origin in the posterior horn of the spinal cord of sensory neurons supplying the heart and these areas. Some patients present with aching in sites of radiated pain as their only symptoms of ischemia. Occasional patients report epigastric distress with ischemic episodes. Less common is radiation to below the umbilicus or to the back.

Stable angina pectoris usually develops gradually with exertion, emotional excitement, or after heavy meals. Rest or treatment with sublingual nitroglycerin typically leads to relief within several minutes. In contrast, pain that is fleeting (lasting only a few seconds) is rarely ischemic in origin. Similarly, pain that lasts for several hours is unlikely to represent angina, particularly if the patient's electrocardiogram (ECG) does not show evidence of ischemia.

Anginal episodes can be precipitated by any physiologic or psychological stress that induces tachycardia. Most myocardial perfusion occurs during diastole, when there is minimal pressure opposing coronary artery flow from within the left ventricle. Since tachycardia decreases the percentage of the time in which the heart is in diastole, it decreases myocardial perfusion.

Unstable Angina and Myocardial Infarction

(See also Chaps. 238 and 239) Patients with these acute ischemic syndromes usually complain of symptoms similar in quality to angina pectoris, but more prolonged and severe. The onset of these syndromes may occur with the patient at rest, or awakened from sleep, and sublingual nitroglycerin may lead to transient or no relief. Accompanying symptoms may include diaphoresis, dyspnea, nausea, and light-headedness.

Script 4

Cluster Headache

Mumbai 2011.

Inspector Suman Tripati and Inspector Zomi Tripati are discussing a case of serial thefts.

ST – This is such a tricky case – its affecting 1 in every 1000 people. Never seen anything like this before.

ZT – hmm… actually that's not true. A couple of years back; we had a series of attacks very similar in nature. But before we could solve the case, the murderers went into hibernation and it remained unsolved. There is a definite pattern and I have a feeling it's the same attackers.

ST – That's interesting. Why do you think the culprits are the same?

ZT – Periodicity. The attacks have happened precisely at 8pm in the nights in half the attacks. The attacks generally took a short time of 30 minutes to an hour. Even in the earlier case about half the attacks were nocturnal.

ST – That's not much for a coincidence!

ZT – It's not just that. That attackers come in clusters on motor bikes. They generally attack men walking alone and seldom women. They attack with tasers usually on the head of the guy.

ST – What happens?

ZT – Results in a deep, excruciating, almost explosive headache. But only on one half of his head – the same side where he was attacked. And victim starts rubbing their head and pacing around in the pain.

ST - Meanwhile, the attackers act as if they are helping him out and pocket his belongings.

ZT – Right on! The victim struggles parasympathetic activation i.e nasal congestion, tears, red eyes and drooping eyelids. That's when the attackers capitalize.

ST – So whats our plan of action?

ZT – We need to identify the central pacemaker of the group who has the taser – and remove the cluster leader. We know the next attack is probably going to be at 8pm tomorrow – lets close in on them rapidly.

COMMENTS FROM THE TEAM

An observational discussion between two cops discussing patterns of a robbery is demonstrated well here. Since cluster headache forms a particular pattern, it fits well into the thief police scenario. The discussion is very straight forwards with subtle cues to remember the features. Nicely done.

Cluster Headache

Cluster headache is a rare form of primary headache with a population frequency of 0.1%. The pain is deep, usually retroorbital, often excruciating in intensity, nonfluctuating, and explosive in quality. A core feature of cluster headache is periodicity. At least one of the daily attacks of pain recurs at about the same hour each day for the duration of a cluster bout. The typical cluster headache patient has daily bouts of one to two attacks of relatively short-duration unilateral pain for 8–10 weeks a year; this is usually followed by a pain-free interval that averages 1 year. Cluster headache is characterized as chronic when there is no period of sustained remission. Patients are generally perfectly well between episodes. Onset is nocturnal in about 50% of patients, and men are affected three times more often than women. Patients with cluster headache tend to move about during attacks, pacing, rocking, or rubbing their head for relief; some may even become aggressive during attacks. This is in sharp contrast to patients with migraine, who prefer to remain motionless during attacks.Cluster headache is associated with ipsilateral symptoms of cranial parasympathetic autonomic activation: conjunctival injection or lacrimation, rhinorrhea or nasal congestion, or cranial sympathetic dysfunction such as ptosis. The sympathetic deficit is peripheral and likely to be due to parasympathetic activation with injury to ascending sympathetic fibers surrounding a dilated carotid artery as it passes into the cranial cavity. When present, photophobia and phonophobia are far more likely to be unilateral and on the same side of the pain, rather than bilateral, as is seen in migraine. This phenomenon of unilateral photophobia/phonophobia is characteristic of TACs. Cluster headache is likely to be a disorder involving central pacemaker neurons in the region of the posterior hypothalamus. cluster Headache: Treatment- The most satisfactory treatment is the administration of drugs to prevent cluster attacks until the bout is over. However, treatment of acute attacks is required for all cluster headache patients at some time.

Acute Attack Treatment - Cluster headache attacks peak rapidly, and thus a treatment with quick onset is required. Many patients with acute cluster headache respond very well to oxygen inhalation. This should be given as 100% oxygen at 10–12 L/min for 15–20 min. It appears that high flow and high oxygen content are important. Sumatriptan 6 mg subcutaneously is rapid in onset and will usually shorten an attack to 10–15 min; there is no evidence of tachyphylaxis. Sumatriptan (20 mg) and zolmitriptan (5 mg) nasal sprays are both effective in acute cluster headache, offering a useful option for patients who may not wish to self-inject daily. Oral sumatriptan is not effective for prevention or for acute treatment of cluster headache.

Preventive Treatments- The choice of a preventive treatment in cluster headache depends in part on the length of the bout. Patients with long bouts or those with chronic cluster headache require medicines that are safe when taken for long periods. For patients with relatively short bouts, limited courses of oral glucocorticoids or methysergide (not available in the United States) can be very useful. A 10-day course of prednisone, beginning at 60 mg daily for 7 days and followed by a rapid taper, may interrupt the pain bout for many patients. When ergotamine (1–2 mg) is used, it is most effective when given 1–2 h before an expected attack. Patients who use ergotamine daily must be educated regarding the early symptoms of ergotism, which may include vomiting, numbness, tingling, pain, and cyanosis of the limbs; a weekly limit of 14 mg should be adhered to. Lithium (600–900 mg qd) appears to be particularly useful for the chronic form of the disorder.

Script 5

Small Pox – clinical features

Varicella major and Varicella minor were born to Mr. and Mrs. Poxvirus of orthopox family. They are very close buddies but are rouges by profession. V minor being small was unable to do much apart from causing some mild injuries and low mortalities in his "enemies", but his big brother V major was a major "bio weapon" by himself.

If anyone happens to lock horns with V major he creates some lesions which are characteristically his with pustules and scabbing. It is also his signature style to hit the persons first from the extremities and face to immobilise and disfigure or blind the enemy and then move on to the trunk. The local police jokingly terms it as various centripetal attacks usually if the person survives even after 2 weeks (12- 14 weeks), the victim runs on a high fever, malaise, vomiting (due to attack on head) and maculo papular rashes (following beating up). Such traumatising was experience that it is generally known among locals that once scabs are formed, the victim becomes " non-contagious" or unable to cause any problem to anyone around in the future. He at times does induce fatal attacks too in which the victim dies within 5-6 days succumbing to the severe injuries.

COMMENTS FROM THE TEAM

A small topic and put in a nice and simple narration. The relation between major and minor along with the potency were well depicted. The remaining points were not very different from the ten pointers but some form of continuity was maintained.

Ten pointers - Small Pox – Microbiology and clinical features

1. Small pox is caused by one of two closely related viruses, Varicella major and Varicella Minor.
2. Both are double stranded DNA Viruses.
3. They are members of the orthopox virus genus of the pox viridae family.
4. Infections with V minor are generally less severe with milder constitutional symptoms and lower mortality rates.
5. Thus V. Major Is the only major bioweapon.
6. Infections with varicella major typically occurs following contact with an infected person from the time that a maculo papular rash occurs through the appearance of scabbing of the pustular lesions.
7. Approximately 12 to 14 days following initial exposure, the patient develops high fever, malaise, vomiting, head ache and maculo popular rashes.
8. The maculo popular rashes begins on the face and extremities and spreads to the trunk (centripetal)
9. The conventional thinking is that once scabs are formed the patient is no longer contagious.
10. Death usually occurs in the 5th or 6th day.

Script 6

The charitable Strepto.

Impetigo

After visiting different regions of the world and making money, Streptococcus decided it was time to give something back to the society. He decided to take a vacation in the tropics. Since he was an orphan and had no children also, he decided to show his love towards children and shower them with his warmth. He specially took a liking towards those who were poor . even when they fell down or scratched themselves, he used to rush to them immediately to tend to the wounds.

After few days he would take them to the beach side and make them paint with red colour on the sand. He also taught them how to build different objects from the sand. Soon the children started buildings called vesicles and pustules. Their creativity was amazing as when all their buildings used to break down, they used to have a honeycomb in their middle.

Strepto also met his long lost neighbor Staph who had set up a shop by the beach side. Staph had a factory called bullous which used to manufacture thin paper crusts.

After the day at beach side, strepto took the children to a resort named cephalexin for 10 days. Here they used to play cricket atleast 4 times a day and used to manage to score 250 runs everytime. At the end of 10 days, the children could not find strepto. No one knew where he had disappeared to. But they had a feeling that he might return after sometime in search of his love whom they taught was present in the Renalty industry.

COMMENTS FROM THE TEAM

A clever way of showing the nicer side of an otherwise popular disease causing agent. It is a fresh approach with a different way to look at bacteria. One clear objective that was served in this script was the difference morphology of the lesions that were produced by the two different bacteria and also the dosage in the form of a score was very creative. The disappearance of strepto after the treatment was also clear. Due credit to the author for a short compilation of sequences along with an unexpected approach to looking at bacteria. Good work.

Impetigo (Pyoderma)

Impetigo, a superficial infection of the skin, is caused primarily by GAS and occasionally by other streptococci or *Staphylococcus aureus*. Impetigo is seen most often in young children, tends to occur during warmer months, and is more common in semitropical or tropical climates than in cooler regions. Infection is more common among children living under conditions of poor hygiene. Prospective studies have shown that colonization of unbroken skin with GAS precedes clinical infection. Minor trauma, such as a scratch or an insect bite, may then serve to inoculate organisms into the skin. Impetigo is best prevented, therefore, by attention to adequate hygiene. The usual sites of involvement are the face (particularly around the nose and mouth) and the legs, although lesions may occur at other locations. Individual lesions begin as red papules, which evolve quickly into vesicular and then pustular lesions that break down and coalesce to form characteristic honeycomb-like crusts Lesions are generally not painful, and patients do not appear ill. Fever is not a feature of impetigo and, if present, suggests either infection extending to deeper tissues or another diagnosis.

The classic presentation of impetigo usually poses little diagnostic difficulty. Cultures of impetiginous lesions often yield *S. aureus* as well as GAS. In almost all cases, streptococci are isolated initially and staphylococci appear later, presumably as secondary colonizing flora. In the past, penicillin was nearly always effective against these infections. However, an increasing frequency of penicillin treatment failure suggests that *S. aureus* may have become more prominent as a cause of impetigo. *Bullous impetigo* due to *S. aureus* is distinguished from typical streptococcal infection by more extensive, bullous lesions that break down and leave thin paper-like crusts instead of the thick amber crusts of streptococcal impetigo. Other skin lesions that may be confused with impetigo include herpetic lesions—either those of orolabial herpes simplex or those of chickenpox or zoster. Herpetic lesions can generally be distinguished by their appearance as more discrete, grouped vesicles and by a positive Tzanck test. In difficult cases, cultures of vesicular fluid should yield GAS in impetigo and the responsible virus in *Herpesvirus* infections.

Streptococcal Impetigo: Treatment

Treatment of streptococcal impetigo is the same as that for streptococcal pharyngitis. In view of evidence that *S. aureus* has become a relatively frequent cause of impetigo, empirical regimens should cover both streptococci and *S. aureus*. For example, either dicloxacillin or cephalexin can be given at a dose of 250 mg four times daily for 10 days. Topical mupirocin ointment is also effective. ARF is not a sequela to streptococcal skin infections, although PSGN may follow either skin or throat infection. The reason for this difference is not known. One hypothesis is that the immune response necessary for development of ARF occurs only after infection of the pharyngeal mucosa. In addition, the strains of GAS that cause pharyngitis are generally of different M protein types than those associated with skin infections; thus the strains that cause pharyngitis may have rheumatogenic potential, while the skin-infecting strains may not.

Script 7

The Monlogue of Mr. Healing

A warm good afternoon to everyone. I am a very natural person and I lead a very lonely life sometimes. People remember me only when things go wrong . sometimes when I am slow, they offer prayers for me to have a fast performance . if that doesn't work, they try stimulating me through various means such as trying to keep me young, destroying those pesky little microorganisms, giving me lots of vitamins as incentives.

I have my assistants belonging to different groups. Each of these groups have their own functions and they make their appearance at different times.

I recently had to work on a minor repair on a gap that had developed in the road that was damaged due to some construction work in the posh area of kormangala

The defect in the road was in a clean area. The gap looked like a straight cut , something like a surgeon would have done or in this case a professional road cutter. . There was not much loss of the cement, tar that made up the road. The edges were still hanging close to each other. Piece of cake really to tidy up this. Sometimes to attract me to this place and start doing the work, they use huge magic threads called sutures. **(characteristics of primary intention**)

I found out from the people surrounding the area that soon after the gap developed, they poured a fluid called Rakhtha which then hardened. They said that this technique was learnt from generations as this prevented the hidden treasures under the road from drying up or being attacked by vicious dreadful microorganisms.**(initial clot formation)**

Within 24 hours, I deployed my neutorphils to start working on the margins. They tire out very fast. Hence by the third day, I replaced them with the larger, more fit group of people called macrophages.**(inflammation)**

We made some infiltrations and cut in the hardened layer and we reached the base layer. From there my trusted lieutenants Epithelials set about their work sincerely and toiled day in and day out . now these epthelials have migrated from a country called epidermis. cell by cell, they managed to build a whole new level/layer that was looking beautiful and healthy and viable that after thinking for 5 days. people decided to remove the initial ugly looking hardened material called scab **(epitheliasiation)**

But this is when the showoffs and those who want to hog the limelight arrive: after the hardwork is done, on the 3rd day, these people from the fibro company arrive and they just blast around anyone who come in their way and on the 5th day they are joined by their sidekicks , the Collagens who dominate till the process of repair is complete. **(organization)**

At the end of it, u don't c how much hard work me and my people have put. Inface there is no trace of any of us except the surface built by epithelium which lasts long as a tribute to our work. The most difficult of repairs is done in places like kalasipaplyam where the roads are terribly damaged b y heavy rains,lack of maintenance by the local municipal board . here the gap in the road is too big to repair, extensive loss of the materials making up the road and even those magic threads cant be used to tie up.**(secondary intention)**

The initial steps are more or less the same. But here I have to take the help of those people from the fibroblast company. They have amazing creatures called granulation. Now they start their work to help my epithelials . now this granulation is incredible . it is basically shaped like vessels lined up side to side which are hard first but once it lose its shyness, it opens up and carries all the magic potion-rakht that we fill in it. Furthermore these granulation work better when they are shown photos of hot girls called VEGF and PDGF. Also these fibroblasts send muscular people along with them to reduce and contract the size . but as is the case, the useless show offs collagen come late and dominate and you can see only them towards the end as they start expanding their family and settle down in the new atmosphere.

Well now you know why I lead such a hard yet simple life.

COMMENTS FROM THE TEAM

Another creative way to convey a message is in the form of a monologue which this script demonstrates quite perfectly. In fact the start has elements of stand up comedy and then it builds into an association of road repair work with skin healing. The steps are correlated nicely and the process could be described in this manner.

One interesting aspect of this script was the humble hardworking character given to healing and the style of narration in the monologues keeps that character consistently throughout. A good example for a monologue.

Script 8

Just a little Delhi Rum

Delirium

Delhi Rum : This is the story of how I helped the Babur bring down the great empire of Bheja Suri.

Babur was a young ambitious cunning king who desperately wanted to rule Delhi. He waged many wars with Bheja Suri, the Emperor of Delhi, always ended up losing. Then he hatched a plan. He sent Bheja 25000 bottles of Rum and its (my) recipe as a gift to prove his allegiance. Bheja Suri was thrilled and he ordered more Rum be continuously made. And thus I was christened – Delhi Rum.

In a few years, Bheja Suri had 2 big problems in running his court – one his sons, Princes of Delhi, who were besotted with me (the world famous in Delhi, Delhi Rum) and lost interest in governance resulting in automatic instability. Sometimes he caught them agitated making plans and hallucinating that they were fighting the enemy. Other times they were hyper aroused in a state of stupor at the royal harem. Whenever he questioned the ministers they said – 'Father, we were hyper-active this week making war plans'.

His second problem were his ministers - They were intoxicated bunch of lazy obese old men – they kept piling up gold coins and were addicted to opium and of course me – Delhi Rum, showing total apathy towards the people of Delhi. They were quiet and withdrawn – hypoactive, allowing Shehanshah Suri to take all the decisions.

Within a few years, Bheja Suri's empire slowly started going into a state of decay – thanks to the ministers and princes – but basically thanks to me! Things started going wrong when we first found the emperor asleep during war, almost killing himself. Bheja Suri himself then started losing cognition – losing memory, losing his visuo-spatial skills as an archer and the worst thing that can happened to Delhi - started mixing our famous Delhi Urdu with Persian. And thus Delhi went into utter confusion.

That's when Babur came to Delhi.

COMMENTS FROM THE TEAM

A very wise and witty monologue from a non living object – the Rum. I really enjoyed the way history was mixed with delirium here and how actual facts – such as when Babur ruled Delhi were shown to result as an effect of the rum. The language and vocabulary is remarkable in this script. Again demonstrating another variety to the methodology.

I would like to also emphasize that this script was written by a non medical and non theatrical person- can you believe it?

Clinical Features of Delirium

A multitude of terms are used to describe delirium, including encephalopathy, acute brain failure, acute confusional state, and postoperative or intensive care unit (ICU) psychosis. Delirium has many clinical manifestations, but essentially it is defined as a relatively acute decline in cognition that fluctuates over hours or days. The hallmark of delirium is a deficit of attention, although all cognitive domains—including memory, executive function, visuospatial tasks, and language—are variably involved. Associated symptoms may include altered sleep-wake cycles, perceptual disturbances such as hallucinations or delusions, affect changes, and autonomic findings including heart rate and blood pressure instability.

Delirium is a clinical diagnosis that can only be made at the bedside. Two broad clinical categories of delirium have been described, hyperactive and hypoactive subtypes, based on differential psychomotor features. The cognitive syndrome associated with severe alcohol withdrawal remains the classic example of the hyperactive subtype, featuring prominent hallucinations, agitation, and hyperarousal, often accompanied by life-threatening autonomic instability. In striking contrast is the hypoactive subtype of delirium, exemplified by opiate intoxication, in which patients are withdrawn and quiet, with prominent apathy and psychomotor slowing.

This dichotomy between subtypes of delirium is a useful construct, but patients often fall somewhere along a spectrum between the hyperactive and hypoactive extremes, sometimes fluctuating from one to the other within minutes. Therefore, clinicians must recognize the broad range of presentations of delirium in order to identify all patients with this potentially reversible cognitive disturbance. Hyperactive patients, such as those with delirium tremens, are easily recognized by their characteristic severe agitation, tremor, hallucinations, and autonomic instability. Patients who are quietly disturbed are more often overlooked on the medical wards and in the ICU, yet multiple studies suggest that this under-recognized hypoactive subtype is associated with worse outcomes.

The reversibility of delirium is emphasized because many etiologies, such as systemic infection and medication effects, can be easily treated. However, the long-term cognitive effects of delirium remain largely unknown and understudied. Some episodes of delirium continue for weeks, months, or even years. The persistence of delirium in some patients and its high recurrence rate may be due to inadequate treatment of the underlying etiology for the syndrome. In some instances, delirium does not disappear because there is underlying permanent neuronal damage. Even after an episode of delirium resolves, there may still be lingering effects of the disorder. A patient's recall of events after delirium varies widely, ranging from complete amnesia to repeated reexperiencing of the frightening period of confusion in a disturbing manner, similar to what is seen in patients with posttraumatic stress disorder.

Script 9

Saving My Grain

Once upon a time in a tiny village in Andhra Pradesh lived a farmer, Bheja Rao who lived with his family. He cultivated paddy for a living. One day things started going badly wrong.

Scene 1:

One fine day his eldest son, Tirupathi Ganesh Mallenapalli more popularly known as Trigem who lived with his wife as a nuclear family, rushed to him.

Trigem : Papa, I'm livid today.

BR : Why son, whats giving you a headache?

Trigem : There is a stork which comes to the farm daily and is destroying our grains. Already the terminal edges of the farm are gone.

BR : Thats is a simply old problem. Just put a scarecrow.

Trigem : No Papa, this stork is pretty smart. It Comes, Gapes to get Rich Pickings. Only when no one is around it swoons down.

BR : Hmmm... let me think abt this.

Scene 2:

Bheja Rao was rich and owned a few horses. The stable manager Sirus Tony came next with a worried expression.

Bheja Rao : Whats causing you a headache, Sirot Tony?

ST : There has been some Horse Trading. 5 Horses are missing. Especially, two of your favorite horses Bhim and Danush.

Bheja Rao : hmmm, Let me see what i can do.

Scene 3:

The farmer had a brother who was a Ganja addict. He was called Dopey Mind all over the village and this was serious problem for the farmer whose reputation was at stake. Farmer's neighbor came running.

Neighbor – puff puff, Do something Bhejaji – My daughter and friends were coming home from college. Dont know what stimulated Dopey Mind, he has gone nuts. He is really bullying them and no one is able to stop him.

Bheja Rao – Thats bad news; I will try my best to leash him.

Scene 4:

Just when he was pondering over the issues, his wife came rushing.

Wife - The water channel to the farm has burst behind our home. Now all the water is flowing into our home. The membrane of our home is cracking.

Bheja Rao – What!

Wife – Not only that. First the calcium deposits from the recent whitewash is coming off. Second, the salts are depositing all over – the TV, fridge, fan. And third the voltage is fluctuating.

Bheja Rao – Feel like my head is going to explode.

Scene 5

After a couple of hours Bheja Rao suddenly looks very relieved like he has solved multiple issues with one stone. Lets find out what he did.

Bheja Rao – Luckily for me, my friend Antony called just then. His right hand man Tripati was in the market and he saw my 5 Horses being Traded. Tripati immediately rescued them. Then Antony and I discussed and my brother Dopey Mind is marrying his daughter. They were apparently in love! On hearing this, Dopey Mind (I should stop calling him that) calmed down immediately. Antony gave me his world famous stork fertilizer which put an end to CGRP. And shhhh! this is a my little secret – the fertilizer was so sticky, not only did it piss off CGRP, we also used it to block the leak in the channel – keeping all the sodium, potassium and calcium just where they belong.

COMMENTS FROM THE TEAM

This script is extremely interesting for two main reasons – Firstly , an attempt was made to show how 4 problems were solved in a simple solution and at the same time covering the subject and secondly, this script describes the pathogenesis of migraine by relating it to a very farfetched farming background. Both these indicate how different a script can be for various different topics. Very nicely done indeed.

Pathogenesis of Migraine

The sensory sensitivity that is characteristic of migraine is probably due to dysfunction of monoaminergic sensory control systems located in the brainstem and thalamus (Fig. 15-1).

Activation of cells in the trigeminal nucleus results in the release of vasoactive neuropeptides, particularly calcitonin gene-related peptide (CGRP), at vascular terminations of the trigeminal nerve. Recently, antagonists of CGRP have shown some early promise in the therapy of migraine. Centrally, the second-order trigeminal neurons cross the midline and project to ventrobasal and posterior nuclei of the thalamus for further processing. Additionally, there are projections to the periaqueductal gray and hypothalamus, from which reciprocal descending systems have established anti-nociceptive effects. Other brainstem regions likely to be involved in descending modulation of trigeminal pain include the nucleus locus coeruleus in the pons and the rostroventromedial medulla.

Pharmacologic and other data point to the involvement of the neurotransmitter 5-hydroxytryptamine (5-HT; also known as serotonin) in migraine. Approximately 50 years ago, methysergide was found to antagonize certain peripheral actions of 5-HT and was introduced as the first drug capable of preventing migraine attacks. The triptans are designed to selectively stimulate subpopulations of 5-HT receptors; at least 14 different 5-HT receptors exist in humans. The triptans are potent agonists of 5-HT_{1B}, 5-HT_{1D}, and 5-HT_{1F} receptors and are less potent at the 5-HT_{1A} receptor. A growing body of data indicates that the antimigraine efficacy of the triptans relates to their ability to stimulate $5\text{-HT}_{1B/1D}$ receptors, which are located on both blood vessels and nerve terminals.

Data also support a role for dopamine in the pathophysiology of certain subtypes of migraine. Most migraine symptoms can be induced by dopaminergic stimulation. Moreover, there is dopamine receptor hypersensitivity in migraineurs, as demonstrated by the induction of yawning, nausea, vomiting, hypotension, and other symptoms of a migraine attack by dopaminergic agonists at doses that do not affect nonmigraineurs. Dopamine receptor antagonists are effective therapeutic agents in migraine, especially when given parenterally or concurrently with other antimigraine agents.

Migraine genes identified by studying families with familial hemiplegic migraine (FHM) reveal involvement of ion channels, suggesting that alterations in membrane excitability can predispose to migraine. Mutations involving the $Ca_v2.1$ (P/Q) type voltage-gated calcium channel *CACNA1A* gene are now known to cause FHM 1; this mutation is responsible for about 50% of FHM. Mutations in the Na^+-K^+ATPase *ATP1A2* gene, designated FHM 2, are responsible for about 20% of FHM. Mutations in the neuronal voltage-gated sodium channel *SCN1A* cause FHM 3. Functional neuroimaging has suggested that brainstem regions in migraine (Fig. 15-2) and the posterior hypothalamic gray matter region close to the human circadian pacemaker cells of the suprachiasmatic nucleus in cluster headache (Fig. 15-3) are good candidates for specific involvement in primary headache.

Script 10

Bandit Queen comes home

In the jungles of Chattisgarh, there once lived a Bandit Queen. She was called Bandit Queen because she wore a band around her head for at least half the days each month. No one knew why! Also, she was a bandit. She looted the evil moneylenders to make a living.

One day, she got one of her couriers came to her.

Courier : Moneylender Nauser Sethy was looting the people of the nearby village. We need to teach him a lesson.

BQ: Im still wearing my band but the situation seems serious. Lets launch.

Sycophant: Didi, you seem to be struggling to ride your horse. You are holding the two sides of your head so tight and seem to be in severe discomfort. I can take care of it and you can stay here and take rest.

BQ : (*Suddenly feeling a little better*) Shut up! I'm perfectly ok, just a small headache and can't take rest. I'm born to bring justice to the people. Lets go.

(chronic head-pain syndrome bilateral tight, bandlike discomfort. builds slowly,fluctuates in severity, and may persist more or less continuously for many days. may be episodic or chronic (present >15 days per month). Completely without accompanying features such as nausea, vomiting, photophobia, phonophobia, osmophobia, throbbing, and aggravation with movement.)

They reach the moneylenders home at 6 in the evening. Everyone seems busy with something or the other. Sycophant watches for sometime for the scene to clear.

Sycophant: Didi, the moneylender is sleeping – but his bed is on top of the money chest. Funny thing he is also wearing a band around his head like you. And he is also holding his head tight on both sides.

BQ : Shut up! What do you mean 'also'?! ...Does he leave his bed ever?

Sycophant : Every 5 minutes he runs to wash basin to vomit; screams at his wife if anyone switches on the light or if the phone rings. She is reading a paper with "International Headache Society and Migraine" written on it. Seems like the Seth has some serious headache tensions.

(International Headache Society's definition of TTH allows an admixture of nausea, photophobia, or phonophobia in various combinations, illustrating the difficulties in distinguishing these two clinical entities. Patients whose headaches fit the TTH phenotype and who have migraine at other times, along with a family history of migraine, migrainous illnesses of childhood, or typical migraine triggers to their migraine attacks, may be biologically different from those who have TTH headache with none of the features.)

BQ : (*face lights up!*) We should then probably take that paper she is holding also with the loot.

Sycophant : Yes, Yes! We can give him a headache, literally then!!

BQ : Shut up! Will you?! You are full of nervous tension. What is wrong with you today??

(pain is a product of nervous tension)

BQ : You wait here. Let me go in for the loot.

Sycophant takes the opportunity to go for a smoke. As Bandit Queen enters the room, she faints in pain.

Seth's wife (who watches all the sitcoms) comes into the room and sees a lady fallen with a band similar to Seth's. Seth is also sleeping.

Seth's wife : (*thinking aloud*) Oh my god! Seth's sister went missing when he was child. Could this be her!

She immediately calls her neighbour Dr Chaturvedi.

Seth's wife : Doctor, I think we have got Seth's little sister back after 30 years. What do you say??

Dr Chaturvedi immediately recognises Bandit Queen, but he puts his duty as a doctor first.

Dr Chaturvedi : Let me take a look at both of them. They both seem to be suffering from Tension type Headache.

Seth's wife confidence in her theory increases.

Dr Chaturvedi : Mrs Sethy, do you have aspirin at home?? That will bring some immediate relief.

Seth's wife : I will just get you a couple. I have told Sethji to do Yoga so many times. He just doesn't listen.

Dr Chaturvedi : Very good idea Mrs Sethy. How do you know so much?

Seth's wife : (*blushes*) Seen it on TV, doctor.

(analgesics such as acetaminophen, aspirin, or NSAIDs. Behavioral approaches including relaxation can also be effective.)

Dr Chaturvedi : You really need to get them both – AMI.TRIP.TYLINE.

Seth's wife : Now tell me doctor, do you think she is Sethji's sister.

Dr Chaturvedi : I wont deny that it is likely. (*winks*)

(For chronic TTH, amitriptyline is the only proven treatment (Table 15-7); other tricyclics, selective serotonin reuptake inhibitors, and the benzodiazepines have not been shown to be effective.)

COMMENTS FROM THE TEAM

A standard and clear script with multiple short lines, ideal for a good performance. Even though in a short span, all aspects of a story build up was covered. The concept was simplified and a very clinical topic was made into a story. Well done.

Tension-Type Headache

Clinical Features

The term *tension-type headache* (TTH) is commonly used to describe a chronic head-pain syndrome characterized by bilateral tight, bandlike discomfort. The pain typically builds slowly, fluctuates in severity, and may persist more or less continuously for many days. The headache may be episodic or chronic (present >15 days per month).

A useful clinical approach is to diagnose TTH in patients whose headaches are completely without accompanying features such as nausea, vomiting, photophobia, phonophobia, osmophobia, throbbing, and aggravation with movement. Such an approach neatly separates migraine, which has one or more of these features and is the main differential diagnosis, from TTH. However, the International Headache Society's definition of TTH allows an admixture of nausea, photophobia, or phonophobia in various combinations, illustrating the difficulties in distinguishing these two clinical entities. Patients whose headaches fit the TTH phenotype and who have migraine at other times, along with a family history of migraine, migrainous illnesses of childhood, or typical migraine triggers to their migraine attacks, may be biologically different from those who have TTH headache with none of the features.

Pathophysiology

The pathophysiology of TTH is incompletely understood. It seems likely that TTH is due to a primary disorder of CNS pain modulation alone, unlike migraine, which involves a more generalized disturbance of sensory modulation. Data suggest a genetic contribution to TTH, but this may not be a valid finding: given the current diagnostic criteria, the studies undoubtedly included many migraine patients. The name *tension-type headache* implies that pain is a product of *nervous tension*, but there is no clear evidence for tension as an etiology. Muscle contraction has been considered to be a feature that distinguishes TTH from migraine, but there appear to be no differences in contraction between the two headache types.

Tension-Type Headache: Treatment

The pain of TTH can generally be managed with simple analgesics such as acetaminophen, aspirin, or NSAIDs. Behavioral approaches including relaxation can also be effective. Clinical studies have demonstrated that triptans in pure TTH are not helpful, although triptans are effective in TTH when the patient also has migraine. For chronic TTH, amitriptyline is the only proven treatment (Table 15-7); other tricyclics, selective serotonin reuptake inhibitors, and the benzodiazepines have not been shown to be effective. There is no evidence for the efficacy of acupuncture. Placebo controlled trials of botulinum toxin type A in chronic TTH have not shown benefit.

Script 11

The Curious Case of Mr. Paget

Medical Theatre Script on Paget's disease of Bone
(Covers – Etiology and Clinical Features; Ref: Harrisons, 17th Ed.)

Characters

1. Mr. Paget - An Architect
2. Judge - An English Court Judge
3. Prosecutor - Young Man
4. Defendant - Middle Aged Man
5. Secretary - Court Assistant
6. Mr. Nucleo -Witness
7. Extra Artists - Cops, court Staff etc.

N.B The parts of script in italics are related to stage settings, order of appearance, movements etc. The part in **Bold Font** *indicates the technical Medical Data, for better understanding the topic, which can be omitted during performances of the script.*

(Scene of a court, trial room. Secretary, prosecutor and defendant seated. The Judge enters. Everybody rises, greets and sits back)

Judge: So Gentlemen, let us proceed.

Secretary: My Lord, the first case of the day. Case no. 234/B, The Curious case of Mr. Paget. Compensation case by the Western Builders, Mr. Paget, Mr. Paget !

(Mr. Paget walks into the trial room, stands in the witness box.)

Secretary: Mr. Paget, now the oath

Mr. Paget: Oh Yeah ! *(displeased)*I will speak the truth, the whole truth and nothing but the truth ! Is that Ok ??

Secretary: Yeah, kind of !

Judge: Mr. Prosecutor, you may question the accused!

Prosecutor: Thank you your honor. Mr. Paget, back again .... What do you do these days?

Mr. Paget: I work as the Architectural consultant for buildings older than 40 years, mainly in the Western Europe.

(Paget's Disease More common in Western Europe and people aged 40 +)

Prosecutor: Oh, Nice. What exactly is the nature of your work?

Paget: I remodel these buildings

(Basic pathology - BONE REMODELLING)

Prosecutor: Remodeling??? Aaahhh !! So razing down the random buildings of your choice and constructing some wired shaped buildings at the same site is remodeling according to you, Mr. Paget ??

Paget: It's not as simple as you say it.

Prosecutor: Oh No, No way... It's complicated!!! Mr. Paget??? May be not as complicated as your previous case, where you were held up in the same trial room for your vulgar fantasies and remodeling of Breasts, Penis, Vulva and what not??

(SITE OF OTHER PAGET's PATHOLOGIES: BREAST, VULVA, PENIS)

Defendant: Objection! Your Honor! He is deviating from the case !

Prosecutor: My Lord, to understand this case I need to establish the character and other activities of Mr. Paget

Judge: Objection Overruled! You may proceed!

Prosecutor: Thank you my Lord ! So, Mr. Paget... you told about remodeling those buildings? Can we know what your qualification to do so is?

Paget: I'm a graduate in TN FRSF 2 and completed SQSTM 1

Prosecutor: Oh I see, but unfortunately Mr. Paget the University records with us show that your name was homozygously deleted from the TNFRSF2 list and the SQSTM 1 certificate of yours read, that you are a mutant.

My Lord, here are the certificates for your reference.

(TNFRSF2 homozygous deletion or SQSTM1 mutation)

(Papers handed over to the Judge through the Secretary.)

Defendant: Objection, My Lord ! My client Mr. Paget is coming from the family of well known Architects. We need to consider his family history in the business when you doubt his qualification.

(FAMILY HISTORY)

Prosecutor: Yes Indeed, Strong Family History. But here see his computer. The Engineer Mr. Paget has got the notorious RSV and Measles Virus on his system. My Lord, this proves his working system is corrupted and malfunctioning.

(PAGET'S MIGHT OCCUR FOLLOWING VIRUS ATTACKS OF RSV AND MEASLES)

Judge: Objection Overruled, You may continue Mr. Prosecutor.

Prosecutor: My Lord, I would like to call in a witness. One of the many drivers in Mr. Paget's numerous notorious razing 'Osteoclast' bull dozers

(osteoclasts – DO BONE DESTRUCTION)

Judge: You May proceed

Assistant: Mr.Nucleo, the osteoclast driver

(Mr. Nucleo enters the witness box, on other side)

Prosecutor: Mr. Nucleo, tell us about the nature of work with Mr. Paget

Mr. Nucleo: He hired a large number of Osteoclast bulldozers from our company and we have been put on the demolishing sites, working day and night razing buildings. He in fact recruited multiple drivers per osteoclast to hasten the destruction.

(NUMBER OF OSTEOCLASTS AND number OF nucleus in EACH OSTEOCLAST INCREASED)

Prosecutor: I heard you have some ranking system for the Osteoclast Bulldozers ?

Mr. Nucleo: We do have RANKs. It's based on that we work. But for increasing the speed of destruction, Mr. Paget gave away more RANKS per osteoclast and supplied a lot of fuel in form of InterLeukin 6 etc. to speed up our work.

(EXPRESSION OF MORE RANK LIGANDS AND IL-6 PROMOTES BONE RESORPTION)

Prosecutor: Thank you Mr. Nucleo. Thank you. That is all from my side, my Lord.

Judge: Mr. Defendant, you have something in the defense of your client?

Defendant: My Lord, Prosecutor has conveniently forgotten about the fact that my client in fact not just razes down the buildings, but he builds new ones instead of the razed ones. Mr. Paget, could you explain to us how you do you go about with all this?

Paget: Yes Sir, Mine is a systematic work. I follow a 3 stage procedure in my Remodeling Business. Stage one, is more of destruction with osteoclasts as explained, at the same time I do increase the water supply to the site in view of stage2 work. Stage 2 essentially involves building up brand new buildings at the site of destruction, a very active work. And in the last stage, As the workers burn out, I close the work with some final touches. When the stage 2 or 3 is happening on one site simultaneously I start Stage one work at some other sites. So it is pretty much complex work, I'm doing.

(3 STAGES; 1- STAGE OF RESORPTION, VASCULARISATION, STAGE OF BONE FORAMTION, STAGE BURNED OUT/SCLEROTIC PHASE)

Defendant: My Lord, You can clearly understand it is a systematic and methodical approach that is followed by my client towards his work. That's all, your honor!

Judge: Mr. Prosecutor???

Prosecutor: Yes, My Lord! The plan sounds so perfect. But you will be shocked to look at these reports. In stage 2 when Mr. Paget reconstructs the buildings it is found he replaces the strong lamellar rods with some weak woven rods. And in the final stage which he calls 'the touch up' he puts in some inappropriate fibrous tissues and cheap mosaic flooring to the building making it awful and weak.

(IN STAGE 2, LAMELLAR BONES ARE REPLACED BY WOVEN BONE, IN STAGE 3 FIBROUS CONNECTIVE TISSUE REPLACEMENT, MAKES NEW BONE BRITTLE)

Judge: Is that over?

Prosecutor: Further Your Honor; look at these photographs of the Northern Tower reconstructed by him. The head of the tower is so huge, with a frontal bossing and look at those bowed and arched foot ends. And all in all, it is no more a tower, look the height is reduced too. This building offer no stability and many cases are already reported of such buildings breaking down. And from the environment point of view, not to forget the Spillage of ALP he causes at work site.

(CLINICAL FEATURES – ENLARGED HEAD, FRONTAL BOSSING, BOWING OF LIMB BONES, SHORTENING, FRACTURES, ALP LEVELS RAISED)

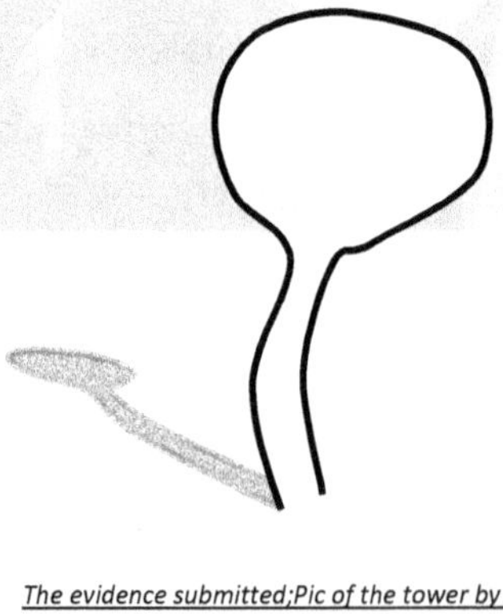

The evidence submitted;Pic of the tower by Mr. Paget

(as a prop, can take the print out of same in A4, for the play)

Judge: Oh, This building looks like a lolypop with a bend stick to it. Mr. Paget, you have anything more to say ?

Paget: I admit, the shapes might look odd for you. But my Lord, Remember, Beauty lies in eyes of Beholder. I wish you all had some X-Ray Eyes so that at least you could appreciate the handcrafted cotton wool spots and the ivory vertebra that I have incorporated in the building.

(X-RAY FINDINGS: COTTON WOOL SPOTS, IVORY VERTEBRA)

Judge: Oh Mr. Paget, I don't think any of those would fancy us. So having found Mr. Paget guilty the court herby declares the termination of all works by him as it is causing a lot of Pain to all of us. Before this passion of Paget becomes a cancerous obsession the court sentences him to lifelong imprisonment and to be fed only with oral Bisphonates and Calcitonin injections.

The Case is closed for now !

(PAIN – IS THE MAIN SYMPTOM, PAGETS DISEASE CAN BECOME MALIGNANT, TREATMENT- ORAL BISPHONATES AND CALCITIONIN INJECTIONS)

COMMENTS FROM THE TEAM

This is an outstanding script. To get a disease like Paget's and incorporate it so perfectly into a real life courtroom is simply the work of perfection. The language, the court etiquette, the attitudes of the prosecutor, Judge and defendant were brought about very nicely. A one act scene which can be performed easily with the pictorial evidence by introducing a prop was simply spectacular.

Not to forget every aspect of Paget's disease was associated with the story very nicely and the interesting technique in script writing makes this a very enjoyable and readable script. Full marks to the author for this one.

Script 12

THE BATTLE OF RHEUMATICUS: a one-act play describing the untold story through the mouth of a soldier.

Cast: 4 soldiers, Lenardo, extra dancing girls, Streptococcus, Army of Pencillin, few extra soldiers.

Play: the Play can be enacted as a monologue delivery on one side of the stage with the other side being used to enact whats being said. Soldiers monologue in black colour. During flashback , the entire stage can be used. Flashback in blue font. Props and lighting setting for stage in green.The learning concept has been highlighted in red.

The untold story of the battle of RHEUMATIC: a soldiers account

(Enter a soldier: in prison clothes, unshaven, dark circles under eyes. Spotlight on him)

Hello everyone. My name is Antibodicus. I am one of the few survivors to tell the tale of the battle in the land of Rheumatic. Well to give you a better understanding of who I am, I was one of the soldiers of the castle of Vitruvius. We were loyal protectors of the man in charge of the castle whose name was Lenardo.

I joined the castle security force just 5 years after its construction was complete and served for 15 years (rheumatic fever occurs only after age of 5 upto a maximum of 15 years.)

I still clearly remember that day when from the neigbouring territory of Haemolyticus, Streptococcus decided to attack us.

Flashback (spotlight fades, focus lights on the other half of the stage, backdrop of castle, enter streptococcus with 4 soldiers and 4 soldiers from another side with weapons as props, smoke)

Streptococcus: group A, its tym to attack the castle of Vitruvius. We will aim for the top part of the castle and try to bring it down.

Army major: haha. U bet loser. We are gonna drive you out in no time you mangy mutt. .

The battle waged on silently ***(subclinical infection)*** for a few days and streptococcus returned back defeated. (choreographed fight scene, with all of streptococcus soldiers defeated while 1 killed on the other side, flashing lights, background mortal combat music)

(spotlight on, other part of stage darkens)

However after 3 weeks, when we soldiers met, we started to find something fishy. It was a hot day and temperatures were soaring high. *(rheumatic fever develops 3 weeks after a streptococcal infection).*

Flashback(spotlight off, orange hue light, table , chairs and silver cups as props, smoke)

Soldier 1: I find lenardo in the company of some unscrupulous ppl in the core floor of the castle.

Soldier 2: Yes they bear resemblance to those ppl from the neigbouring village of Haemolyticus from where streptococcus attacked us. The look so alike

Soldier3: very true. And lenardo and these relatives of his are not treating us properly. They are not paying us well. I think its time we started stealing treasures and attacking them.

Soldier 1: I think all this has happened cause our mind has been freaked out by the war with streptococcus.

(the pathophysiology of Rheumatic fever is 1) cross reactivity with cardiac antigens as they bear close resemblance to the M proteins of Steptococcus and 2) microbe initiated autoimmune reactivity)

(spotlight on, other part of stage dark)

So After two weeks we decided to attack the core treasure room in the castle. We chalked out a plan where most of the soldiers would be deployed near the core treasure room while we would station only a few in the corridors *(bites the heart and licks the joints)*

(spotlight continues to be on, other part of stage, dim light, door as prop or formation door, soldiers enacting stealing treasure)

For those of you not familiar with the land of rheumaticus, daytime here is called systole and night time is called diastole. So it was decided that in the middle of night , we would steal the treasure from the core room. However we had to be careful as the door to this room was a creaky one and made lots of noise in the middle of the night whenever we went to steal the treasure*(Mid Diastolic Murmur)*

But then we received some bad news during our review meeting

(spotlight off, smoke , orange hue light, tables, chairs, golden cups)

FLASHBACK

Soldier 1: u know what. When I was keeping watch over the large corridors, I saw lenardo's men at different places. They are very quick and fleeting. And they are not present in just one corridor and keep on changing their positions.

Soldier 2: hmmm. We need to move more quickly. I have an idea. Y don't we distract all of them in a way and keep them busy so we can steal more treasures.

Soldier 3: I have some MINOR report. I counted our treasures and the stocks of our ESR and CRP have increased tremendously. Plus the temperatures of the day are getting hot and its increasingly uncomfortable to stay in the corridors *(Minor criteria of jones: fever, arthralgia, raised esr n CRP titres)*

Soldier monologue:And so after that meeting we decided to celebrate the festival of St.vitus and distract them

We first painted around the stage with red colour and left the central area clear

(erythema marginatum).

We got some girls and made them dance to a choreography that was purposeful, rapid to keep the tempo going and more importantly we ensured that none of the steps were repeated

(Sydenham chorea: purposeful involuntary rapid non repetitive movements) .

there were lots of prominent people who looked so bony admiring the different body parts of the girls like elbow, shin, back of the occiput. Crazy people admiring weird body part!!!!! *(subcutaneous nodules)*

(spotlight off , flashing lights for dance, smoke)

FLASHBACK

(Dance sequence to a bollywood number happening with scene description as above few people sprawled on the floor, dance area to have been painted with red colour)

suddenly enter lenardo

(Strobe lighting (lightning and sound of galloping horses in background), lenardo and pencillin enter)

Lenardo: ur time is up my boys. Surrender or face the wrath of the mighty new army

Soldiers: we will not give up. We will fight till we achieve our goal or martyrdom

Lenardo : fine . have it your way. Pencillin ATTACK.

(flashing lights, mortal komabt background music, chorpgraphed fight scene like in gladiators)

(spotlight on)

Soldier monologue:And thus we waged a battle against pencillin and we were captured while a few died. And those that were captured had to face punishment

(smoke, spotlight off)

Flashback

Lenardo: those of you guilty of infiltrating the corridor, I punish you for a period in jail under the watchful eyes of pencillin for 5 years or till you turn 21.

And those of you guilty of entering the core room, I I punish you for a period in jail under the watchful eyes of pencillin for 10 years or till you rot into adulthood.

(Treatment regimen)

(fading music, only spotlight, spotlight fades, curtain falls)

And hence ends my story which I narrate from the prison cells

THE END

COMMENTS FROM THE TEAM

This script in particular is extremely unique. Very different from the style commonly adapted. It takes a monologue and couples it with lots of acting, dialogues, fight sequences, music and light effects. The author has demonstrated the fantastic ability to make this otherwise very detailed topic into an extremely colourful, musical , action packed theatrical marvel. If this script is enacted with all the right costumes, light effects, sound and props, there is no doubt that this will make a very powerful and high impact show.

The other important aspect to note in the style of writing is the concept of a simultaneous flashback happening in the background on the stage. This would leave the audience with a very unique experience and keep them gripped to their seats. To write a script with a concept of simultaneous events on stage simply is a demonstration of intellectual brilliance. A very valuable contribution indeed. Hats off.

Script 13

3 Greedy Queens

Medical Theatre Script on Cushing 's disease / Syndrome
(Covers – Etiology and Clinical Features; Ref: Harrisons, 17th Ed.)

Characters

6. King Cortical Simha - An Old Indian King
7. Swami Ramdev - A Sage/ God Man
8. Pitutrisa - Youngest Queen
9. Ectopica - Middle Aged Queen
10. Exogenousa - Oldest Queen
11. Extra Artists - Soldiers, Disciples of Sage, Maids of Ectopica etc.

N.B The parts of script in italics are related to stage settings, order of appearance, movements etc. The part in **Bold Font** *indicates the technical Medical Data, for better understanding the topic, which can be omitted during performances of the script.*

Narrator: Long long ago, how long ago even we don't know, the old but strong King of **Adrenal** Land Shri **Cortical** Simha desperate to have a heir to the throne approached Swami Ram Dev for a solution

(Swami's Ashram, King enters with queens, Swami Meditating, Opens the eyes as the king approaches)

Swami: King Cortical Simha, Glad to see you at my ashram son!!! Don't panic! I understand the problem and your heart's desire to have a successor at this age. Blessed are you for I may have a solution with me today.

Cortical Simha: Glad to hear your words God Man. I married thrice... tried a lot .....But all went in vain ....as I have no kids still (sighs) Oh! Holy one, bless me with strong kids... who will keep up the name of my dynasty. Here are my reports from the Court Physician (hands over the scrolls)

Swami: Hmmm..... (Looks the scrolls) Son, I fear to say this, (pause) but you doesn't seem to get excited enough these days and your secretion levels are pretty low – I mean the magical **cortisol** produced by you is not enough as per the analysis to give you healthy kids. Probably your dear queens can help you with this.

Simha: I understand Swaami!!! Here I will call the queens one by one!!! Help them to help me ....

Guards, Bring in Yuvarani Pituitrisa – my youngest and most beautiful wife!!!!

(Enter Pituitrisa)

Pituitrisa: Here I am, my Lord!

Swami: Yuvarani Piturisia... Dear, Listen carefully......you look still young and active ...Understand your capabilities, you are different from other queens... think about the good days where you could charm and stimulate the king with your hormones... Yes yuvarani... it is time to produce more of those magical ACTH hormones, and get our Raja aroused.

(EXCESSIVE PRODUCTION OF ACTH FROM PITUITAY GLAND STIMULATE ADRENAL CORTEX TO PRODUCE MORE CORTISOL)

Pitutrisa: Oh my king, my beloved.....now wait for the magic of my hormones!

Simha: Oh, I simply love those hormones, my beautiful princess. Gurads, Now Bring in the 2nd queen of mine, Queen Ectopia

(Enter Ectopia, leave pitutrisa)

Swami: Ectopia, my dear things doesn't look promising with you, at this age. For producing magical ACTH It looks like you will have to seek the help of your maids, I mean those Carcinoid tumor maids ... So my child, seek the help of you pancree, ovarii and thymus maids for some ACTH. Or In fact you may also ask the smoking maid – Bronchogenic Carcionma to spare a little of her ACTH! May Good happen to you!!

(ECTOPIC ACTH PRODUCTION FROM BRONCHOGENIC CARCIONOMA, CARCINOID TUMOURS OF THYMUS, PANCREAS, OVARY ETC.)

Ectopia: I will ensure my king gets enough of hormones to get the arousal. I'm ready for anything.

Simha: Go Ectopia... Get your (winks) beautiful maids ready and steady! Now bring in my 3rd queen Exogenousa, the oldest of all, but the dearest to me

(enter Exogenousa, leave Ectopia)

Exogenousa: Oh Swami; Pardon me. I overheard all your conversations... I'm too old to produce the magical ACTH, and I have no maids with me to help me.... I don't think I can help my King.

Swami: Oh Dear, I understand. I suggest you use this wonder drug I've got for you - the Steroids.... Use it on him!!! He will love you for this. And your wishes and dreams will be fulfilled!

(STEROID THERAPY/OVERUSE)

Swami: Son, Please Call in the other two Ranis also

(others also enter)

Swami: My dear Children beware! Beware!!!! Do not over use the magical hormones or wonder drugs on the King . Don't you also My King, succumb to the temptations. Remember, over dosage can lead to devastating effects. I shall visit you all at your palace in 9 months from now to hear the good news.

Narrator: Nine Months later, Swami Ramdev in the palace of King Cortical Simha

(On scene the queens, enter Swami)

Swami: Oh Queens, I'm here.... Eager to see your newborns. Heard the palace is blessed with 2 girls and one boy!! Where are the little ones? **(PREVALNCE IN FEMALES MORE)**

Pitutrisa: Oh you are here.... Was waiting for you ...Look here, look at my daughter the princess.... a fat ,lazy girl with purplish lines on the abdomen who already has got Hypertension doing nothing. **(Truncal Obesity, Lethargic, Purple Abdominal Striae, Hypertension)**

Ectopica: Swamiji......How about this one...My daughter with a Moon Face !!! Oh, look at the hair on her face !!! And the little one has got all broken bones already. **(MOON FACE, HIRSUTISM,OSTEOPOROSIS)**

Exogenousa: Swami!!! Nothing can beat this!!!! Have a look at my dear son... he has got a hump like a buffalo on his back, his is body is swollen up and his pee smells of glucose. Why did you do this to us ??? **(BUFALLO HUMP, EDEMA, GLYCOSURIA)**

Swami: Oh ! God ! I knew this would happen!!!!! I had warned you all in fact!!!! Where is the King??

Ectopica: Oh... Don't even think of meeting him!! He got a huge nodular mass on his body. Ashamed to come out he sits all day in his room. **(NODULAR HYPERPLASIA OF ADRENAL CORTEX)**

Swami: Hmmm.... I understand the poor old king could not take all the overstimulation from all of you!!! I had warned you ladies not to overdo the regimen with the Old husband of yours, now look what you have caused to your man and the babies. For your greediness this is the penance! On your cushion shaped babies, the kingdom shall be known from now as **Cushings** Kingdom. The next time I come I will tell you the cure for it!!!!! Till then keep praying .... And nothing else!!!!

COMMENTS FROM THE TEAM

This script is a complete package with a humorous story line, a strong plot, with a nice build up and a moral to conclude. A disorder like Cushing's disease is brilliantly converted into a story of a king with 3 wives, and how he wants a solution to the problem adds the comedy to the scenario. The attempt made here is different as in the features of the disease appear in the babies, which is quite a smart way to put it. The usual tendency is to have the features suffered by the main characters of the play, but this method of description made the story more enjoyable. Many times while converting topics into a script, the conclusion is usually abrupt, but here the story ended powerfully along with some witty lines towards the end. An extremely cool creation indeed.

Script 14

TROY Adaptation of infective endocarditis

Characters :

Aureus: Strongest bacterium attacking Troy (Staph. aureus)

HACEK: Weaker set of bacteria (Haemopilus Actinobacillus Cardiobacter Eikinella Kingella)

StrepA: one of HACEK's friends. Helped in preparing battle ground for HACEK

Virid: Another weaker bacteria (Strep. viridans)

Bacter: King of Bacteria

Priam: Trojan King

MSCRAMMs: Aureus' warriors (Microbial Surface Components Recognising Adhesin Matrix Molecules)

Mito: One of Trojan's best warriors (Mitral valve)

Aorto: Mito's brother (Aortic valve)

Aorti: Mito's sister

Scene 1: Before approaching Troy. Aureus resting.

Enter: HACEK

HACEK: King Bacter sent me. He needs to see you. They are waiting for you.

Aureus: Meet me outside in a moment.

HACEK: Are the stories about you true? They say you are very virulent. (S. aureus- virulent bacteria. Can invade even with intact endothelium)

Aureus: Wouldn't be bothering with building resistance against serum & peptides then? Would I? (serum and peptides in blood counter act bacteria)

HACEK: Troy is the strongest heart I've ever seen. I wouldn't want to fight him.

Aureus: That's why no one will remember your name.

HACEK: I would rather attack them when they are vulnerable my lord. When the Trojan WBC warriors are lesser (neutropenia) or their antibodies are less (immunosuppressed).

Virid: My lord, I heard some of them are drug abusers (more susceptible). It would be easy to attack. And my good friend StrepA has already helped me weaken my victim's valve (RHD).

Aureus: Your business is war boy. Do it right.

Scene 2: Bacter's army approaching Troy. In front of the great Trojan endothelial gates.

Priam: All my life I've lived by a code. And the code is simple. Use your serum, bring out your peptides and defend your country. Troy is mother to us all. Fight for her! (defense mechanism)

Scene 3: Aureus and his warriors are fast approaching the shores of Troy.

Aureus: MSCRAMMS. My brothers of the sword. I'd rather fight by you than any army of thousands. Let no Trojan forget how menacing we are! We are adhesive lions. Help me stick to the Trojan endothelial gates. And when we enter my brothers. do not forget that we use their own fibrin n platelets to form are thrombus. Our vegetation to dwell and multiply. (help to adhere to vessel wall)

MSCRAMMS: What do we do next?

Aureus: Be thou patient. Some bacteria attack while others rest inside. For we are vulnerable when we are out in foreign territory. It is not possible for Trojan medication to attack when we hide (medication only effective against active bacteria). Brothers! You know whats waiting beyond the shores? Immortality! Take it. Its yours! Fire Septic emboli!

After a gruesome battle.... Troy was left injured

Mito: If killing is your only talent, then that's your curse! How many brothers of mine have you damaged? How many sons and father and husbands? How many brave Aureus? You have weakened me and now i regurge (mitral regurgitation).

Aureus: You are the best valve I've ever fought.

Mito: My brother Aorto... You have injured him too and left him just like me (Aortic regurgitation). My sister Aorti is so scared. She has closed herself inside so tight (Aortic Stenosis). I cannot change what has happened. If they ever tell my story, let them say I walked with Bacter. Fungi rise and fall like the winter wheat. But these names will never die (very rarely affect valves). Let them say, "I lived in the time of Aureus".

COMMENTS FROM THE TEAM

This script is a very clever adaptation of the movie troy with pathogenesis of infective endocarditis. The usual tendency is to plot around the subject while making a story but here the author has very nicely demonstrated how the subject is catered to the plot. The characters, the language, the style of sentences are very consistent with the epic war movie throughout the script with extreme resemblance to dialogues and details in the context of infective endocarditis. Associating such a complicated topic to almost every sequence of the parts of the movies is simply the work of brilliance. Terrific work on the adaptation.

Pathogenesis of infective carditis

Unless it is injured, the endothelium is resistant to infection by most bacteria and to thrombus formation. Endothelial injury (e.g., at the site of impact of high-velocity blood jets or on the low-pressure side of a cardiac structural lesion) causes aberrant flow and allows either direct infection by virulent organisms or the development of an uninfected platelet-fibrin thrombus—a condition called *nonbacterial thrombotic endocarditis* (NBTE). The thrombus subsequently serves as a site of bacterial attachment during transient bacteremia. The cardiac conditions most commonly resulting in NBTE are mitral regurgitation, aortic stenosis, aortic regurgitation, ventricular septal defects, and complex congenital heart disease. These conditions result from rheumatic heart disease (particularly in the developing world, where rheumatic fever remains prevalent), mitral valve prolapse, degenerative heart disease, and congenital malformations. NBTE also arises as a result of a hypercoagulable state; this phenomenon gives rise to the clinical entity of *marantic endocarditis* (uninfected vegetations seen in patients with malignancy and chronic diseases) and to bland vegetations complicating systemic lupus erythematosus and the antiphospholipid antibody syndrome.

Organisms that cause endocarditis generally enter the bloodstream from mucosal surfaces, the skin, or sites of focal infection. Except for more virulent bacteria (e.g., *S. aureus*) that can adhere directly to intact endothelium or exposed subendothelial tissue, microorganisms in the blood adhere to sites at NBTE. If resistant to the bactericidal activity of serum and the microbicidal peptides released locally by platelets, the organisms proliferate and induce a procoagulant state at the site by eliciting tissue factor from adherent monocytes or, in the case of *S. aureus*, from monocytes and from intact endothelium. Fibrin deposition combines with platelet aggregation, stimulated by tissue factor and independently by proliferating microorganisms, to generate an infected vegetation. The organisms that commonly cause endocarditis have surface adhesin molecules, collectively called microbial surface components recognizing adhesin matrix molecules (MSCRAMMs), that mediate adherence to NBTE sites or injured endothelium. Fibronectin-binding proteins present on many gram-positive bacteria, clumping factor (a fibrinogen- and fibrin-binding surface protein) on *S. aureus*, and glucans or FimA (a member of the family of oral mucosal adhesins) on streptococci facilitate adherence. Fibronectin-binding proteins are required for *S. aureus* invasion of intact endothelium; thus these surface proteins may facilitate infection of previously normal valves. In the absence of host defenses, organisms enmeshed in the growing platelet-fibrin vegetation proliferate to form dense microcolonies. Organisms deep in vegetations are metabolically inactive (nongrowing) and relatively resistant to killing by antimicrobial agents. Proliferating surface organisms are shed into the bloodstream continuously.

The pathophysiologic consequences and clinical manifestations of endocarditis—other than constitutional symptoms, which probably result from cytokine production—arise from damage to intracardiac structures; embolization of vegetation fragments, leading to infection or infarction of remote tissues; hematogenous infection of sites during bacteremia; and tissue injury due to the deposition of circulating immune complexes or immune responses to deposited bacterial antigens.

Script 15

The opioid RAP

Opioid analgesics

Man – Lying by my side

Clutching my leg

'oh God take this pain away from my leg,

Cant move, cant eat, cant sleep

Pain surely is the son of the beast,

Popping hundred of these tablets all day

But none of them take the pain away,

These silly pills just aren't efficient

The increase in pain is quite significant.

(Enter opioid)

Opioid - Whose your daddy, whose your daddy

Pills- you are my daddy

Of pain relief we don't have a clue.

Opioid – I am the most reliable, most effective

Don't cry all you other pills don't look dejected

Where severe pain relief is expected

Forget the cause and pop me irrespective.

Pain Inhibiting neurons is how I act

That is the nervous system but you already know that

pop me in your mouth or straight in the vein

I may act slower but effective all the same

Whose your daddy

Whose your daddy

Pills – you are my daddy

We are less potent than stew

You are my daddy

Of your greatness we never knew!!

Opioid – But for my greatness a dark side I do possess

I can make you itch, put you to sleep or your respiratory system I will depress

Pop too many and you can't evacuate stools

Don't tell nobody but I'm not very cool

Even the sound of my name, makes hippies drool!

Now go out confident and get yourself a dame

Sorry I can't help libido, I'm useful only for pain.

COMMENTS FROM THE TEAM

Full marks for creativity here for the musical, poetic rap depiction of opioid analgesics as it really adds a huge variety to the ways a topic can be converted into an entertaining fun to read material, form a complex group of sentences in a text book. All aspects of opioid analgesics along with the patient's agony due to pain are emphasized very well with usage of medically relevant words to rhyme with. An excellent method of presentation.

Opioid Analgesics

Opioids are the most potent pain-relieving drugs currently available. Furthermore, of all analgesics, they have the broadest range of efficacy, providing the most reliable and effective method for rapid pain relief. Although side effects are common, they are usually not serious except for respiratory depression and can be reversed rapidly with the narcotic antagonist naloxone. The physician should not hesitate to use opioid analgesics in patients with acute severe pain. Table 12-1 lists the most commonly used opioid analgesics.

Opioids produce analgesia by actions in the central nervous system. They activate pain-inhibitory neurons and directly inhibit pain-transmission neurons. Most of the commercially available opioid analgesics act at the same opioid receptor (-receptor), differing mainly in potency, speed of onset, duration of action, and optimal route of administration. Although the dose-related side effects (sedation, respiratory depression, pruritus, constipation) are similar among the different opioids, some side effects are due to accumulation of nonopioid metabolites that are unique to individual drugs. One striking example of this is normeperidine, a metabolite of meperidine. Normeperidine produces hyperexcitability and seizures that are not reversible with naloxone. Normeperidine accumulation is increased in patients with renal failure.

The most rapid relief with opioids is obtained by intravenous administration; relief with oral administration is significantly slower. Common side effects include nausea, vomiting, constipation, and sedation. The most serious side effect is respiratory depression. Patients with any form of respiratory compromise must be kept under close observation following opioid administration; an oxygen saturation monitor may be useful. The opioid antagonist naloxone should be readily available. Opioid effects are dose-related, and there is great variability among patients in the doses that relieve pain and produce side effects. Because of this, initiation of therapy requires titration to optimal dose and interval. The most important principle is to provide adequate pain relief. This requires determining whether the drug has adequately relieved the pain and the duration of the relief. *The most common error made by physicians in managing severe pain with opioids is to prescribe an inadequate dose. Since many patients are reluctant to complain, this practice leads to needless suffering*. In the absence of sedation at the expected time of peak effect, a physician should not hesitate to repeat the initial dose to achieve satisfactory pain relief.

An innovative approach to the problem of achieving adequate pain relief is the use of patient-controlled analgesia (PCA). PCA requires a device that can deliver a baseline continuous dose of an opioid drug, as well as preprogrammed additional doses whenever the patient pushes a button. The patient can then titrate the dose to the optimal level. This approach is used most extensively for the management of postoperative pain, but there is no reason why it should not be used for any hospitalized patient with persistent severe pain. PCA is also used for short-term home care of patients with intractable pain, such as that caused by metastatic cancer.

Script 16

SHOULDER MOVEMENTS

The groovy shoulder

When one flexes the arm at the shoulder joint

There is one small joint

Which you must remember

Whether it is July or November

There is a gamble of two muscles

Pectoralis major and anterior deltoid in the tussels.

To Teres major, Latissimus dorsi was happily married

But while extending, these got joined with posterior deltoid

In adduction of course,

The joint decided a better course

It went with two majors (pectoralis major and Teres major)

On the way they stopped for some gazers

The two majors danced with subscapularis during medial rotation

Even anterior deltoid and latissimus dorsi soon joined the happy flirtation.

If one wants the joints to laterally rotate

Then there is difference in the mate

Posterior deltoid dances with infraspinatus

Even teres minor comes and triangulates

When just abduction is desired

Supraspinatus and mid-deltoid are required

But if kapil dev has to do the bowling

Come trapezius and serratus anterior following

Small muscles provide stability

Large ones give it mobility

And shoulder joint dances

Dances and dances.

COMMENTS FROM THE TEAM

Full credit to the author for attempting to make a poem on an anatomical topic. Anatomy is one of the challenging areas when it comes to understanding it and due to the large levels of facts and details it is very difficult to usually carve it into some form of creativity. A poem is a nice method to convey this message.

Script 17

Primary insomnia

There was a girl called insomnia who lived in Bangalore. People believed that she used to love in a world of her own creation. She was always thinking and imagining about some or the other thing.

People called her crazy because she never took rest. Her parents were fed up of her because of her abnormal behavior. She barely closed nor eyes, she couldn't sleep at all. Even if she tried to sleep, she would wake up. Sometimes she used to get up in the middle of night and stare at her roof. Her parents took her to a psychiatrist, but all in vain.

Then she was taken to a man called Mr. Benzo. He was the assistant doctor in a famous hospital called DRUGS. He helped her to recover a lot. She slowly started falling in love with Mr. Benzo. She almost got addicted towards him. After 2 months Mr. Benzo died in a car accident. She lost her mind. She couldn't live a moment without him. After a month she started feeling better when she met with the chief Dr. Trazodone. He loved and cared about her. Slowly she started recovering from her sleeping problems. She started feeling better and slept well.

COMMENTS FROM THE TEAM

A simple and short way to convey the topic. Very straight forward.

Primary Insomnia

Many patients with chronic insomnia have no clear, single identifiable underlying cause for their difficulties with sleep. Rather, such patients often have multiple etiologies for their insomnia, which may evolve over the years. In addition, the chief sleep complaint may change over time, with initial insomnia predominating at one point, and multiple awakenings or nonrestorative sleep occurring at other times. Subsyndromal psychiatric disorders (e.g., anxiety and mood complaints), negative conditioning to the sleep environment (psychophysiologic insomnia, see below), amplification of the time spent awake (paradoxical insomnia), physiologic hyperarousal, and poor sleep hygiene (see above) may all be present. As these processes may be both causes and consequences of chronic insomnia, many individuals will have a progressive course to their symptoms in which the severity is proportional to the chronicity, and much of the complaint may persist even after effective treatment of the initial inciting etiology. Treatment of insomnia is often directed to each of the putative contributing factors: behavior therapies for anxiety and negative conditioning (see below), pharmacotherapy and/or psychotherapy for mood/anxiety disorders, and an emphasis on maintenance of good sleep hygiene.

If insomnia persists after treatment of these contributing factors, empirical pharmacotherapy is often used on a nightly or intermittent basis. A variety of sedative compounds are used for this purpose. Alcohol and antihistamines are the most commonly used nonprescription sleep aids. The former may help with sleep onset but is associated with sleep disruption during the night and can escalate into abuse, dependence, and withdrawal in the predisposed individual. Antihistamines may be of benefit when used intermittently but often produce rapid tolerance and may have multiple side effects (especially anticholinergic), which limit their use, particularly in the elderly. Benzodiazepine-receptor agonists are the most effective and well-tolerated class of medications for insomnia. The broad range of half-lives allows flexibility in the duration of sedative action. The most commonly prescribed agents in this family are zaleplon (5–20 mg), with a half-life of 1–2 h; zolpidem (5–10 mg) and triazolam (0.125–0.25 mg), with half-lives of 2–3 h; eszopiclone (1–3 mg), with a half-life of 5.5–8 h; and temazepam (15–30 mg) and lorazepam (0.5–2 mg), with half-lives of 6–12 h. Generally, side effects are minimal when the dose is kept low and the serum concentration is minimized during the waking hours (by using the shortest-acting, effective agent). Recent data suggest that at least one benzodiazepine receptor agonist (eszopiclone) continues to be effective for 6 months of nightly use. However, longer durations of use have not been evaluated, and it is unclear whether this is true of other agents in this class. Moreover, with even brief continuous use of benzodiazepine-receptor agonists, rebound insomnia can occur upon discontinuation. The likelihood of rebound insomnia and tolerance can be minimized by short durations of treatment, intermittent use, or gradual tapering of the dose. For acute insomnia, nightly use of a benzodiazepine receptor agonist for a maximum of 2–4 weeks is advisable. Benzodiazepine receptor agonists should be avoided, or used very judiciously, in patients with a history of substance or alcohol abuse. The heterocyclic antidepressants (trazodone, amitriptyline, and doxepin) are the most commonly prescribed alternatives to benzodiazepine receptor agonists due to their lack of abuse potential and lower cost. Trazodone (25–100 mg) is used more commonly than the tricyclic antidepressants as it has a much shorter half-life (5–9 h), has much less anticholinergic activity (sparing patients, particularly the elderly, constipation, urinary retention, and tachycardia), is associated with less weight gain, and is much safer in overdose. The risk of priapism is small (~1 in 10,000).

Script 18

SPASOTOXICALLY CORRECT....?..!

Amidst the unstable political scenario of the nation, yoga guru, Baba Tetadev (Clostridium tetani) was looking to create mayhem(contamination of wounds) since people were just taking him for granted. However, he knew he wouldn't be able to stir the public by himself and needed some assistance. (C.tetani doesn't itself evoke inflammation unless co-infection with other organisms present) So he resorted to taking help from a dear friend, Mr. Spasotox (toxin- Tetanospasmin) who was a very shrewd person.

Together they schemed a scandalous plot. They carefully chose their targets of which the first and most vulnerable were Baba's followers comprising of the poor (devitalized tissue) with low standards of living (low redox potential) and a few foreigners (foreign bodies). Baba Tetadev immediately set out to bend his followers not just physically but also mentally. (toxin enters devitalized tissue)

Meanwhile the conniving Mr.Spasotox was out to hit the next target. Soon he was at the portals of CNS Bhavan where on being denied entry he slickly bribed the security guards at the peripheral terminals. (toxin binds to peripheral motor neuron terminals)

He then entered the vast axonal compound (toxin enters axon) and made his way to the parliament house (nerve cell body) where he saw the members and the Ruling party-Sonial cord. (spinal cord-CNS). Witty Mr.Spasotox managed to befriend the party members (pre-synaptic terminals) and misled them into believing that Baba Tetadev was upto no good. In this way he blocked their ability to think clearly.(toxin blocks release of inhibitory neuro-transmitters) He even succeeded in suppressing the resistance put up by some level headed ministers such as Sri. Gali Sin (inhi. NT-glycine) and Sri. Gabar Singh (inhi. NT-GABA). Thus Mr.Spasotox achieved the most crucial step of the plan- disrupting the functioning of Sonial cord.(MOA-Tetanospasmin: disruption of proper function of synaptic vesicle release apparatus)

The ruling party now became *rigid and in order to prevent the spread of Baba Tetadev's propaganda called Tetanus (limit polysynaptic spread of impulses) they recruited the police (agonists and antagonist recruited rather than inhibited) and ordered them to arrest Baba.

Meanwhile in the NaamDila grounds ,the effects of the yoga guru were seen to manifest as his followers blindly followed him despite experiencing *stiffness and pain in the neck, shoulder,back and proximal limb muscles. Atleast their hands and feet were relatively spared. As they clenched their teeth in pain a few developed a *lockjaw (Trismus). A few others were practicing a peculiar yoga posture with *arched backs (opisthotonos) and they *grimaced with pride and sneered at the struggling amateurs (risus sardonicus).

All of a sudden the police entered the grounds and there was violence! Some were *spasmodically attacked (Spasm attack) while those who put up strong resistance were threatened to be *choked (reduced ventilation, apnea, laryngospasm). In a matter of 7 hours

(median time of onset after injury-7 days) operation "Tetanus" had ravaged throughout the nation injuring lay men (generalized tetanus), infants (neonatal tetanus) and led to several local uprisings (local tetanus).

However it didn't end at this, now that the Ruling party was intoxicated, astute Spasotox wanted to also agitate the Opposition, PGP- Preganglionic Party (toxin also affects preganglionic sympathetic neurons). They immensely sympathized (sympathetic hyperactivity) with Baba Tetadev and upheld his movement. Some over-enthusiastic members danced with joy and their hearts beat fast ^(tachycardia) because Sonial cord's name was now tarnished. They began to ^sweat profusely and the summer heat made them warm ^(hyperpyrexia).Not realizing the vigor with which they danced some sustained ^fractures and muscle ruptures. Hence this outrageous gesture only led to further complications making matters worse.

Thus goes on… a typical desi saga…

One that is led by a Baba

In an attempt to be 'politically correct',

Simple principles one tends to forget .

When one uses corruption (toxin tetanospasmin a.k.a Mr.Spasotox) to curb corruption (So called objective of Baba Tetadev)…

Truly and surely there is no solution!

Is this what our nation has come to..? Is there any respite from such atrocious happenings..?!...Our only hope for recovery would be the sprouting of new politicians rather reformers (recovery requires sprouting of new nerve terminals) to put things into perspective and work towards building an "anti-spasotox" future…!

*clinical manifestations
^complications

COMMENTS FROM THE TEAM

We have so far come across all sorts of stories, associations, movie adaptations, poems and monologues for various medical topics. Here is a script adapted to real life politics ending with a social message. The incorporation of almost every aspect of tetanus along with subtle elements of humour makes this script an extremely interesting and enjoyable reading. The description of baba tetadev and the yoga formations depicting the clinical features were the best highlights of this story. The fact that every second word has been associated with some aspect of tetanus clearly demonstrates the ability of the author to be extremely creative and imaginative.

Clinical Manifestations of Tetanus

Generalized tetanus, the most common form of the disease, is characterized by increased muscle tone and generalized spasms. The median time of onset after injury is 7 days; 15% of cases occur within 3 days and 10% after 14 days.

Typically, the patient first notices increased tone in the masseter muscles (trismus, or lockjaw). Dysphagia or stiffness or pain in the neck, shoulder, and back muscles appears concurrently or soon thereafter. The subsequent involvement of other muscles produces a rigid abdomen and stiff proximal limb muscles; the hands and feet are relatively spared. Sustained contraction of the facial muscles results in a grimace or sneer (risus sardonicus), and contraction of the back muscles produces an arched back (opisthotonos). Some patients develop paroxysmal, violent, painful, generalized muscle spasms that may cause cyanosis and threaten ventilation. These spasms occur repetitively and may be spontaneous or provoked by even the slightest stimulation. A constant threat during generalized spasms is reduced ventilation or apnea or laryngospasm. The severity of illness may be mild (muscle rigidity and few or no spasms), moderate (trismus, dysphagia, rigidity, and spasms), or severe (frequent explosive paroxysms). The patient may be febrile, although many patients have no fever; mentation is unimpaired. Deep tendon reflexes may be increased. Dysphagia or ileus may preclude oral feeding.

Autonomic dysfunction commonly complicates severe cases and is characterized by labile or sustained hypertension, tachycardia, dysrhythmia, hyperpyrexia, profuse sweating, peripheral vasoconstriction, and increased plasma and urinary catecholamine levels. Periods of bradycardia and hypotension may also be documented. Sudden cardiac arrest sometimes occurs, but its basis is unknown. Other complications include aspiration pneumonia, fractures, muscle rupture, deep-vein thrombophlebitis, pulmonary emboli, decubitus ulcer, and rhabdomyolysis.

Neonatal tetanus usually occurs as the generalized form and is usually fatal if left untreated. It develops in children born to inadequately immunized mothers, frequently after unsterile treatment of the umbilical cord stump. Its onset generally comes during the first 2 weeks of life.

Local tetanus is an uncommon form in which manifestations are restricted to muscles near the wound. The prognosis is excellent. *Cephalic tetanus*, a rare form of local tetanus, follows head injury or ear infection and involves one or more facial cranial nerves. The incubation period is a few days and mortality is high.

Script 19

HEART BROKEN

(J seen working frantically, answering phone calls, very stressed!)

Narrator: Ah, J, look at him...it's true, that people are anything but relaxed nowadays; working to support one's family, working to keep one's job, working to save one's own life! His industry has just recently experienced a boom; but what's really been blown to smithereens is...

(McD's theme... "It's finger lickin' good!" J orders fast food, eats and rests back, tired)

Narrator: Whatever happened to the good old times, when food was actually edible, and actually...healthy? All this finger lickin' has caused a quiet turmoil within the city of J...let's have a look inside...

(Inside J)

Narrator: In a secret meeting of the illegal Dyslipidemic Gang...

Leader: Ok, we've got enough fatty members now to cause real damage. We have three choices of location: the Left Anterior Descending Highway; The Right Coronary Highway, and the Left Circumflex Highway. We want to damage the whole thickness of that office complex. We shall set up camp in the LAD...infiltration starts tomorrow.

Narrator: And the Gang sets to work...constructing their hideout, which they called... the 'Plaque'...time passes; and the finger lickin' just doesn't stop!

News reporter: It's the third year of growing unrest in the city of J; an illegal establishment has just been set up by the members of the rogue Dyslipidemic Gang, nearly blocking one of the crucial highways supplying the central Sanguine Transport Office. So far, there has been no disturbance caused within the office complex that has been particularly affected; although damage is thought to be imminent. This was a surveillance report for the National Homeostatic News Channel.

Pathologist: Here, we see that a plaque has been formed due to atherosclerosis. This may be caused due to various reasons, mainly an unhealthy diet, sedentary lifestyle as seen here. The dyslipidemic state in J has lead to the partial blockage in the Left Anterior Descending Artery of J's heart. What will happen next?

News reporter: Some members were seen to escape from the Plaque due to its rupture; ThromboRoad Repair Team's work has left the highway completely blocked. Homeostatic experts are worried now. The further road now shows stagnation in traffic; which may affect the functioning of the affected Office Unit.

(6 Hours Post Blockage)

Pathologist: Due to the sluggish blood flow, fluid extravasates and interstitial edema occurs in the myocardium. Myocytolysis is seen.

(12 hours Post Blockage)

Narrator: At the office of the Leukocyte Active Protection Dept. (LAPD)...

Officer: We've got trouble around the Apex of the Left Office; this part of the office is starting to shut down. I'm sending my forces down there for damage control. Sergeant, get your neutrophils ready. (Sergeant leaves)

Pathologist: Now, coagulative necrosis sets in, seen as intensely eosinophilic muscle fibers with loss of striations. Fibers show nuclear degenerative changes and appear hyaline. Grossly, the affected region of the heart is pale, and drier. Neutrophilic infiltration is marginal.

(24 Hours Post Blockage)

News Reporter: Breaking news just coming in; larger parts of the affected office are shutting down, and the office structures are progressively disintegrating. The Plaque hasn't been demolished yet, and traffic is still stagnant. Vehicles are starting to park off the road. This is catastrophe! Reporting for the National Homeostatic News Channel.

Pathologist: Grossly, the area shows cyanosis, and hemorrhagic spots.

(48 – 72 Hours Post Catastrophe)

News Reporter: The situation seems to be growing grimmer. There is complete chaos and disarray in the affected Office; structural disintegration is advanced. The LAPD has successfully infiltrated the damaged Office, and is attempting to clear out debris.

Pathologist: The necrosis complete with loss of nuclei. Neutrophilic infiltrate extends into the affected area. Grossly, the infarct develops a defining yellow border. Thus, the Myocardial infarction has occurred. It is transmural, in that it affects the entire thickness of the ventricular wall.

Narrator: The Neutrophil forces of the LAPD work hard to clear debris; and by the 7th day...

(End of 1 Week)

News Reporter: After initial recovery work by the neutrophils, the LAPD has recruited its special MacroP team to aid final clearing of debris... BREAKING NEWS: The Criss-Cross Collagen Constructing Company has offered to send its finest Fibroblast workers to the damaged site to aid in the repair of the Office buildings. Going over to our on-site correspondent...

C: We have with us the CEO of the CCCCC. Sir, why this decision?

CEO: Well, the damage caused by the Dyslipidemic Gang was pretty devastating. It's sent shockwaves in the form of reduced traffic flow along highways and roads all over the city. I felt morally obliged to aid in the construction of support framework for the damaged Office. It will aid the Sanguine Transport Office to function.

C: Thank you! The fibroblast workers have arrived, and have started putting up scaffoldings of collagen and other supportive framework where there was once a well-functioning office. New roads are being constructed to aid supply the workers with raw material. The effects of the Plaque are truly saddening!

(2 Weeks post Catastrophe)

News reporter: The last of the Office debris have been cleared, and the collagen framework is under progress. MacroP has cleared the wayward traffic.

Pathologist: Macrophages show lipofuscin and hemosiderin accumulation due to hemorrhage-clearing.

(4-6 Weeks after Catastrophe)

News Reporter: The CCCCC workers have completed the structural reinforcement of the affected Office. Sadly, however, the Office will never be the same again!

(Clear scene, J seen rubbing his chest in discomfort)

Narrator: And, J, too, will never be the same again. He was in severe, sudden pain when the blockage occurred. He was terrified, and urinated less frequently. He had low grade fever, and breathing was difficult. His ECG showed ST segment elevation, characteristic of MI. His serum cardiac markers, creatine kinase and cardiac troponins were elevated. If only, he'd stopped finger lickin' earlier, and got off his chair often...how prosperous the City of J would've been!

Xx_______________xX

COMMENTS FROM THE TEAM

This script gives the inside story at a molecular level describing the process of myocardial ischemia associating it with in house news channels and reporters from within the body. Some very smart creativity like for LAPD and CCCC was quite spectacular. The language and style was consistent throughout and depicted the news reporting from the homeostatic channel consistently. Smart writing.

Script 20

THE TALE OF TWO BROTHERS

(story on the clinical features and treatment of leprosy)

Long long ago in **MYCOBACTRICITY**, there ruled a king named **Hansen.** He was a just king, and ruled **Mycobactricity** with an **iron ROD** . The king two sons, **prince Tuberan** and **prince Lepran** .The latter being the elder of the two. (LL : ELDER)

Now, the king did not favour young prince Tuberan and had a soft corner for prince Lepran .

The king had his reasons, prince lepran was :-

- Well known throughout the kingdom
- He had a lion like countenance
- And smooth shiny skin with lots of hair ,(although it would fall eventually)
- He was also sensitive compared to his younger brother
- And was progressive in thought

But he had certain drawbacks , he had:-

- Poor smell
- Very bad eyesight
- Weak bones and muscles
- And was know to be weak in the balls (bacilli invading the upper respiratory tract mucosa , eyes, bones, muscles, testis)

While prince Tuberan was dramatically opposite :-

- He was known only at one or few places in the entire city
- Was copper red in colour with dry scaly skin and very less hair
- And was known to be an insensitive man
- Moreover, he was always nervous and was very self limiting in his thoughts ,and had none of the other good features of his brother.

But he had two important qualities :-

- Clarity of thought
- Marked height

And so he always considered himself to be no:1 (cell mediated immunity) and his elder brother as no:2(immune complex mediated)

Now, the king had three ministers

RIFA

DAP

And CHLOE

Who were always on the look out for the throne. They read the mind of prince Tuberan and realized that he hated his brother , so they offered to help him by eliminating prince Lepran forever .

Now it was definitely not an easy task to eliminate prince Lepran ;he was strong and resistant to most of the poisons they sort to give him .It took 12 months of poisoning by the ministers to kill him.

Every month, RIFA and CHLOE would supervise the poisoning of the prince with 600 mg and 300 mg of strong poisons , while CHLOE and DAP would daily give 50mg and 100 mg of homemade poison to the prince.

Finally the prince died .

Prince Tuberan was overjoyed, but little did he know that he was to last no more than 6 months more.

The clever RIFA would administer 600 mg of strong poison every month, while DAP would make the prince take his own concoction of 100 mg. Finally, the young prince Tuberan too met his brothers fate.

King Hansen was heartbroken....:(

COMMENTS FROM THE TEAM

This is a simple and straightforward script depicting the clinical features of leprosy using associations with a royal story of jealousy and conspiracy. The description of the characteristics of the prince with the type of lesion especially with the mention of one of them to be an "insensitive man", was very witty humor. The characterization for the ministers was well shown along with the doses into the story. That makes it easy to revise along with reading an entertaining story. Very impressive work over all.

CURRICULUM VITAE

PERSONAL PARTICULARS

Name : Dr. Jagdish Chaturvedi

Date of birth : 27-3-1984

Address : No.15, Type 5, NIMHANS Quarters,
Dairy circle Campus,
Bengaluru.560029.

Email : hemoglobin84@yahoo.com
hemoglobin84@gmail.com

Phone Number : Residence 08026564843
Mobile +919740076668

ACADEMIC QUALIFICATIONS-

February 2009-February 2012 – DNB ENT St John's Medical College Hospital, Bangalore.
National Board of Examinations, New Delhi.

September 2001 – August 2006: M.B.B.S Sree Siddhartha Medical College, Tumkur.
Rajiv Gandhi University of health sciences, Karnataka.

March 1999 – March 2001: 12th Standard Sudarshan Vidya Mandir, Jayanagar, Bengaluru.
Karnataka Pre University Course (PUC)

March 1989 – March 1999: 10th Standard Vivekenanda Educational Centre, Bengaluru.
Indian School of Secondary Education (I.C.S.E)

REGISTRATION-

1. **Karnataka Medical Council :** KMC Registration no: 78403
   Date of registration : 24-10-2007

2. **Director Identification Number**: 0334575 **Government of India**
   Founder and Director : Yukti Medical Innovations Private Limited

MEMBERSHIP IN PROFESSIONAL BODIES-

1. Association of Otolaryngologists of India – Member
2. Association of Otolaryngologists of India, Bangalore Branch - Member
3. International Association for the Study of Pain (IASP) –Member #55814.
4. Foundation for Head and Neck Oncology – FHNO

INTELLECTUAL PROPERTY-

PATENT PENDING :

Indian Patent: Ear Nose Throat Multi-scope Recorder
Application Number- 1293/CHE/2010 A
Date of Filing- 10/5/2010
Publication Date- 13/8/2010
Status – Awaiting Examination (No Oppositions)

: An Educational System and Improved Teaching Methodology Using Dramatics and Theatrical Art Forms.
Application Number- 3020/CHE/2011
Date of Filing- 03/9/2011
Publication Date- 13/8/2010
Status – Awaiting publication

PCT: Ear Nose Throat Multi-scope recorder
Application Number- PCT/IN2010/000756
Filing Date- 19/11/2010
Status- Awaiting Publication

CLINICAL EXPERIENCE-

IN INDIA :

16 /2/2009 – to date : **St John's Medical College and Hospital**, Bangalore
Department of Otorhinolaryngology and
Head and Neck Surgery

Status : Post graduate resident,
3rd year DNB (Diplomat of National Board)

Resposibilities : OPD and Ward duties

- Minor OT procedures
- Emergency duties
- Case presentations
- Weekly seminar presentations
- Journal club
- Research projects
- Training in Basic surgical skills
- Training in Endoscopic surgeries
- Training in Head and Neck surgeries
- Training in office endoscopies
- Assisting and supervising junior house officers and interns for minor surgeries as well as ward and emergency
- Participation in weekly clinical and inter departmental

ENT – Anatomy meet
ENT- Radiology meet
ENT- Pathology meet

Active participation in Teaching programmes
Audit meetings and ENT Journal Clubs.

At ST.JOHNS MEDICAL COLLEGE HOSPITAL, INDIA, a 1200 bedded tertiary care teaching hospital. During my **first year of post graduate training** my work included initial management of patients on a routine and emergency basis, attending ward rounds, clinics and grand rounds, participating in journal club, teaching sessions and interdepartmental meetings. In this period I was taught basic surgical skills. I was also trained to organize Continuation of Medical Education [CME] and Conferences for my Department.

During the **second year of my post graduate training** I consolidated my basic skills. I have been actively involved in research activities of the department. I have developed basic competencies in ear and nasal surgery. My duties have included assisting the consultants in surgeries of the ear, nose and head & neck malignancies. Other duties included attending regular teaching sessions and Weekly clinics. (Vertigo/Allergy/Speech/OSA Clinics)

During the **third year of my postgraduate training** I was given more responsibilities such as teaching undergraduates during clinics and tutorials. I was also given the task to teach and guide my junior colleagues during minor OT and emergency procedures. I was trained to organize /conduct workshops and also be a part of many clinical research papers. Surgical opportunities were extended to assisting more complicated and prolonged cases. Basics on application of biostatistics were also taught during this period. I was scheduled to attend rotational clinical training in the following fields-

Head and neck Onco-surgery Training – 1 month (January-February 2011)
TATA Memorial cancer hospital, Mumbai.

Neurosurgery clinical training - 2 weeks (March 2011)
NIMHANS, Bangalore.

General Surgery and plastic surgery - 6 weeks (March- April 2011)
St John's Medical College hospital, Bangalore.

15/9/2010- 15/3/2011: **Health Informatics Group, St John's Research Institute**
Bangalore

Status: Specialist Consultant Validator

Responsibilities: Validating health fact sheets related to Ear, Nose and Throat
Checking validity of content
Making necessary corrections
Cross checking references

During these 6 months I validated health sheets that were written by Junior doctors/ dentists who work in the health informatics group using resources such as text books, journals, internet references, Pub med and evidence based databases such as the Cochrane database. I validated the entire syllabus of both clinical and basic information that came under Ear, Nose and Throat. This post was assigned during my post-graduation by the institute and the head of the department.

15/2/2008 – 14/8/2008: **National Institute of Mental Health and Neuro Sciences**
(NIMHANS), Bangalore.

Status: Non PG Junior Resident in Psychiatry

Responsibilities: History taking and clinical examination

- Ward and short stay duties
- Emergency duties
- Attending teaching program's
- Rural visits
- Weekly OCD Clinics
- Forensic and criminal psychiatry duties

The National Institute of Mental Health and Neuro Sciences is a multidisciplinary Institute for patient care and academic pursuit in the frontier area of Mental Health and Neuro Sciences. During my 6 months here I received training in basic psychiatry skills (History taking, Psychological assessment, Clinical examination, Emergency management of psychiatric conditions) and attended new patient and follow up patient OPD'S. I also attended weekly OCD Clinics under Dr. YC Janardhan Reddy where I learnt to perform Exposure and Response prevention [ERP] on 5 patients and Behavioral Therapy on 7 patients.

12/9/2007 - 31/1/2008: **Sagar Apollo Hospital**, Bangalore.

Status : Resident Doctor

Responsibilities : History and clinical examination

- Intrpretation of investigations
- Ward night duties
- Attended teaching progam's

Sagar Apollo Hospital is a 665 bedded multi-speciality hospital located in Jayanagar, Bangalore. During my 3 months here I learnt to manage basic ward and emergency skills [History taking, clinical examination, interpreting lab reports]. My role was to perform night duties on 15 days/ month where I was incharge of examining patients admitted in private wards in the hospital. I also attended classes conducted there for DNB candidates in various departments.

17/10/2006 – 16/10/2007: District Hospital Tumkur
Compulsory Rotatory Internship for 1 year.

Status : Intern

Responsibilities : History taking and clinical presentation
Sending and collecting investigation repots
Training in basic clinical skills
Night duties

The District Hospital in Tumkur, Karnataka is a 500 bedded multidisciplinary hospital. I underwent rotatory clinical internship for a period of 12 months where I attained exposure to clinical skills in department of medicine, surgery , OBG and community medicine for 2 months each. ENT, Ophthalmology,Radiology and psychiatry for 15 days each and pediatrics and Emergency medicine for 1 month and 15 days in Psychiatry.

OVERSEAS :

Observer posts-

1/5/2004 – 31/5/2004 : **University Hospital North Staffordshire,**
Department of Histopathology
Stoke –on-Trent, U.K.

Status : Student Elective
Supervison -Dr. B Latifaj

Responsibilities : Assist in Biopsy sectioning

- Assist in Autopsies
- Learn examination of Histopathological slides
- Attend interdepartmental meetings

University Hospital North Staffordshire is a multi speciality hospital under NHS in stoke on trent. I was a student elective in the department of histopathology where I assisted in making histopathological sections and biopsies. I learnt how to examine slides and gross pathology. I also assisted in autopsies two days a week and attended inter departmental meets.

WORKSHOPS/CONFERENCES –

PRESENTED-

August 2011: **Event: NATIONAL WORKSHOP ON RESEARCH METHODOLOGY**

Presented: "Hypothesis and framing of objectives" using an innovative methodology of Medical theatre. The 1 hour talk was presented to an audience of 500 participants using Stand up comedy and dramatic narration. This unique presentation was well accepted by participants and teachers.

July 2011: **Event: NSRCEL ECOSYSTEM SUMMIT 2011**

Presentation: Presented the updates and development of the Ear Nose Throat Multiscope recorder or The Yentroscope. Demonstrated the basic prototype with an overview of the advanced versions along with details of current status.

October 2010: **Event : ENGINEER 2010, NITK, Suratkal. E- PITCH Contest** category BIZZ.

Presentation : Presented an idea and Business plan of an invention- Ear nose throat Multiscope recorder – Selected in finals from 150 ideas submitted in India.

Category – Award for best entrepreneurship business plan.

October 2010: **Event : AOI 6th Southzone and 28th Karnataka state ENT Conference.**MMC and RI, JK grounds, Mysore.

Presentation : Factors effecting post tonsillectomy pain.
Jagdish C, Hemanth V, Kapila M, Shilpa C, Nayar RC.

Category : Free paper

September 2010: **Event : 13th International Congress on Pain**
Palais De Congress, Montreal, Canada.

Presentation : An assessment of factors effecting post operative pain in patients undergoing Tonsillectomy.

Category : Poster presentation. (Finacial aid granted)

Febuary 2010 : **Event : <u>SINOCON 2010 – CME on Sinonasal disease</u>**
Command Hospital, Air force. Bangalore.

Presentation : Correlation of CT PNS findings versus endoscopic findings in chronic rhino sinusitis with specific reference to soft tissue details.

Jagdish C, Ravi CN.

Category : Post Graduate Award Paper.

October 2009 : **Event : <u>AOI – 5TH South zone and 13th TNCON</u>.**
Chennai , Tamil Naidu.

Presentation : Correlation of CT PNS findings versus endoscopic findings in chronic rhino sinusitis with specific reference to soft tissue details.

Category : Free paper

May 2009 : **Event : <u>27TH Karnataka state conference</u>**, M.R.M.C. Gulbarga.

Presentation : Neonatal Hearing Screening Device.
Jagdish C, Ramesh A.

Category : Poster presentation

ATTENDED-

May 2011: **Facilitators Training program for Theatre –in- Education**
Christ University, Bangalore
A 3 day workshop to help train participants become facilitators for theatre workshop.
As an internship to the training and as a requirement for completion of the course, I was made to facilitate 3 workshops for 240 Business school students / workshop for 8 hours.

April 2011: **Medical Education Technology Workshop**
Cardinal Gracias Hall, St John's Medical College,
Conducted by – Medical Education Cell
Contents- Learning objectives, Micro Teaching, Presentation Skills.

October 2010: **10th FHNO-IFHNOS 2010 Conference**
Hotel Taj Residency, MG Road. Bangalore.
A 3 day conference on recent advances in head and neck surgery.
Attended live surgeries and panel discussions.

April 2010 : **National Board of Examinations workshop on theory examination**
S. Rly Head quarters Hospital, Chennai-23. 2010
Workshop on Theory examination module – Actively participated in a mock theory exam and was taught self evaluation as per DNB regulations.

March 2010 : **CME on Recent Advances in ORL and OSCE Workshop**.
St John's Medical College, Hospital. Bangalore.
Participated in an Objective Structured Clinical Exam which taught principles of answering OSCE questions in DNB examinations.

December 2009 : **Workshop on Endoscopic sinus surgery.**
Delhi Metro Hospital. New Delhi.
V P SOOD'S Nasal Foundation.
Workshop comprising of latest advances in endoscopic sinus surgeries with presentations in basic anatomy and live surgical and cadaveric demonstration.

November 2009 : **National Accrited Board of Hospitals (NABH)Workshop.**
ZANDIG Knowledge Management Pvt Ltd. Bangalore.
A 3 day Work shop on Basic hospital accridition skills and auditing of a Hospital for NABH accridition.

May 2009 : **PG Dissertation Workshop and SPSS Practicum.**
St John's Medical College and Hospital.Bangalore.
A 7 days workshop on how to conduct a study, analyze data and use Statistical Package for Social Sciences (SPSS).

AWARDS-

1. **ENGINEER 2010, NITK, Suratkal. E- PITCH Contest category BIZZ.**
   Presented an idea and Business plan of an invention- Ear nose throat Multiscope recorder – Awarded 1st prize.

3. **H. Kishore Chandra Prasad Quiz Competition for postgraduates at AOI – 5TH** South zone and 13th TNCON 09. 2nd Place.

4. **Gold medal awarded for Best Poster presentation.**
   Poster: Neonatal Hearing Screening Device
   27TH Karnataka state conference 2009, M.R.M.C. Gulbarga.

5. **Paper presentation in CME on sinonasal disease – SINOCON 2010**
   Title – "Correlation of CT PNS findings versus Endoscopic findings in chronic rhino sinusitis with specific reference to soft tissue details."
   SINOCON 2010.Command Hospital, Bangalore.- 3rd position.

6. **Best out going student- 12th std. Sudarshan vidya mandir college.**

For overall development and excellence.

RESEARCH PROJECTS-

BIODESIGN AND MEDICAL INNOVATIONS-

1. **Title** : **Ear , Nose, Throat Multiscope and Recorder [EMR].**

Status : Principal Inventor

Responsibilities : Inventor of device.
Founder of company – Yukti medical innovations pvt ltd.
Finalizing and manufacture of prototype.
Marketing of product
Conducting studies/Field trials using this product

Description : A device which enables otoscopy, nasal endoscopy and 90 degree video laryngoscopy using the same portable instrument and recording of information into a memory card from which images can be stored and later on transmitted for teaching purposes and telemedical consultation. Patent application filed in may 2010. Abstract for patent published in the "**Offical Journal of Patents**" by the government of India.

Current status- Incubation - Presently formalities in process for an incubation program by IIMB, Bangalore with NSRCEL(Nadathur S Raghavan Center for Enterpreneurial Learning).

Grants – Received Techno Entrepreneur promotional Program grant for 14 lakhs for the phase 1 development of prototype for Ear Nose Throat Multi-scope Recorder.

Mentor : Dr. Ravi C Nayar. Professor and Head, Department of ENT.
Head of Biomedical innovations program. St John's Medical College.
Mr. Vasudev Avadhani. Advisor.NSRCEL. IIM – Bangalore.
Mr. Suryanarayana A , COO. NSRCEL. IIM-Bangalore

2. **Title : Electromagnetic forceps**

Status : Research assistant
Principal Inventor : Dr. Ravi C Nayar
Responsibilities : Assistance in the prototype development of the forceps
Addition of valuable inputs in the design and function
Being part of the development process of the product
Business development and market analysis
Assistance in carrying out field trials
Value addition for future models

Description : The electromagnetic forceps evolves from the idea that a surgical foreign body removal forceps can be made charged electromagnetically such that the removal of metallic foreign bodies from hollow organs/ tissues can be removed safely, quickly and with minimal surrounding damage. This capability to charge these instruments can help attain better grip on the metallic foreign body and facilitate removal.

Current status: Patent filed , Prototype development in progress.
Mentor : Dr. Ravi C Nayar. Professor and Head, Department of ENT.
Head of Biomedical innovations program. St John's Medical College.

3. **Title : Vertigo Rotatory Chair to assess function of all semi-circular canals**.

Status : Research Assistant

Principal Inventor : Dr. Ravi C Nayar

Responsibilities : Assistance in the prototype development of the chair
Addition of valuable inputs in the design and function
Being part of the development process of the product
Business development and market analysis
Assistance in carrying out field trials
Value addition for future models

Description : The current vertigo chair limits its assessment to only the lateral semi-circular canal. This is due to the uni-axial pendular motion of the chair. The concept of developing a vertigo rotatory chair that can assess the functions of all the three semi-circular canals can be utilized to diagnose problems related to any of the three canals and also assist considerably in the management.

Current status: Patent filed
Prototype development in progress.

Mentor : Dr. Ravi C Nayar. Professor and Head, Department of ENT.
Head of Biomedical innovations program.

4. **Title : Development of indeginous tracheostomy tubes with special functions**

Status : Research Assistant
Principal Inventor : Dr. Ravi C Nayar
Responsibilities : Assistance in the prototype development of the tubes
Addition of valuable inputs in the design and function
Being part of the development process of the product
Business development and market analysis
Assistance in carrying out field trials
Value addition for future models

Description : The tracheostomy tubes available in the Indian market are expensive and not readily available to patients in many parts of India. By developing indigenous, cost effective tracheostomy tubes , the availability of these tubes for Indian health care can be significantly increased. Further special needs for an Indian patient can be catered to by modifying these tubes and customizing it to the Indian setting, such as to overcome problems such as tube blockage and increased collection of secretions which are commonly faced due to poor tube care.

Current status: Prototype development of various tubes in progress.

Mentor : Dr. Ravi C Nayar. Professor and Head, Department of ENT.
Head of Biomedical innovations program.

ACADEMIC RESEARCH-

1. **Title : Factors affecting post operative pain in patients undergoing Tonsillectomy.**

Status : Principal Investigator

Responsibilities : Data collection
Analysis of data
Writing the article

Description : A prospective observational study at a tertiary referral centre using the Wong Baker's Facial Pain scale for measurement of pain at 6 hrs, 24 hrs and 48 hrs post operatively. Results show a positive correlation with 5 factors responsible for pain in tonsillectomy patients.

Guide- Dr Ravi C Nayar. Professor and Head, Department of ENT.

2. **Title** : **The effects of Personality traits on patients with Allergic Rhinitis**.

Status : Co- investigator

Responsibilities : Analysis of ICD 10 IPDE screening questionnaire
Writing the psychiatric correlation aspects in article.
Data collection

Description : A prospective observational study using IPDE screening questionnaire to identify personality traits in patients proven to have allergic rhinitis with skin prick tests.
A positive correlation has been established with two main personality traits.

Guide- Dr. Ravi C Nayar. Professor and Head, Department of ENT.

3. **Title : Vertigo clinic- St John's Medical College Hospital.**

Status : Training assistant

Responsibilities : Complete clinical assessment and examination of patients with vertigo.
Perform Electro Nystagmography
Perform VEMP – Vestibular evoked myogenic potential.
Entry into Proforma for quarterly audits.

Description : Assistant as a part of the weekly Vertigo clinic at St John's medical college. Expertise in Electronystagmography [ENG] and Vestibular Evoked Myogenic Potential [VEMP] . [2009-Present]. Recently participated in the first audit of the clinic with results correlating with ENG and VEMP.

Team : Dr. Ravi C Nayar. Head of Department ENT. Chief supervisor- Vertigo clinic.

Dr. Athira R. Serior resident.
Dr. Alex Mathew. Senior resident.
Dr. Jagdish Chaturvedi. Junior Resident.
Dr. Dali Chandran. Junior resident.

4. **Title : 5 Year retrospective study on Tracheal stenosis**.

Status : Co investigator

Responsibilities : Data collection
Writing of the article
Analysis of data.

Description : Analysis of our experiences with resection and anastomosis at St John's medical college hospital and problems encountered with T- Tube insertion were assessed by retrospectively reviewing patient data. Factors responsible for outcome was analysed.

Publications from data : Nandakumar R, **Jagdish C**, Prathibha CB, Shilpa C , Sreenivas V ,Balasubramanya A M , Nayar RC .Tracheal Resection with end to end anastomosis for post intubation cervical tracheal stenosis – A retrospective five year analysis

Guide- Dr Ravi C Nayar. Professor and Head, Department of ENT.

5. **Title : Correlation of CT PNS findings versus Endoscopic findings in chronic rhino sinusitis with specific reference to soft tissue details**.

Status : Principle investigator

Responsibilities : Data collection
Analysis of data
Drafting of article

Description : A prospective observational study at a tertiary referral centre correlating soft tissue findings in CT scan with endoscopic findings intraoperatively. Factors responsible for correlation are assessed.

Guide- Dr Ravi C Nayar. Professor and Head, Department of ENT.

6. **Title : Measurement of internal diameter of trachea, a comparative study between 4 different methods**.

Status – Principle investigator (Thesis)

Responsibilities – Data Collection
Performing tracheoscopies in cadaveric trachea
Measurement recording in CT and using Vernier's Calipers.

Review of literature
Writing of the article.

Description : Prospective analytical study comparing internal diameter of trachea by measurements obtained using Vernier Calipers (in resected specimens), Endoscopic software, Endoscopic formula and CT measurements.

Guide- Dr Ravi C Nayar. Professor and Head, Department of ENT.

PUBLICATIONS-

PUBLISHED-

J. Chaturvedi, V. Shobha, P. Rout, B. Alexander & R. Nayar : Diagnostic Dilemma Of A Midline Destructive Disease - A Case Report. *The Internet Journal of Otorhinolaryngology.* 2010 Volume 12 Number 1.

Category – Case Report

Jagdish C, K Hegde, A Ananthamurthy, RC Nayar.Submental Epidermoid cyst- A Case Report. *The Internet Journal of head and neck surgery.* 2010 Volume 4 Number 2.

Category- Case Report

Jagdish C. Mystery of the Vanishing Cyanosis. Student British Medical Journal.
Student BMJ 2010;18:c6181

Category- Review

Jagdish C, Hemanth V**,** Kapila M**,** Shilpa C**,** Nayar RC. An assessment of post operative pain in patients undergoing tonsillectomy**.** Minerva Otorhinolaryngologica.

Category- Original Article

Jagdish C, N Sinha, K V Pawan, A M Balasubramanya. A case of fluid in the right occipital condyle secondary to persistant increase in intracerebral pressure.Internet Journal of Radiology
.

Category- Case report

A B Nair, **Jagdish C**, BV Manjula, R Nandakumar, M Correa, A Sawkar. A case of isolated laryngeal Candidiasis in an immunocompetant individual with an unusual presentation mimicking laryngeal carcinoma. Malaysian Journal of Otorhinolaryngology

Category – Case Report.

Jagdish C, Nandakumar R, Prathibha CB, Shilpa C , Sreenivas V ,Balasubramanya A M , Nayar RC .Tracheal Resection with end to end anastomosis for post intubation cervical tracheal stenosis – A retrospective five year analysis. Journal of laryngology and otology.

Category- Original article.

A Ramesh, **Jagdish C**, M Nagapoornima, P N Suman Rao. Validation of a low cost calibrated mechanical noisemaker for hearing screening of neonates. Indian Journal of Medical Research.

Category– Original Article

Jagdish C, N Shankar, RC Nayar. Near complete tracheal ring in a cadaver of Indian origin. International Journal of Anatomical Variations.

Category- Case Report

ACCEPTED-

Jagdish C, Hemanth V, Balasubramanya AM.Unilateral orbital proptosis secondary to a carotico-cavernous fistula.American Journal of Otorhinolaryngology.

Category- Clinical radiology- Case report.

Jagdish C, Sreenivas V, Hemanth V, Nandakumar R. Management of Adult Recurrent Respiratory Papillomatosis with oral Acyclovir following Micro Laryngeal Surgery–A Case series. Journal of laryngology and otology.

Category- Case series

Jagdish C.Cleanliness can be injurious to health. Post Graduate Medical Journal.

Category- Letter to Editor.

Hemanth V, Sreenivas V, **Jagdish C**. Meningioma presenting as a mass in the External auditory canal. Indian Journal of Radiology and Imaging.

Category – Case report

Hemanth V, **Jagdish C**, Anitha R, Kapila M, Galgali A, Nayar RC. Personality traits in Allergic Rhinitis: A case series. Journal of Allergy and Clinical Immunology.

Category- Original Article.

PROFESSIONAL EXPERIENCE –

Routine Clinical Skills

- Insertion of intravenous canulae and venepuncture.
- Urinary Catheterisation.
- Suturing of wounds.
- Providing Basic Life support.

Accident and Emergency Experience:

Involved in the management of:

- Compromised airway, Stridor
- Foreign bodies in the ear, nose, esophagus, trachea and bronchus
- Intracranial complications of Chronic Suppurative Otitis Media
- Epistaxis and Ear Bleed
- Orbital Trauma and Optic nerve injury
- CSF Rhinorhoea and Otorhoea
- Facial and Recurrent Nerve Injury
- Sudden Sensory neural hearing loss
- Acute Episode of vertigo
- Head injury and Laryngeal tracheal Trauma

Emergency Procedures (assisted and performed under consultant supervision)

- CPR
- Tracheotomy
- I&D of abscesses (Ludwig's, Peritonsillar etc.)
- Rigid Esophagoscopy / Bronchoscopy for foreign body removal
- Emergency Facial Nerve Decompression
- Management of cut throat injuries with emergency reconstruction
- Reduction and surgical management of nasal fractures and facial trauma

Experience in Head and Neck Surgery

Surgery	Independently	Assisted	Observed
Thyroidectomy		Yes	
Submandibular Gland Excision		Yes	
Thyroglossal cyst excision		Yes	
Ranula excision		Yes	
Lymph Node Biopsy			
Dermoid excision	Yes		
Carotid artery ligation			Yes
Tracheostomy	Yes		
Incision and drainage of deep neck abscess		Yes	
Modified neck dissection		Yes	
Total Laryngectomy		Yes	
Tracheal stenosis stenting		Yes	
Pectoralis major myocutaneous flap			Yes
Tracheal anastomosis		Yes	
Tracheal stenosis excision		Yes	
Tonsillectomy	Yes		
Adenoidectomy	Yes		
Tongue tie release	Yes		
Salivary Duct Calculus extraction		Yes	
Inter Maxillary Fixation			Yes
Palatal Mass Excision		Yes	
Mandibular Fracture Plating			Yes
Lingual Thyroid excision		Yes	
Drainage of Peritonsillar Abscess	Yes		
Drainage of retropharyngeal abscess		Yes	
Buccal Biopsy	Yes		
Foreign body removal (Fish Bone)	Yes		
Esophagoscopy		Yes	
Bronchoscopy		Yes	
Laryngoscopy (Rigid + Fibreoptic)	Yes		
Microlaryngoscopy		Yes	
Laser excision of subglottic stenosis		Yes	
Glossectomy		Yes	
Madibulectomy		Yes	
Cleft palate repair			Yes
Uvulopalatopharyngoplasty		Yes	
Buccal carcinoma excision		Yes	

Experience in Ear Surgeries

Surgery	Independently	Assisted	Observed
Myringotomy/ Grommet insertion	Yes		
Polypectomy		Yes	
Wax Granuloma / foreign body removal	Yes		
Tympanoplasty/ mastoidectomy		Yes	
Stapedotomy		Yes	
Ossiculoplasty		Yes	
Mastoid abscess drainage	Yes		
Ear Lobe Repair	Yes		
Glomus Tympanicum excision		Yes	
Facial nerve decompression		Yes	
CSF Otorhoea closure		Yes	
Temporal bone resection			Yes
Facial nerve grafting		Yes	
External Auditory canal reconstruction		Yes	
Acoustic neuroma excision			Yes
Cochlear implant			Yes

Exposure to FESS and other Nasal Endoscopic Surgeries

Endoscopic surgeries under supervision, or assisting them.

- Diagnostic nasal endoscopies
- Endoscopic polypectomy
- Endoscopic dacryocystorhinostomy
- FESS for Sino nasal Polyposis
- FESS for Chronic sinusitis
- FESS for allergic fungal sinusitis (AFS)
- Endoscopic CSF Rhinorhoea Closure
- Endoscopic orbital decompression
- Endoscopic nasal biopsies

Laser Surgeries in ENT, assisted or observed

- Partial turbinectomy
- Microlaryngoscopy for vocal nodules
- Microlaryngoscopy for respiratory papillomatosis
- Tracheal stenosis
- Wide excision Glossectomy for carcinoma tongue

Nasal Surgery

Surgery	Independently	Assisted	Observed
Septoplasty		Yes	
Submucous Resection		Yes	
Septorhinoplasty		Yes	
Fracture Nasal bone reduction		Yes	
Lateral Rhinotomy		Yes	
Caldwell Luc's surgery		Yes	
Rhinosporidiosis excision		Yes	
Septal perforation closure		Yes	
Foreign Body removal	Yes		
Intranasal biopsy		Yes	
Septal abscess drainage	Yes		
Medial Maxillectomy		Yes	
External frontoethmoidectomy		Yes	
Maxillary artery ligation		Yes	
Maxillectomy		Yes	

TEACHING EXPERIENCE:

- Lecturing undergraduate doctors and junior nursing staff during post Graduation.

Topics – Foreign Body in the Nose and Ear- 3rd Year MBBS- July 2010
Nasal Packing procedure – 3rd Year MBBS Students – June 2010
Approach to Tonsils – 3rd Year MBBS Students - May 2010
Examination of the Ear – 3rd Year MBBS Students. – March 2010
Anatomy of Larynx – 3rd Year MBBS Students. – June 2010
Anatomy of Nasal Cavity – 3rd Year MBBS Students. – August 2010
Anatomy of Ear – 3rd Year MBBS Students. – October 2010
Physiology of Hearing – 3rd Year MBBS Students. – November 2010
Sinusitis – 3rd Year MBBS Students. – April 2011

INNOVATIVE TEACHING:

MEDICAL THEATRE- The art of learning and teaching medical topics by utilizing theatrical skills and converting them into entertaining skits/plays.Topics taught using medical theatre –

1. Peadriatric Milestones – 4th Year MBBS
2. Systemic Lupus Erythromatosis – 2nd Year MBBS

Medical Theatre Workshops- Organizing secretary and director of the medical theatre workshop where a professional group of theatre artists (Script People's Theatre, Bangalore) and senior medical faculty join together to train 30 medical students difficult to understand medical topics with the help of theatre and dramatics. This Innovative teaching has a huge potential to provide alternate teaching and learning methodology for medical students and teachers.

Workshops Conducted

April 2011 - **Medical Theatre Workshop- St John's Medical College**
Facilitators- Sibu Vaz, Rajesh PI (Script People Theatre)
Dr. Anil Abraham – HOD, Dermatology
Dr. Arvind Kasturi- HOD, Community Medicine
Dr. Jagdish Chaturvedi- PG ENT (Organizer)
Topics- Meniere's disease, Mechanism of action of Penicillin.
Participants- 35 Medical Students

May 2011- **Medical Theatre Workshop- St John's Medical College**
Facilitators- Sibu Vaz, Rajesh PI (Script People Theatre)

Dr. Arvind Kasturi- HOD, Community Medicine
Dr. Jagdish Chaturvedi- PG ENT (Organizer)
Topics- Broca's aphasia, Temporal arteritis, Parathyroid hormone, Sympathetic and parasympathetic system, Rheumatic fever and Cushing's syndrome.
Participants- 25 Nursing students + 10 Medical Students

June 2011 - **Medical Theatre Workshop- Astrazaneca**
Facilitators- Sibu Vaz, Rajesh PI (Script People Theatre)
Dr. Jagdish Chaturvedi- PG ENT (Organizer)
Topics- Soft skills for medical representatives – Negotiator , strategists, multi-taskers, alliance builders, Influencers, New business acquirers
Participants- 85 Medical Representatives/ Special task force members

August 2011- **Medical Theatre Workshop- Vasan Eye Hospital, Jayanagar**
Facilitators- Sibu Vaz, Rajesh PI (Script People Theatre)
Dr. Jagdish Chaturvedi- PG ENT (Organizer)
Topics- Soft skills for hospital NABH team and NABH content on Continuous Quality Improvement.
Participants- 85 NABH hospital team comprising of doctors, nurses, customer care and housekeeping staff.

September 2011- **Medical Theatre workshop for Nursing faculty of St. John's Hospital.**
Facilitators- Sibu Vaz, Rajesh PI (Script People Theatre)
Dr. Jagdish Chaturvedi- PG ENT (Organizer)
Topics- Introduction and implementation of Medical Theatre.
Participants- 65 nursing college teaching faculty and clinical instructors.

EQUIPMENT EXPERIENCE:

Use of all basic instrumentation on outdoor basis and within the operation theatre.

COMPUTER SKILLS

- I am proficient in the use of MS Office XP (word, access, excel and PowerPoint) and using the Internet for academics and research.
- Competent in digital photography and multimedia
- Acquired training in applications of MAYA 3D Software.
- Competent in utilization of Patent searching software
- Competent in development of Pretzi presentation skills
- Competent in utilization of Statistical Package for Social Sciences (SPSS)

OTHER EXPERIENCES –

1/5/2011- 1/7/2011- **Preliminary Tabla**
Rhythms institute of fine Arts,
Jayanagar. Teacher- Rajesh

1/11/2006 – 1/2/2007 : **Preliminary drumming**
Rhythms institute of fine Arts,
Jayanagar. Teacher- Asis Ghosh

1/11/2002- 7/11/2002 : **Course in Councelling skills**
Banjara Academy, Bangaluru.

1/11/2001- 1/6/2002 : **Course - Design a dot com.**
Skills in html, java, photo Shop, office, win 98, win Xp.
BOSTON computer institute. Jayanagar, Bangalore.

MISSION STATEMENT:

- My goal is to improve health care system in India either by developing low cost devices that provide better diagnosis and management or by incorporating better systems into the current clinical set up in neglected or under researched areas for betterment of the Indian health care system.

AREA OF INTEREST :

- **Title : To study psychiatric co-morbidities that are associated with chronic ENT conditions**

To study psychiatric co-morbidities that are associated with chronic ENT conditions causing prolonged morbidity and affects quality of life. I would also aim to open doors to a new field called Psychiatric-otolaryngology where in patients with chronic ENT conditions would be analyzed to detect any psychiatric co- morbidities that are predisposing or delaying recovery so as to prevent morbidity and improve quality of life.

- **Title : To research in the field of pain and the factors that cause pain in ENT diseases and factors that affect post operative pain.**

Studying in these areas would help to develop better management strategies to reduce post operative pain by incorporating standardized protocols and check lists prior to surgery so that pain avoiding steps can be carried out uniformly.

➢ **Title : Innovative Teaching**. – Medical Theatre

I am extremely enthusiastic when it comes to teaching and would like to make teaching and learning ENT as fun and exciting as possible by using attractive and innovative techniques that would help build interest in the minds of aspiring medical students/Junior residents in the field of ENT. I would also like to incorporate theatre, acting, role playing and gaming in the methods for teaching ENT. Currently I have been organizing monthly workshops where in medical topics are taught by using the tools of theatre and dramatics.

➢ **Title : Endoscopic surgeries**. (Functional Endoscopic Sinus Surgery).

Endoscopic surgeries and its innovative applications need further exploration in order to enable surgeries safely without effecting cosmesis.
PERSONAL STATEMENT –

- I have worked in the most austere and the most extravagant conditions and have learned to appreciate that good clinical skill is the backbone of a department.

- I understand the need for and appreciate the role of AUDIT in clinical practice. Given an opportunity would like to participate in clinical research.

- I am able to discuss treatment options with patients in a way they can understand.

- I am hardworking, focused and sincere and am easy to get along with as a part of a team and can maintain good friendly relations with my colleagues and other paramedical staff.

EXTRA CURRICULAR/HOBBIES/INTERESTS-

1. **Drama and theatre-**

Acted in Script Peoples Theatre Annual production VISION 2020- The Apricot Jar
Played the role of an Auctioneer. Directed by Upendra. 3 shows performed on 18,19,20th August
2011 for 1500 people/ show at the Christ University Auditorium.

Participated in numerous one act plays throughout School and college.
Won NIE playfest 99 most popular play and best actor Awards.

Worked with **Bangalore Little theatre in 1999**. With Sibu B Vaz and Rajesh P I. **Workshop "spot 99", "playback"theatre**.

Specialize in Comedy – Stand-up, slapstick , Satire and mimicry.

First Prize- Skit competition- Campus day 2009. St John's medical college Hospital.

First Prize- Skit competition- Campus day 2010. St John's medical college Hospital.

2. **Music** - Qualified Drumer and Guitarist.

3. **Sports** - Cricket, Table tennis, Badminton , swimming, Snooker, foot ball, volley Ball, chess.

AWARDS :

First Place- Tug of war competition – Campus day 2009. St John's medical college Hospital.
Winner- Cricket Tournament- Campus day 2009.St John's medical college Hospital.
2nd Runners- T10 Cricket tournament for Doctors, May 2010. Bangalore.

NEWS –

- An article titled " **Taking a run at better health**" published in "**The GAZETTE**" Montreal,Canada on September 1st 2010 describing my run on a treadmill at the pain congress for charity to developing nations .For every Kilometer jogged an a aum of 50 euros would go for charity. I totaled to 25 Kilometers and contributed 1250 euros to charity.

  http://www.montrealgazette.com/business/Taking+better+health/3467834/story.html

- Semifinalist – Clean and clear **Face of Bangalore** contest courtesy Bangalore times. Article and photo published on 10/10/2010 for performing stand up comedy.

http://lite.epaper.timesofindia.com/mobile.aspx?article=yes&pageid=36&sectid=edid=&edlabel=TOIBG&mydateHid=26-09-2010&pubname=&edname=&articleid=Ar03601&publabel=TOI

REFEREES:

1. **Dr Ravi C Nayar, MS DNBE DLORCS DCCF**
   Professor & Head
   Department of ENT & head and neck surgery,
   St Johns medical college.
   e-mail-ravi-nayar-ent@hotmail.com

2. **Dr Balasubramanya MS**
   Professor,
   Department of ENT & head and neck surgery,
   St Johns medical college.
   e-mail-drbalaam@gmail.com

3. **Dr. C.R. Chandrashekar**
   Professor of Psychiatry
   NIMHANS,Bangalore-560029
   e-mail-psychiatry@nimhans.kar.nic.in

4. **Dr.Y.C.Janardhana Reddy**
   Addl. Professor of Psychiatry
   NIMHANS,Bangalore-560029
   e-mail-jreddy@nimhans.kar.nic.in

5. **Suryanarayanan A**
   Chief executive officer
   NSRCEL
   IIMB, Bangalore
   surya@iimb.net

6. **Dr. Mohan Rao**
   Chief Operating Officer
   E Health TBI
   PESIT, Bangalore
   repakalliraork@rediffmail.com

"IT IS IMPORTANT TO DO WELL

WHEN IT IS IMPORTANT TO DO WELL"

- Dr. Jagdish Chaturvedi

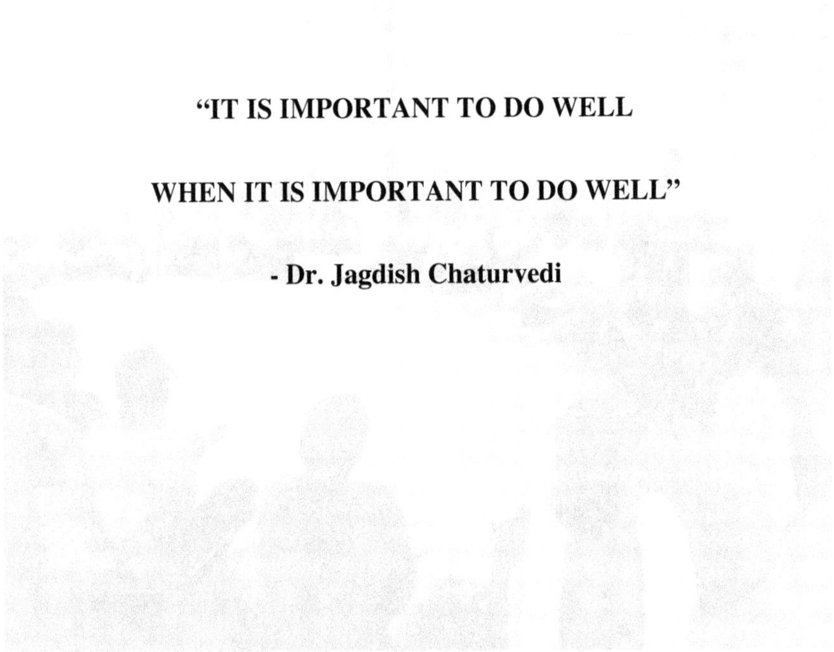

www.ingramcontent.com/pod-product-compliance
Lightning Source LLC
Chambersburg PA
CBHW051443170526
45166CB00001B/101
* 9 7 8 1 5 1 9 4 5 8 0 9 4 *